Riccardo Poli   Wolfgang Banzhaf
William B. Langdon   Julian Miller
Peter Nordin   Terence C. Fogarty (Eds.)

Genetic Programming

European Conference, EuroGP 2000
Edinburgh, Scotland, UK, April 15-16, 2000
Proceedings

 Springer

Volume Editors

Riccardo Poli
Julian Miller
The University of Birmingham, School of Computer Science
Edgbaston, Birmingham B15 2TT, UK
E-mail: {R.Poli,J.Miller}@cs.bham.ac.uk

Wolfgang Banzhaf
University of Dortmund, Department of Computer Science
44221 Dortmund, Germany
E-mail: banzhaf@cs.uni-dortmund.de

William B. Langdon
Center for Information Science
Kruislaan 413, 1098 SJ Amsterdam
E-mail: W.B.Langdon@cwi.nl

Peter Nordin
Chalmers University of Technology, Dept. of Physical Rescource Theory
412 96 Göteborg, Sweden
E-mail: nordin@fy.chalmers.se

Terence C. Fogarty
South Bank University, School of Computing
Information Systems and Mathematics
London SE1 0AA, UK
E-mail: fogarttc@sbu.ac.uk

Cataloging-in-Publication Data applied for

Die Deutsche Bibliothek - CIP-Einheitsaufnahme

Genetic programming : European conference ; proceedings / EuroGP 2000,
Edinburgh, Scotland, UK, April 15 - 16, 2000. Riccardo Poli ... (ed.).
- Berlin ; Heidelberg ; New York ; Barcelona ; Hong Kong ; London ;
Milan ; Paris ; Singapore ; Tokyo : Springer, 2000
  (Lecture notes in computer science ; Vol. 1802)
  ISBN 3-540-67339-3

CR Subject Classification (1998): D.1, F.1, I.5, I.2, J.3

ISSN 0302-9743
ISBN 3-540-67339-3 Springer-Verlag Berlin Heidelberg New York

Springer-Verlag is a company in the BertelsmannSpringer publishing group.
© Springer-Verlag Berlin Heidelberg 2000
Printed in Germany

Typesetting: Camera-ready by author
Printed on acid-free paper   SPIN: 10720165   06/3142   5 4 3 2 1 0

Lecture Notes in Computer Science 1802

Edited by G. Goos, J. Hartmanis and J. van Leeuwen

DATE DUE FOR RETURN

Springer
Berlin
Heidelberg
New York
Barcelona
Hong Kong
London
Milan
Paris
Singapore
Tokyo

Preface

This volume contains the proceedings of EuroGP 2000, the European Conference on Genetic Programming, held in Edinburgh on the 15th and 16th April 2000. This event was the third in a series which started with the two European workshops: EuroGP'98, held in Paris in April 1998, and EuroGP'99, held in Gothenburg in May 1999. EuroGP 2000 was held in conjunction with EvoWorkshops 2000 (17th April) and ICES 2000 (17th-19th April).

Genetic Programming (GP) is a growing branch of Evolutionary Computation in which the structures in the population being evolved are computer programs. GP has been applied successfully to a large number of difficult problems like automatic design, pattern recognition, robotic control, synthesis of neural networks, symbolic regression, music and picture generation, biomedical applications, etc. In recent years, even human-competitive results have been achieved by a number of groups.

EuroGP 2000, the first evolutionary computation conference of the new millennium, was the biggest event devoted to genetic programming to be held in Europe in 2000. It was a high quality conference where state-of-the-art work on the theory of GP and applications of GP to real world problems was presented. Like the two previous workshops, EuroGP 2000 was organized by EvoGP, the genetic programming working group of EvoNET, the Network of Excellence in Evolutionary Computing. The aims are to give European and non-European researchers in the area of genetic programming as well as people from industry an opportunity to present their latest research and discuss current developments and applications. The conference was sponsored by EvoNET and Marconi Telecommunications.

In total 39 papers were submitted to the conference by researchers in 23 different countries. Through a rigorous double blind review process 27 papers where accepted for publication in this volume and for presentation at the conference (14 for oral presentation and 13 as posters). Many of these are by internationally recognised researchers in genetic programming and evolutionary computation and all are of a high quality. This has been ensured by an international program committee including the best GP experts worldwide. We would like to thank them for their thorough work and dedication.

We would also like to thank John Koza, the father of the field of genetic programming, who gave a tutorial at the conference, and David Goldberg, one of the most influential researchers in the field of evolutionary computation, who gave an invited speech.

April 2000   Riccardo Poli, Wolfgang Banzhaf, William B. Langton
Julian Miller, Peter Nordin, and Terence C. Fogarty

Organization

EuroGP 2000 was organized by EvoGP, the EvoNet Working Group on Genetic Programming.

Organizing Committee

Program co-chair:	Riccardo Poli (University of Birmingham, UK)
Program co-chair:	Wolfgang Banzhaf (University of Dortmund, Germany)
Publicity chair:	William B. Langdon (CWI, The Netherlands)
Local chair:	Julian Miller (University of Birmingham, UK)
Publication co-chair:	Peter Nordin (Chalmers University of Technology, Sweden)
Publication co-chair:	Terence C. Fogarty (South Bank University, UK)

Program Committee

Peter Angeline, Natural Selection, New York, USA
Wolfgang Banzhaf, University of Dortmund, Germany
Marco Dorigo, Free University of Brussels, Belgium
Gusz Eiben, University of Leiden, The Netherlands
Terence C. Fogarty, South Bank University, UK
James A. Foster, University of Idaho, USA
Chris Gathercote, Harlequin, UK
Thomas D. Haynes, Wichita State University, USA
Hitoshi Iba, University of Tokyo, Japan
Christian Jacob, University of Calgary
Robert E. Keller, University of Dortmund, Germany
William B. Langdon, CWI, The Netherlands
Evelyne Lutton, INRIA, France
Nic McPhee, University of Minnesota, USA
Jean-Arcady Meyer, Ecole Normale Superieure, France
Julian Miller, University of Birmingham, UK
Peter Nordin, Chalmers University of Technology, Sweden
Riccardo Poli, The University of Birmingham, UK
Joao C. F. Pujol, UFMG - Pampulha, Brazil
Conor Ryan, University of Limerick, Ireland
Marc Schoenauer, Ecole Polytechnique, France
Michele Sebag, Ecole Polytechnique, France
Terry Soule, St. Cloud State University, USA
Andrea G. B. Tettamanzi, Genetica, Italy
Adrian Thomson, University of Sussex, UK

Marco Tomassini, University of Lausanne, Switzerland
Hans-Michael Voigt, GFaI Berlin, Germany
Peter Whigham, University of Otago, New Zealand
Yin Yao, University of Birmingham, UK

Sponsoring Institutions

Napier University, Edinburgh, UK.
Marconi Communications Limited.
EvoNet: the Network of Excellence in Evolutionary Computing.

Table of Contents

Posters

On the Impact of the Representation on Fitness Landscapes

Paul Albuquerque[1], Bastien Chopard[1], Christian Mazza[2], and Marco Tomassini[3]

[1] CUI, Université de Genève, 24 rue Général-Dufour, CH-1211 Genève 4, Switzerland.
Paul.Albuquerque@cui.unige.ch, Bastien.Chopard@cui.unige.ch
[2] Labo de Probabilités, UFR de Mathématiques, Université Claude-Bernard Lyon–I,
43 Bd du 11-Novembre-1918, 69622 Villeurbanne Cedex France.
mazza@probas1.univ-lyon1.fr
[3] ISS, Université de Lausanne, CH-1015 Lausanne, Switzerland.
Marco.Tomassini@iismail.unil.ch

Abstract. In this paper we study the role of program representation on the properties of a type of Genetic Programming (GP) algorithm. In a specific case, which we believe to be generic of standard GP, we show that the way individuals are coded is an essential concept which impacts the fitness landscape. We give evidence that the ruggedness of the landscape affects the behavior of the algorithm and we find that, below a critical population, whose size is representation-dependent, premature convergence occurs.

1 Introduction

Genetic programming (GP) is a fast developing field which aims at evolving a population of computer codes towards a program that solves a given problem [Koz92]. This approach, based on the principle of natural selection of the fittest, has proved successful in several practical situations.

However, GP is applied in a rather empirical way and little is known about the convergence properties of the evolution algorithms. Also, the role of several parameters in the mathematical implementation of the GP remains unclear [GO98]. of a theory

Traditionally, genetic programs are coded with a tree representation, although some studies exist in which a linear coding has been proposed ([WO94], [Nor94] and [KB96]). Linear program representations are in some sense equivalent to tree representations if they span the same program space once decoded. However, the dynamical evolutionary process, i.e. the way in which the genotypic search space is traversed, can vary widely depending on the representation and on the genetic operators.

In this paper we would like to show that the chosen representation for the computer programs is likely to affect crucially the properties of the standard GP algorithm. The reason is that the coding shapes the fitness landscape over the search space quite differently as a function of the representation. For instance, we believe that there is an increasing complexity in the landscape structure as we go from a fixed length linear to a tree representation.

We consider the class of Boolean functions to gain some insight into this matter. This class of problems often provides test cases. In this setting we compare two simple linear representations of a program, one of fixed and the other of variable length. The latter representation, although simpler than the tree representation, already contains some of its key features.

For the sake of simplicity and due to the fact that some theoretical results are available in [Cer96] and [Cer98], we focus on a particular evolution algorithm introduced by Goldberg [Gol90], and known as Generalized Simulated Annealing (GSA). The GSA has a generation-dependent mutation-selection mechanism, but no crossover. Yet it enables us to clearly distinguish between the fitness landscapes induced by both representations and examine their influence on the convergence properties of the algorithm.

In a further study we shall apply the same approach to the tree-based GP and identify the role of crossover.

The paper is organized as follows. In section 2 we informally discuss the various steps leading to the implementation of an evolutionary algorithm. Section 3 is devoted to the description of the GSA. We then apply in section 4 the GSA to the case of Boolean functions represented respectively as strings of bits and of integers. In section 5 we give our simulation results concerning the behavior of the mean fitness and fitness variance for different population size. We also empirically verify the existence of a critical population size with respect to the convergence of the GSA towards a global optimum, and measure stopping times. We finally draw some conclusions in section 6.

2 Strategy

2.1 Statement of the Problem

Given an optimization problem, for example the Traveling Salesman Problem (TSP), we first define the notion of potential solution, namely for the TSP a path passing exactly once through each city. Each potential solution is then assigned a fitness reflecting the quality of the solution, e.g. the length of the path in the TSP case. The goal is to find the best or a good approximation of the best solution (i.e. with highest fitness).

2.2 Choice of the Representation

The application of an optimization algorithm to a given problem requires a preliminary encoding of the potential solutions. A common encoding is of course the binary encoding. However, the TSP solutions are usually represented as strings of integers: assuming the cities are numbered, a string of integers simply gives the visiting order.

Often the choice of the representation is imposed. Nevertheless, we believe that some time and imagination should be devoted to seek alternate representations which might prove better. Prior knowledge about the fitness function must be exploited. For example, if the fitness function is a function from $[-3, +3]^2$ to $\mathbb{R}^+$ which is rotationally invariant, then it is wise to use polar coordinates on $[-3, +3]^2$. We give another illustration of the importance of the representation (see [Bäc96]). A k-ary alphabet requires strings of length $l' = l \log 2 / \log k$ to encode 2^l search points, and for this length l', there are $(k+1)^{l'}$ schemata. Note that $(k+1)^{l'} < 3^l$, for $k > 2$ and l large enough. Therefore a binary alphabet offers the maximum number of schemata, 3^l, for any discrete code.

2.3 Geometry

The choice of a representation defines a geometry by specifying the notion of neighbor. Two solutions are neighbors if one is obtained from the other through an elementary change at the representation level. For example, with the binary encoding, the underlying geometry is that of the hypercube endowed with the Hamming distance; here an elementary change consists in flipping one bit. More generally, the choice of a representation determines a graph structure: a node represents a solution and there are links between neighbors. It carries a natural distance: the minimal path distance on the graph. We also have the notion of fitness landscape on this graph. The ruggedness of the fitness landscape determines the difficulty of the search. The presence of several sharp local optima hinders the search, while the chances of success increase with the number of global optima. The difficulty of a path is reflected in the total hill descent along the path oriented towards the fitter solution. Some statistics on the total hill descent over a sample of paths that remain outside the set of maximum fitness solutions except at their endpoint, gives an indication on the difficulty of the fitness landscape. If fitness fluctuates little with distance, loosely speaking if geometry and fitness are correlated, then the fitness landscape is relatively smooth, and hence the search is easier.

2.4 Variation Operator

The variation operator is responsible for the exploration of the search space and should be linked to the geometry. Short jumps should occur with comparatively high probability, so as to ensure an efficient local search, while long jumps ought to be rare, but likely enough to allow escape from a local optimum. The variation operator should also be biased towards higher fitness solution. This operator in standard GP consists in the combination of the mutation, crossover and selection operators. Unlike mutation, which is fitness independent and ensures ergodicity of the Markov chain, crossover is expected to increase the mean fitness of the population and thus contribute to speed up convergence.

3 Mutation-Selection Algorithm

In this section, we describe the stochastic optimization algorithm proposed by Goldberg in [Gol90]. A formal description of the algorithm can be found in [Cer96] and [Cer98]. Following Cerf, we call it Generalized Simulated Annealing (GSA). The GSA will allow us to emphasize the importance for an evolutionary algorithm of the representation used to encode the problem.

The search space S is a finite set of large cardinality. A fitness function $F : S \rightarrow \mathbb{R}$ assigns to each individual of S a numerical value reflecting the quality of the solution it represents. The goal is to maximize the fitness function, i.e. to find an optimal individual in S. The algorithm given below evolves an initially random population towards a population containing only individuals with maximal fitness. Evolution arises through the action of the genetic operators: mutation and selection.

Consider a population of size p. The mutation-selection algorithm takes place in the product space S^p. Mutation acts independently on each individual in the population. The mutation probability is controlled by a parameter τ, and vanishes as $\tau \rightarrow 0$. The mutation probability usually decreases with the distance in S. The mutation operator must be irreducible, i.e. any individual can be reached, starting from an arbitrary individual, within a finite number of mutations. The minimal number of mutations required to go from one individual to another, also reflects in some sense the distance separating them. Note that mutation, unlike selection, is fitness independent.

Contrarily to mutation, selection really acts on S^p because it can not be decomposed into independent processes. It is also highly reducible. It favors probabilistically the fitter individuals in a population. An individual z_i in a population $z \in S^p$ has a probability

$$\frac{e^{\beta F(z_i)}}{\sum_{i=1}^{p} e^{\beta F(z_i)}}, \tag{1}$$

of surviving selection. Here β is the analog of an inverse temperature controlling the selection strength. The selective pressure increases with β. In the limit $\beta \rightarrow \infty$, the limiting distribution is the uniform distribution over the fittest individuals in the current population.

Classical mutation-selection algorithms are not usually described as the random perturbation of a simple selection process, i.e. selection with uniform distribution over the fittest individuals in the current population. For a fixed intensity level of the random perturbations, this model evolves exactly as a standard mutation-selection algorithm. For the GSA, mutation and selection are, as $\tau \rightarrow 0$ and $\beta \rightarrow \infty$, asymptotically vanishing perturbations of the simple selection process. The underlying idea is that the disruptive action of mutation should disappear with time as the selective pressure becomes stronger and the average fitness in the population increases. However, the mutation probability must remain high enough to prevent the random walk from being trapped in a local optimum (premature convergence), while the selective pressure allows anti-selections (meaning selection of individuals with lower fitness) with diminishing probability.

Assume that β controls the mutation in the sense that $\tau = e^{-\beta}$. For large populations, Cerf proved that the cooling schedule $\beta \sim \log n$ ensures convergence of the GSA towards an optimal solution. More precisely, let R be the minimal number of mutations needed to join two arbitrary individuals in S, Δ the maximal fitness difference, and δ the minimal fitness difference over pairs of individuals with different fitness. Cerf showed the existence of a critical population size p_c for the convergence of the algorithm. For $p > p_c$, there is a cooling schedule $\beta = \beta_0 \log n$ that guarantees convergence to a population containing only individuals with maximum fitness. The critical population size p_c satisfies the bound

$$p_c < \frac{R\Delta}{\min(1, \delta)}. \tag{2}$$

In analogy to physics, the constant β_0 is an initial inverse temperature. Its value should be low enough to ensure that thermal agitation (personified here by mutation) will be sufficient through the run to allow escape from any local optimum. Hence, β_0 depends on the number and size of hill descents along the paths followed by the GSA towards an optimal solution. Neither β_0, nor p_c, are easily computable or approximable.

4 Application to Boolean Functions

4.1 Representation

We apply the GSA to the following toy problem. We consider Boolean functions on 5 variables. There are $2^{2^5} \approx 4.2 \times 10^9$ such functions, since these are the mappings from $\{0, 1\}^5$ to $\{0, 1\}$. We choose two representations.

Assuming some ordering of $\{0, 1\}^5$, a standard representation of a Boolean function is given by a string of bits, where each bit corresponds to the image of an element of $\{0, 1\}^5$, that is a Boolean function is represented by its table of values. Thus, each function is uniquely represented and the search space is $S = \{0, 1\}^{32}$. We will call this representation the binary representation.

A Boolean function can be represented by a polynomial over $\mathbb{Z}_2$. We write $+$ for the binary addition (logical XOR) and $*$ (logical AND) for the binary multiplication. Note that there are many polynomials representing the same function, since $x_k * x_k = x_k, x_k + x_k = 0$. We first represent a Boolean function by a polynomial whose expression is completely developed (e.g. $(x_1 + x_4) * x_2 = x_1 * x_2 + x_2 * x_4$). We then use the correspondence rules: $+0 \leftrightarrow 0, *0 \leftrightarrow 1, +1 \leftrightarrow 2, *1 \leftrightarrow 3, +x_1 \leftrightarrow 4, \ldots, +x_5 \leftrightarrow 13, *x_5 \leftrightarrow 13$, to map a polynomial to a string of integers. For example, the polynomial $x_1 * x_2 * x_2 + x_3 * x_4 * x_5 + 1$ is first rewritten as

$*x_1$	$*x_2$	$*x_2$	$+x_3$	$*x_4$	$*x_5$	$+1$

or

$$\boxed{+x_1}\boxed{*x_2}\boxed{*x_2}\boxed{+x_3}\boxed{*x_4}\boxed{*x_5}\boxed{+1}$$

and then using the correspondence rules, we get the string of integers

$$\boxed{5}\boxed{7}\boxed{7}\boxed{8}\boxed{11}\boxed{13}\boxed{2}$$

or

$$\boxed{4}\boxed{7}\boxed{7}\boxed{8}\boxed{11}\boxed{13}\boxed{2}$$

Thus, Boolean functions can be represented as strings of integers, representation which is of course highly degenerate. This degeneracy is due to the presence of many "introns", i.e. redundant parts of code which come from the identities $x_1 + x_1 = 0, 0*x_5 = 0, 1*x_2 = x_2$, etc.... . Henceforth, we will restrict our attention to $S = \{0, \ldots, 13\}^l$, for some fixed l. If l is too small, not all Boolean functions can be represented by an element of $\{0, \ldots, 13\}^l$. The longest completely developed reduced polynomial expression is obtained by developing $(1 + x_1) * \ldots * (1 + x_5)$. Therefore, the minimal l needed to represent all Boolean functions is 81. This second representation will be called the integer representation.

4.2 Mutation and Geometry

The search space is either $S = \{0, 1\}^{32}$, or $S = \{0, \ldots, 13\}^l$. More generally, we can write both spaces as $S = \{0, \ldots, k\}^l$, for some fixed l. The Hamming distance d on S is defined as the number of positions at which two strings differ. Nearest neighbors in S are elements at distance one from each other. With this metric, $S = \{0, 1\}^{32}$ is the 32-dimensional hypercube.

The probability $q_\tau(x, y)$ that an individual $x \in S$ mutates into an individual $y \in S$ is given by:

$$q_\tau(x, y) = (1 - \tau)^{l - d(x,y)} \left(\frac{\tau}{k}\right)^{d(x,y)}. \tag{3}$$

Note that $q_\tau(x, y) > 0$, for any $x, y \in S$. Asymptotically as $\tau \to 0$,

$$q_\tau(x, y) = \begin{cases} 1 - l\tau + \mathcal{O}(\tau^2), & \text{if } x = y, \\ \dfrac{\tau}{k} + \mathcal{O}(\tau^2), & \text{if } x \sim y, \\ \mathcal{O}(\tau^2), & \text{otherwise} \end{cases} \tag{4}$$

where $x \sim y$ denotes the nearest neighbor relation in S. Therefore, the mutation operator essentially describes a nearest neighbor move with however non-zero probability for longer jumps.

4.3 Fitness

We fix a goal function f_0 that the algorithm must find. The fitness of an arbitrary Boolean function f is given by

$$F(f) = 2^5 - \sum_{k \in \{0,1\}^5} (f_0(k) - f(k))^2, \tag{5}$$

here $f_0(k), f(k)$ are treated as integers. The above sum is the least-square error over the sample set $\{0, 1\}^5$. The algorithm searches for the best approximation of f_0 in S. Recall that, for the integer representation, f_0 itself may not belong to S.

The fitness function for the binary representation is obtained by replacing the above sum by the Hamming distance to f_0. The associated fitness landscape is unimodal and hence very simple. It is identical for every f_0. The interest for the landscape generated by the integer representation lies in the presence of many large plateaus. This fitness landscape depends on f_0 and is comparatively more complicated.

Learning 5-bit Boolean functions is rather uninteresting. We emphasize that our interest resides in the difference in landscape resulting from the choice of the representation, and how this affects the search.

5   Simulation Results

It was only necessary to test one goal function for the binary representation, since, as previously mentioned, all goal functions are equivalent for this representation (i.e. yield exactly the same unimodal fitness landscape). For the integer representation, we tested four different goal functions using various lengths of strings (10, 15, 20) the number of variables being kept constant equal to 5. This provided us with different fitness landscapes.

The four goal functions are:

1. $f_{0,1} = x_1 + x_2 + x_3 + x_4 + x_5$ represented as a string of length 10,

4	6	8	10	12	0	1	1	1	1

2. $f_{0,2} = 1 + x_1 x_2 x_3 + x_3 x_4 x_5$ represented as a string of length 15,

4	7	9	8	11	13	2	0	0	0	0	0	0	0	0

3. $f_{0,3} = 1 + x_1 x_2 x_3 + x_2 x_3 x_4 + x_3 x_4 x_5$ represented as a string of length 15,

4	7	9	8	11	13	2	6	9	11	0	1	1	1	1

4. $f_{0,4} = x_2 + x_4 + x_5 + x_3 x_4 + x_1 x_2 x_3 x_4 x_5$ represented as a string of length 20,

1	2	10	10	9	13	7	9	5	6	10	7	12	10	7	3	10	9	2	3

Our simulations were run sequentially on a Sun Ultra 10 workstation.

5.1   Choice of the Run Parameters

The parameters of the runs, initial inverse temperature β_0 and initial mutation probability τ_0, were chosen by inspection. For a given goal function and different population size, we varied β_0 and τ_0 over some interval and compared performances. Constant parameter values were then selected for each goal function. Theoretical predictions assert that τ_0 influences the speed of convergence, but not the convergence itself, unlike β_0 whose role is crucial for both purposes.

The values of these parameters are given in table 1 for both representations.

Table 1. Parameters of the runs.

goal function	β_0	τ_0
$f_{0,1} = x_1 + x_2 + x_3 + x_4 + x_5$	0.075	0.05
$f_{0,2} = 1 + x_1 x_2 x_3 + x_3 x_4 x_5$	0.07	0.06
$f_{0,3} = 1 + x_1 x_2 x_3 + x_2 x_3 x_4 + x_3 x_4 x_5$	0.08	0.06
$f_{0,4} = x_2 + x_4 + x_5 + x_3 x_4 + x_1 x_2 x_3 x_4 x_5$	0.08	0.04
binary representation	0.5	0.005

5.2 Mean Fitness, Fitness Variance and Critical Population Size

The increase in mean fitness is steep during the beginning of the runs before slowing down and finally stabilizing. The non-zero probability of selecting lower fitness individuals from the current population allows the mean fitness to punctually decrease. The fluctuations inside a population due to mutations are smoothed out by the selection operator and the cooling, while the difficulty of the landscape reinforces them. The smoothing increases with the population size p. It seems that rugged landscapes yield steeper initial increase in mean fitness followed by a much slower regime. Moreover, the mean-fitness fluctuates more during the runs.

In figure 1, we plot for a population size $p = 15$, the mean fitness of the population averaged over 100 runs, as a function of the number of generations.

The fitness variance in the population also measures the diversity of individuals. Rugged landscapes allow a greater fitness variance, because neighboring individuals may have very different fitness and distant ones comparable fitness. We observe from the fitness variance that the landscape of $f_{0,1}$ is relatively easy, while those of $f_{0,2}, f_{0,3}, f_{0,4}$ are of comparable difficulty. The binary representation yields the simplest landscape: unimodal and constant over any sphere centered about the unique global optimum. However, convergence occurs faster, in terms of generations, in the $f_{0,1}$ case, because the size of the global optima set compensates for the structurally more complex landscape.

We plot in figure 2 the fitness variance in the population, averaged over 100 runs, as a function of the number of generations for a population size $p = 15$. Since the fitness variance for the binary representation is comparatively smaller than for the integer one by a factor 10^2, we use a different scale for the sake of legibility.

For the integer representation, we then average the mean-fitness over the last generations after convergence is reached and over 100 runs. We plot this in figure 3 against the population size. Here we observe the existence of a critical population size p_{crit} as predicted by Cerf. It is interesting to point out the relation between the difficulty of the fitness landscape and the critical population size p_{crit}. We indirectly control the difficulty of the landscape through the string length of the integer representation and the choice of the goal functions. The string length is actually equal to the quantity R in equation (2). The choice of the goal function acts on the average difficulty of a path in S^p. A given goal func-

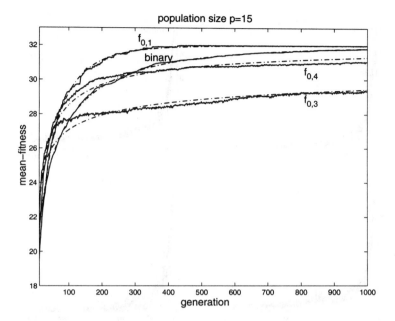

population size p=15

Fig. 1: Average mean fitness vs. generation for a population size $p = 15$. The integer representation curves are labeled $f_{0,1}$, $f_{0,3}$ and $f_{0,4}$. The solid line corresponds to the numerical data. We fit the data (dash-dot line) with the function $32 - \beta_1 \exp(\beta_2 x^{\beta_3})$ (see §5.3).

tion may be represented by several strings, because the integer representation is degenerate. Hence, the size of the global optima set in S depends on the goal function. Paths leading to this set will be longer on average if this set is small, and more difficult in rugged landscapes. We think here of the difficulty of a path as the total amount of anti-selections (i.e. hill descents) required to travel along it. The ratio of the size of the global optima set to the size of S^p increases with the ratio of the minimal length necessary for representing the goal function, to the length of the actual representation.

For the binary representation, the critical population size p_{crit} equals 2. Notice that for $p = 1$, the algorithm is a blind random walk with jump probabilities decreasing through the run. For any $p \geq 2$, the algorithm converges with probability 1 to the global optimum, since there are no local optima and the GSA is asymptotically similar to a Simulated Annealing (see [AM]).

From figure 3, we extract values for p_{crit} and compare them with the theoretical bound in Cerf's formula (2) with $\Delta = 2^5 = 32$, $\delta = 1$ and $R = 10, 15, 20$ respectively. The comparison in table 2 shows that this bound, which was in no way meant to be optimal, proves very bad, as could have been guessed from its simplicity. It is however easily computable or at least approximable.

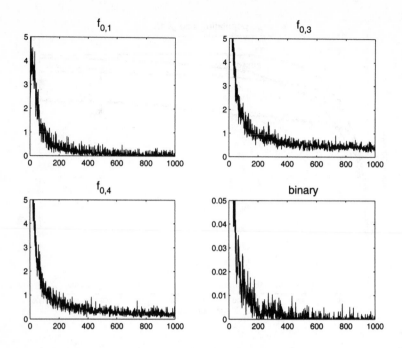

Fig. 2: Average fitness variance (y-axis) vs. generation (x-axis) for a population size $p = 15$. Notice the different scale in the binary case (see text).

5.3 Fit of Mean Fitness

We fit the mean fitness of the population, averaged over 100 runs, for 10 population size above the critical value, with the function

$$32 - \beta_1 \exp(\beta_2 x^{\beta_3}). \tag{6}$$

Recall that in our application 32 is the maximal fitness value. The correlation between fit and data is above 99% for the goal functions $f_{0,1}$ and $f_{0,2}$, 96% for $f_{0,3}$, 97% for $f_{0,4}$ and near 100% for the binary representation. A fit of the data for a population size $p = 15$ is plotted in figure 1.

The initial fitness growth is controlled by β_1. The speed of convergence to maximum mean fitness depends essentially on β_3, whereas β_2 controls the transition from the sharp initial increase to the slower regime.

Except for the goal function $f_{0,1}$, there was no clear behavior of the parameters as a function of the population size. However, we can compare $f_{0,1}$ through $f_{0,4}$ from the order of magnitude of β_3 (see table 3), the smaller values yielding slower convergence. Hence, we infer that in increasing order of landscape difficulty we have $f_{0,1}$, $f_{0,2}$, $f_{0,4}$ and $f_{0,3}$, the binary representation scoring between $f_{0,1}$ and $f_{0,2}$.

We sample the search space by independently drawing individuals from a uniform distribution and computing their fitness. The variance of the fitness

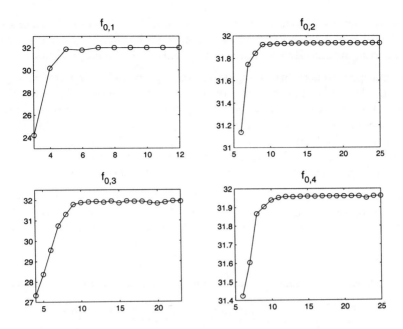

Fig. 3: Critical population size for $f_{0,1}, f_{0,2}, f_{0,3}, f_{0,4}$. Population size (x-axis) vs. average mean fitness after convergence (y-axis).

Table 2. Value of critical population size p_{crit} and theoretical bound (2).

goal function	p_{crit}	bound
$f_{0,1} = x_1 + x_2 + x_3 + x_4 + x_5$	5	320
$f_{0,2} = 1 + x_1 x_2 x_3 + x_3 x_4 x_5$	9	480
$f_{0,3} = 1 + x_1 x_2 x_3 + x_2 x_3 x_4 + x_3 x_4 x_5$	9 or 10	480
$f_{0,4} = x_2 + x_4 + x_5 + x_3 x_4 + x_1 x_2 x_3 x_4 x_5$	10 or 11	640

Table 3. Order of magnitude of β_3.

goal function	β_3
$f_{0,1} = x_1 + x_2 + x_3 + x_4 + x_5$	0.7
$f_{0,2} = 1 + x_1 x_2 x_3 + x_3 x_4 x_5$	0.4
$f_{0,3} = 1 + x_1 x_2 x_3 + x_2 x_3 x_4 + x_3 x_4 x_5$	0.02
$f_{0,4} = x_2 + x_4 + x_5 + x_3 x_4 + x_1 x_2 x_3 x_4 x_5$	0.2
binary representation	0.5

distribution obtained from this sample, gives some indication on the ruggedness of the landscape. A large standard deviation could be caused by the presence of multiple local optima. Hence, more rugged landscapes should yield noisier data.

With respect to the critical population size, one would be tempted to replace Δ in formula (2) by the standard deviation of the sample. The heuristics is based on the observations that the initial population is drawn from a uniform distribution and the fitness variance in the population decreases during the run. Hence it is reasonable to believe that the fitness fluctuations along a path leading to a global optimum are bounded by the standard deviation of the sample. In table 4 we display the mean and variance of a sample of size 10^6.

For the binary representation, there are C_k^{32} individuals at Hamming distance k from the goal function. This corresponds to a binomial distribution $Bin(32, 1/2)$, whose mean 16 and variance 8 are easily computable.

Table 4. Mean and variance of a fitness sample drawn from a uniform distribution.

goal function	mean	variance
$f_{0,1} = x_1 + x_2 + x_3 + x_4 + x_5$	16.01	2.55
$f_{0,2} = 1 + x_1 x_2 x_3 + x_3 x_4 x_5$	16.12	5.94
$f_{0,3} = 1 + x_1 x_2 x_3 + x_2 x_3 x_4 + x_3 x_4 x_5$	15.61	12.2
$f_{0,4} = x_2 + x_4 + x_5 + x_3 x_4 + x_1 x_2 x_3 x_4 x_5$	16.08	7.42

5.4 Stopping Times

We record the following stopping time: the first generation at which the mean fitness reaches the value 31.9. We average the stopping time over 1000 runs and plot this as a function of the population size p in figure 4. We then fit the data with the function

$$\frac{1}{\gamma_1 + \gamma_2 p}. \tag{7}$$

The correlation between fit and data is above 99% for $f_{0,1}, f_{0,2}, f_{0,3}$, 97% for $f_{0,4}$ and 99% for the binary representation. In table 5, we display the values of the parameters of the fit.

Table 5. Parameters for the fit by $\frac{1}{\gamma_1 + \gamma_2 p}$ of the average stopping times.

goal function	$\gamma_1 \quad [\times 10^{-3}]$	$\gamma_2 \quad [\times 10^{-3}]$
$f_{0,1} = x_1 + x_2 + x_3 + x_4 + x_5$	-0.8317	0.9117
$f_{0,2} = 1 + x_1 x_2 x_3 + x_3 x_4 x_5$	0.4815	0.2889
$f_{0,3} = 1 + x_1 x_2 x_3 + x_2 x_3 x_4 + x_3 x_4 x_5$	0.0764	0.0158
$f_{0,4} = x_2 + x_4 + x_5 + x_3 x_4 + x_1 x_2 x_3 x_4 x_5$	0.2247	0.0190
binary representation	-0.0742	0.0523

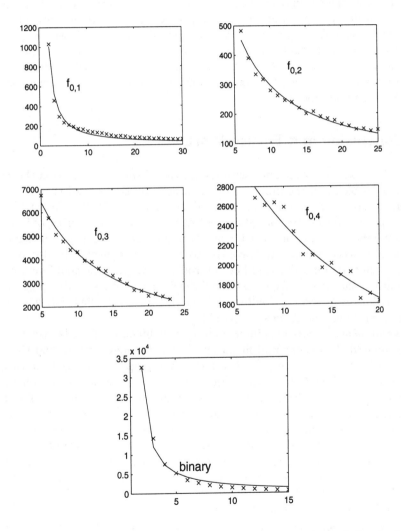

Fig. 4: Average stopping times (y-axis) vs. population size (x-axis) for the integer (labels $f_{0,1}$ to $f_{0,4}$) and binary representation. The solid line is a fit of the data by $\frac{1}{\gamma_1 + \gamma_2 p}$.

We observe that, for the same population size, the average stopping times are the highest for $f_{0,3}$ and decrease at a similar rate (given by γ_2) for $f_{0,4}$. They decrease at a faster rate for $f_{0,2}$ and much faster for $f_{0,1}$. The binary representation scores somewhere between $f_{0,2}$ and $f_{0,4}$. Once again, the difficulty of entering the global optima set depends on its size and the average difficulty of a path leading to it. Therefore the stopping times give a measure of the difficulty of the fitness landscape.

The fits of the stopping times indicate roughly that the gain in time is proportional to the increase in population size. However, the computation time for our sequential algorithm also increases proportionally to the population size p. Hence, provided $p > p_{crit}$, an increase in p yields no benefits in terms of speed of convergence. This hints strongly towards parallelizing the algorithm.

We must mention that the actual computation time involved in one generation depends strongly on the representation and its parameters. Here we do not care about real time performances, but only about the dependence on the number of generations.

6 Conclusion and Future Work

In this work, we were concerned with the impact of the representation chosen to encode a problem, on the difficulty of the fitness landscape, and thereof on the convergence of a type of GP algorithm. We compared for Boolean functions two linear representations: one as a string of bits and another as a string of integers. The first representation produced a simple unimodular landscape, the same for all goal functions. The second gave rise to a more complicated landscape with large plateaus. Its difficulty could be controlled by varying the goal function and the length of the strings. We investigated the difficulty of the landscape by using an algorithm proposed by Goldberg ([Gol90]), which we called, following Cerf, Generalized Simulated Annealing (GSA).

Several measures provided indications on the difficulty of the landscapes. We first compared the exponential fit parameters of the mean fitness and then the fitness variance during the run. The latter was greater in rugged landscapes. Large variance of the fitness distribution, obtained from a sample of individuals drawn from a uniform law, hints towards the presence of multiple local optima and thus reflects the ruggedness. The appearance of a critical population size for the integer representation, which seemed to increase with the ruggedness of the landscape, constitutes an interesting phenomenon. It implies that the population size must be big enough to ensure convergence. This apparently new result in GP agrees with the theoretical predictions given by Cerf in a more general context ([Cer96] and [Cer98]). There seems to exist no precise bound yielding a useful estimate for the critical population size.

Besides the ruggedness, another important feature of the landscape is the size of the global optima set. This became apparent as we measured stopping times. The binary representation has only one global optimum as opposed to several for the highly degenerate integer representation. This hinders the search and explains why the stopping times for the binary representation are higher than those observed in structurally more complex landscapes. Therefore, the measure of stopping times for entering some neighborhood of the global optima set constitutes a good measure of the difficulty of the landscape.

We are convinced of the necessity of performing some statistical tests before adopting a representation in order to get a picture of the associated fitness landscape, and then ponder the adequacy of a search algorithm. In this sense, the variation operator and the representation must be closely linked. The variation operator is expected to increase the mean fitness. Therefore it must be adapted to the representation. Inadequacy at this level will result in poor performance of the algorithm. This is probably often the case for the standard crossover operator in GP under the tree representation. In future work, we intend to apply the GSA to tree-based GP and later investigate the role of crossover.

Acknowledgements

This work is supported by the Swiss National Science Foundation.

References

[AM]   P. Albuquerque and C. Mazza. Mutation-selection algorithm: a large deviation approach. In preparation.

[Bäc96]   Thomas Bäck. *Evolutionary Algorithms in Theory and Practice*. Oxford University Press, 1996.

[Cer96]   Raphaël Cerf. The dynamics of mutation-selection algorithms with large population sizes. *Annales de l'Institut Henri Poincaré*, 32(4):455–508, 1996.

[Cer98]   Raphaël Cerf. Asymptotic convergence of genetic algorithms. *Advances in Applied Probability*, 30(2):521–550, 1998.

[GO98]   David E. Goldberg and U. O'Reilly. Where does the good stuff go and why? how contextual semantics influences program structure in simple genetic programming. In *Genetic Programming, First European Workshop, EuroGP98*, volume 1391 of *Lecture Notes in Computer Science*, pages 16–36, Heidelberg, 1998. Springer-Verlag.

[Gol90]   David E. Goldberg. A note on Boltzmann tournament selection for genetic algorithms and population-oriented simulated annealing. *Complex Systems*, 4(445-460), 1990.

[KB96]   R.E. Keller and W. Banzhaf. Genetic programming using genotype-phenotype mapping from linear genomes into linear phenotypes. In J. R. Koza, D. E. Goldberg, D. B. Fogel, and R. L. Riolo, editors, *Genetic Programming 1996: Proceedings of the First Annual Conference*, pages 116–122, Cambridge, MA, 1996. The MIT Press.

[Koz92]   John R. Koza. *Genetic Programming: On the Programming of Computers by Means of Natural Selection*. MIT Press, Cambridge, MA, 1992.

[Nor94]   P. Nordin. A compiling genetic programming system that directly manipulates the machine code. In K. E. Kinnear, Jr., editor, *Advances in Genetic Programming*, pages 311–331. The MIT Press, Cambridge, MA, 1994.

[WO94]   M. Wineberg and F. Oppacher. A representation scheme to perform program induction in a canonical genetic algorithm. In *Parallel Problem Solving from Nature- PPSN III*, volume 866 of *Lecture Notes in Computer Science*, pages 292–301, Heidelberg, 1994. Springer-Verlag.

The Legion System: A Novel Approach to Evolving Heterogeneity for Collective Problem Solving

Josh C. Bongard[1]

AI Lab, Computer Science Department
University of Zurich, Winterthurerstrasse 190
CH-8057, Zurich, Switzerland
bongard@ifi.unizh.ch

Abstract. We investigate the dynamics of agent groups evolved to perform a collective task, and in which the behavioural heterogeneity of the group is under evolutionary control. Two task domains are studied: solutions are evolved for the two tasks using an evolutionary algorithm called the Legion system. A new metric of heterogeneity is also introduced, which measures the heterogeneity of any evolved group behaviour. It was found that the amount of heterogeneity evolved in an agent group is dependent of the given problem domain: for the first task, the Legion system evolved heterogeneous groups; for the second task, primarily homogeneous groups evolved. We conclude that the proposed system, in conjunction with the introduced heterogeneity measure, can be used as a tool for investigating various issues concerning redundancy, robustness and division of labour in the context of evolutionary approaches to collective problem solving.

1 Introduction

Investigations into heterogeneous agent groups are only just getting under way. To cite two examples, in [20], morphological heterogeneity is studied, in which physical robots have non-overlapping sets of sensors and effectors; in [3], physical and simulated robots with distinct motor schemata are referred to as behaviourally heterogeneous groups.

These studies stand in contrast to biological models, such as action selection [17] and behaviour thresholds [24, 7], in which the underlying control algorithms of the agents are equivalent, but changes to the parameters of an agent's control algorithm lead to behavioural differentiation. Agents in [22] exhibit large morphological and behavioural variation, but this variation, in the context of collective problem solving, was not addressed.

In biological systems, individual cells in an organism contain (near-) identical genomes; although individual organisms within a species exhibit differing alleles, the actual gene complement across organisms within a species is the same. In contrast, evolutionary algorithms are not limited by this constraint: evolved agent groups can exhibit large behavioural differentiation. To this end, the work presented here is concerned with the dynamics of behaviourally *heterogeneous* groups, in which not only the observed behaviours, but also the underlying control architectures of the agents are differentiated.

In this report, simulated agents are studied. However, there is a growing body of literature dedicated to heterogeneous robot groups. Arkin and Hobbs [1] delineate a number of dimensions along which enlightened design of robot groups should proceed. Mataric *et al* have implemented groups of robots in which heterogeneity is realized through spatial differentiation within the task space in order to minimize physical interference [10, 11], or by implementing a dominance hierarchy, in which inferior robots can only perform a subset of the basic behaviours available to more dominant robots [16].

These studies, however, take a simplistic view of heterogeneity, in that the differences between agents in the group are decided upon by the designers. For example, in the case of territoriality, each agent is assigned its own area prior to execution of the task. It has been pointed out [3] that most of this work is also simplistic in that heterogeneity is treated as a binary property. In a series of studies [3], groups of robots learned to perform a collective task by tuning the heterogeneity of the group to best perform the task. For foraging and cooperative movement tasks, it was shown that groups invariably converge on homogeneous behaviours; in the case of robot soccer, the teams converge on heterogeneous behaviours [3].

Although these studies were concerned with the degree of heterogeneity in a group as a consequence of the task domain, emphasis was placed on exercising a measure for heterogeneity called *social entropy* [5]. Herein it is shown that by using an evolutionary approach to heterogeneous group behaviours, a simplified measure of heterogeneity can be formulated which overcomes some of the drawbacks of social entropy, explained in Sect. 2.

Evolutionary approaches to heterogeneity include the work by Bull and Fogarty [8], who present an island-model genetic algorithm that encodes classifier systems used to control a quadruped robot; in [21], cascade neural networks [9] are evolved for parity computation using an incremental genetic algorithm. In both investigations, however, the behavioural niches of the groups are predetermined.

In [14], a genetic programming approach is introduced in which niche determination is more dynamic: behaviours are evolved for a pride of lions in a predator/prey task domain. Each individual s-expression in the GP population codes for each and all of the behaviours required by members of the pack. The merit of evolving team behaviours, as opposed to evolving individual behaviours which are later combined to form a team, is pointed out in [12]: individual-level evolutionary systems must somehow overcome the credit assignment problem.[1]

---

[1] The credit assignment problem also appears in learning approaches to group heterogeneity. This problem, as noted in [4], prompted the development of a new (and heavily domain specific) type of reinforcement learning heuristic, *shaped reinforcement learning* [15].

In Luke and Spector's model, the behaviour for each individual lion in a pride is represented as a branch in an s-expression which encodes all of the behaviours for the team. This model successfully avoids the credit assignment problem, and allows for emergent problem decomposition: the amount of divergence (and convergence) between the behaviours of the individual lions is shaped by the selection pressure exerted by the predator/prey task domain. However, this model suffers from two serious drawbacks.

First, more diverse groups are implicitly favoured by the system, because each individual agent possesses its own distinct behaviour: in order to obtain a subset of k agents that perform equivalent behaviours, the system must evolve the same behaviour k times in the same s-expression. Second, the system scales with the number of agents performing the task: for n agents, the s-expression must contain n branches.

2 The Model

We now introduce an augmented genetic programming system, called the Legion system, which shares the advantages of the system described in [14], but overcomes its limitations.

2.1 The Legion System

Each individual s-expression in the Legion population encodes behaviours for an entire agent group, and is composed of two or more branch s-expressions. The first branch s-expression is the *partition s-expression*, and dictates how an agent group is to be partitioned into a set of behavioural classes. The partition s-expression is evaluated in depth-first order, in order to determine how many behavioural classes the agent group will contain, and how many agents will be assigned to each behavioural class. When a SPLIT operator is encountered, kf agents are assigned to the next available behavioural class, where k is the number of agents not yet assigned a behavioural class, and f is the floating point value ($0 \leq f \leq 1$) returned by the SPLIT operator's left branch. The remaining $k(1-f)$ agents are further partitioned when the next SPLIT operator is encountered. When the final SPLIT operator is encountered, the remaining agents are placed into the two next behaviour classes. Any remaining behavioural s-expressions are deleted. If the final SPLIT operator is encountered and there remains only one more behaviour s-expression, this last behaviour s-expression is duplicated, and the remaining agents are divided into the two identical behaviour s-expressions. In subsequent generations, mutation and crossover events may differentiate these two branch s-expressions.

It follows from this that, as opposed to the model in [14], the Legion system can dynamically change the number of behavioural classes in an agent group over evolutionary time, as well as modifying the behaviours of members of each class[2]. Moreover, by modifying the number of SPLIT operators in partition s-expressions, selection pressure can increase or decrease the number of behavioural classes—and thus the heterogeneity—of agent groups over evolutionary time.

The remaining branch s-expressions in a Legion s-expression, referred to as *behaviour s-expressions*, are domain-dependent and encode the actions performed by agents assigned to that behavioural class. Fig. 1 presents the architecture of the Legion population in pictorial form.

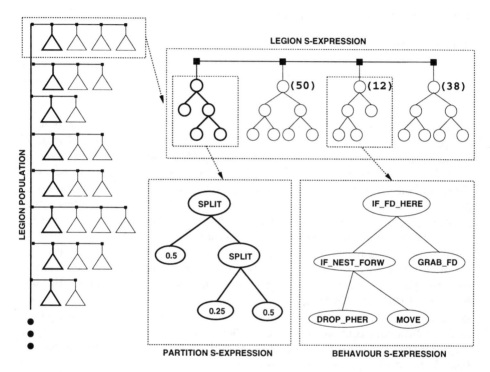

Fig. 1. A pictorial representation of the Legion system The bracketed numbers next to the three behaviour s-expressions denote the percentage of agents from a group that would be assigned to that behavioural class. The percentages, and the number of behavioural classes, are determined by the partition s-expression.

Crossover in the Legion system is accomplished by *restricted breeding*, similar to [14]: given two Legion s-expressions s_1 and s_2 with partition and behaviour s-expressions $\{p_1, b_{1,1}, b_{2,1} \ldots b_{i,1}\}$ and $\{p_2, b_{1,2}, b_{2,2} \ldots b_{j,2}\}$, the partition s-expressions of the two children are created by sub-tree crossover of p_1

---

[2] This process was modelled on the biological concept of gene families, which are produced by gene duplication and differentiation over evolutionary timescales [19], [18].

and p_2, and the behaviour s-expressions are created by the pairwise crossings of $\{(b_{1,1}, b_{1,2}), (b_{2,1}, b_{2,2}), \ldots (b_{i,1}, b_{i,2})\}$, where $i \leq j$.

If it is accepted that the amount of heterogeneity in an agent group is dependent on the number, membership and differentiation of the behavioural classes in a Legion s-expression, and the constitution of behavioural classes in a Legion system is under evolutionary control, then it follows that the amount of heterogeneity in the agent groups evolved by the Legion system is subject to the selection pressure of the task domain.

2.2 The Heterogeneity Measure

In [3] a measure of heterogeneity, social entropy, is presented and defined as

$$H = -\sum_{i=1}^{M} p_i \log_2(p_i), \tag{1}$$

where M is the number of behavioural classes in an agent group, and p_i is the probability that any given agent is a member of the behavioural class i. Social entropy thus takes into account the number and membership sizes of the behavioural classes in a group, but does not consider the differences between agents in different classes. A more complicated measure of social entropy is given in [5] which takes into account inter-class behavioural differences. However, this measure is domain-specific, and relies on details of the capabilities of agents within the group, such as perceptual or internal state.

When evolving behaviours for agent groups, a fitness function is usually formulated which calculates some quantitative measure of the facility of the group to accomplish its assigned task. The fitness function is dependent on the behaviours of the agents within the group; differences in fitness between any two given agent groups imply behavioural differences between those groups. Thus, in an evolutionary context, a measure of heterogeneity can be formulated based solely on the fitness values of agents within the group, and not directly on the behaviours of the agents themselves.

Consider a group of n agents which has been partitioned by the Legion system (or some other evolutionary algorithm) into a set of behavioural classes $B = \{b_1, b_2, \ldots, b_c\}$. Let f be the fitness value of this agent group. Let $P = \{p_1, p_2, \ldots, p_{2^c-1}\} - \emptyset$ be the power set of B. We can then iteratively assign agents in the group to the behavioural classes of p_i, and compute the fitness $f(p_i)$ of the group. Each behavioural class in p_i is assigned $\frac{n}{|p_i|}$ agents. We can now define the heterogeneity measure as

$$H = 1 - \frac{\sum_{p_i \in P}(\sum_{j=1}^{|p_i|} |a_j|)f(p_i)}{(\sum_{p_i \in P} \sum_{j=1}^{|p_i|} |a_j|)f}. \tag{2}$$

From Eqn. 2 it follows that if the groups assigned to all the subsets of P achieve the same fitness value as that attained by the original, heterogeneous group, the heterogeneity value is zero. This indicates that agents in the different classes, as determined by the original partition, do not exhibit distinct behaviours. If the groups all behave differently than the original heterogeneous group, then the heterogeneity measure will differ from zero. This indicates that members in the different classes perform distinct behaviours. Moreover, if the fitness values $f(p_i)$ are lower[3] than the fitness value for the original partition f, then H will approach unity. This is formalized as

$$H = \begin{cases} 0 & : \quad \text{if} \quad \forall p \in P, f(p) = f \\ > 0 & : \quad \text{if} \quad \exists p' \in P, f(p') < f, \quad \text{and} \quad \forall \bar{p} \in P - p', f(\bar{p}) = f \\ 1 & : \quad \text{if} \quad \forall p \in P, f(p) = 0 \end{cases} \quad (3)$$

The advantage of this domain-independent, fitness-based heterogeneity measure is that it explicitly incorporates the concept of division of labour. When all of the agents in the group are forced to perform only a subset of the behaviours evolved for them (chosen from among the behavioural class combinations in P), and then perform poorly (indicated by a lowered fitness value for the chosen combination), this indicates that a range of behaviours have evolved for this group, all of which must be performed in order to successfully solve the collective task.

3 Results

The first task domain studied is synthetic, and was designed in order to test the Legion system on a task domain in which both homogeneous and heterogeneous groups can optimally solve the given task. This task is named the Travelling Mailman Problem, or the TMP.

Consider a city with s streets that produce $\{l_1, l_2, \ldots, l_s\}$ letters each day, which must be collected by a fleet of mailmen. Each mailman can collect one letter each day. The goal of the mailmen is to arrange themselves across the streets in the city so as to minimize the amount of uncollected mail. At the beginning of each simulation, each mailman indicates the street number which will be his mail route for the duration of the simulation. The total amount of uncollected mail at the end of the simulation is given by

$$\sum_{i=1}^{n} \sum_{j=1}^{s} \begin{cases} u_j - m_j & : \quad u_j > m_j \\ 0 & : \quad u_j \leq m_j \end{cases}, \quad (4)$$

where s is the number of streets, n is the number of iterations in the simulation, u_j is the amount of uncollected mail at street j, and m_j is the number of mailmen servicing street j.

In Table 1, the information necessary for applying the Legion system to the TMP is given.

---

[3] We here assume that a high fitness value is desirable; for tasks in which low fitness values are desirable, H is computed by flipping the numerator and denominator given in Eqn. 2.

Fitness Function	Equivalent to Eqn. 4		
Termination Criteria	500 generations are completed		
Non-terminal Nodes	Name	Arity	Description
	IF_ST_CAP	2	j = evaluated left branch
			k = evaluated right branch
			if $u_j > m_j$, move to street j
			else move forward k streets
	PLUS	2	left branch + right branch
Terminal Nodes	The two integer constants zero and unity		
Population Size	500		
Number of Generations	250		
Selection Method	Tournament selection; tournament size = 2		
Maximum Tree Depth	7		
Maximum Behavioural Classes	3		
Mutation Rate	1% chance of node undergoing random replacement		

Table 1. Legion System Parameters for the Travelling Mailman Problem The fitness function is a decreasing function; lower fitness values imply a more fit solution.

The second task studied was food foraging in simulated ant colonies [2, 6, 4]. Twenty ants operating within a 32 by 32 toroidal grid must locate food placed at two food sources, and return as much food as possible to a single nest. Ants may lay and sense pheromones, which can be used by the ant group to increase the rate of food retrieval. At each time step of the simulation, each ant performs one action, based on the state of its local environment.

The fitness function used to evaluate the performance of an ant colony is given by

$$f + r + \sum_{i=1}^{n} t_i. \tag{5}$$

In the fitness function, f stands for functionality. Given an ant colony (a_1, a_2, ..., a_n), f is set to 0 if no ant attempts any behaviour; 1 if at least one ant attempts one of the three behaviours *grab food*, *drop pheromone* or *move*; 2 if at least two ants a_i and a_j attempt one of these three behaviours, and the behaviours of a_i and a_j are distinct; and 3 if at least three ants a_i, a_j and a_k attempt one of the three behaviours, and the behaviours of a_i, a_j and a_k are distinct. The functionality term f is used to motivate initial Legion groups to evolve ant colonies with high functionality.[4]

Ants removing food from the food piles are rewarded by r, the number of food pellets removed by the colony from the food piles. The final term of the fitness

[4] In [6], a similar fitness function to that of Eqn. 5 was employed, but the functionality term f was not used. Because of this, evolved behaviours reported in [6] were produced with a population size of 64000 over 80 generations. These solutions were roughly as fit as the evolved solutions reported in this work, which were generated using a population size of 500 over 250 generations.

function rewards colonies for returning food to the nest as quickly as possible: n is the number of food pellets returned to the nest, and t_i is the number of time steps remaining in the simulation when food pellet i was returned to the nest.

In Table 2, the information necessary for applying the Legion system to the food foraging problem is given.

Fitness Function	See Eqn. 5	
Termination Criteria	250 generations completed, or all food returned to nest	
Non-terminal Nodes	IF_FD_HERE	The ant is standing on a food pellet
	IF_FD_FORW	There is food in front of the ant
	IF_CARRYING_FD	The ant is carrying a food pellet
	IF_NEST_HERE	The ant is standing on the nest
	IF_FACING_NEST	The ant is facing the nest
	IF_SMELL_FOOD	There is a food pellet next to the ant
	IF_SMELL_PHER	There is pheromone next to the ant
	IF_PHER_FORW	There is pheromone in front of the ant
Terminal Nodes	MOVE_FORW	Move one cell forward in current direction
	TURN_RT	Turn 90 degrees clockwise
	TURN_LT	Turn 90 degrees counterclockwise
	MOVE_RAND	Move two cells in a random direction
	GRAB_FD	Pick up a food pellet, if one is here
	DROP_PHER	Drop pheromone at current position
	NO_ACT	Do not perform any action
	MOVE_DROP	Move one cell forward; drop pheromone
Population Size	500	
Number of Generations	250	
Selection Method	Tournament selection; tournament size = 2	
Max Tree Depth	7	
Max Behavioural Classes	3	
Mutation Rate	1% chance of node undergoing random replacement	

Table 2. Legion System Parameters for the Food Foraging Problem The fitness function is an increasing function; higher fitness values indicate a more fit solution.

The thin lines in Fig. 2 report data generated by 30 runs of the Legion system applied to the TMP. Figs. 2 a) and b) plot the heterogeneity (as given in Eqn. 2) and the number of behavioural classes, respectively, of the fittest mailman group at the end of each generation from a typical run of the Legion system. Figs. 2 c) and d) depict the average heterogeneity and number of behavioural classes, respectively, in the Legion population as a whole, averaged over the 30 runs. The parameters for the runs are given in Table 1.

The thick lines in Fig. 2 report data generated by 30 runs of the Legion system applied to the food foraging problem. Figs. 2 a) and b) plot the heterogeneity (as given in Eqn. 2) and the number of behavioural classes of the most fit ant group at the end of each generation in a single run of the Legion system. Figs.

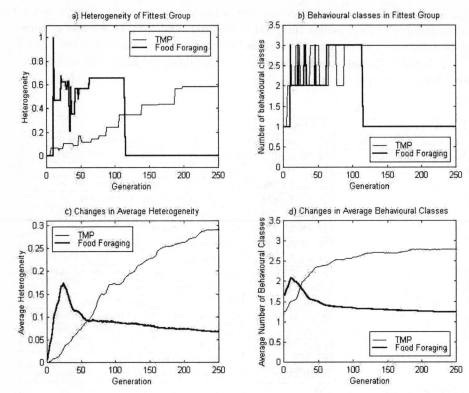

Fig. 2. Changes in heterogeneity for a set of runs of the travelling mailman and the food foraging problems: a) and b) show changes in the heterogeneity and the number of behavioural classes, respectively, of the fittest agent group at each generation during a typical run; c) shows changes in the average heterogeneity of the Legion population; d) shows changes in the average number of behavioural classes for the population. The results in c) and d) are averaged over 30 runs.

2 c) and d) depict the average heterogeneity and number of behavioural classes, respectively, in the Legion population as a whole, averaged over the 30 runs. The parameters for the runs are given in Table 2.

4 Discussion

For the mailman groups evolved for the TMP, Fig. 2 c) shows that the heterogeneity of the groups increases over evolutionary time. Fig. 2 d) shows that mailman groups rapidly approach the asymptote of the maximum possible number of behavioural classes. By comparing the slopes of Figs. 2 c) and d) it becomes clear that even after the Legion population is saturated with agent groups with the maximum number of behavioural classes, new agent groups continue to exhibit increased heterogeneity.

This result is further supported by the data from the sample TMP run shown in Figs. 2 a) and b). In this run, after generation 100, the most fit mailman group always contains three behavioural classes (see Fig. 2 b)). However, subsequent agent groups continue to increase in heterogeneity until the 200th generation (see Fig. 2 a)).

In contrast to these results, the data in Figs. 2 c) and d) show that for the food foraging problem, simulated ant colonies exhibit less heterogeneity over evolutionary time. Our investigations suggest that the initial, rapid increase and subsequent gradual decrease in heterogeneity seen in Figs. 2 c) and d) is due to the generation of a fit behaviour within a single behavioural class of a heterogeneous, ancestral colony. This fit behaviour is then assimilated by a larger fraction of ants in descendant colonies, until eventually all ants in a descendent colony use this behaviour, rendering these descendent colonies completely homogeneous. This hypothesis was supported by studying the lineages of several ant colonies during evolution (data not shown). Note also that the height of the peaks in Figs. 2 c) and d) fall short of the values obtained by corresponding mailman groups in Figs. 2 c) and d).

The tendency of foraging groups to converge on homogeneous solutions, as shown in Fig. 2, supports the findings in [3], in which a set of simulated robots foraging for different coloured pucks converge, via a learning algorithm, on identical members.

In both sets of runs, the maximum number of behavioural classes for any agent group was restricted to three. This was done to minimize computation time: computation of Eqn. 2 increases exponentially with the number of behavioural classes. However, this upper limit was sufficient to demonstrate the convergence to heterogeneous and homogeneous agent groups in the TMP and the food foraging tasks, respectively, and also that group heterogeneity can change even when the number of behavioural classes remains fixed.

These two sets of experiments demonstrate that heterogeneity is neither implicitly nor explicitly affected by the Legion system alone; rather, the amount of heterogeneity is domain-specific. From this it follows that the Legion system serves as a kind of heterogeneity 'divining rod': agent groups that perform better with either differentiated or undifferentiated members are revealed as such by the Legion system.

The Legion system can also be used to artificially exert selection pressure in favour of either homogeneous or heterogeneous groups. For groups that tend to converge on heterogeneous solutions, clamping the maximum number of behavioural classes to one ensures the evolution of only homogeneous groups (this follows from the definition of H in Eqn. 2).

Conversely, by incorporating the heterogeneity measure into fitness functions for problem domains in which agent groups tend to become more homogeneous over time, groups with both a high fitness and high heterogeneity can be generated. This technique was applied to the food foraging problem: the same procedure was used as that summarized in Table 2, but the fitness function used was $h(f + r + \sum_{i=1}^{n} t_i)$, where h is defined in Eqn. 2, and f, r and t_i are explained

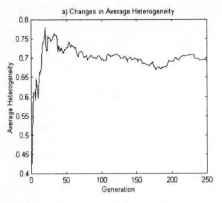
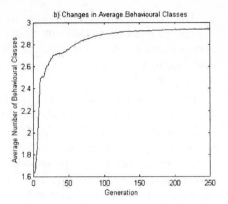

Fig. 3. Artificially evolving heterogeneity for the food foraging task: The Legion system was run for the food foraging task, using the parameters given in Table 2. The fitness function used was $h(f + r + \sum_{i=1}^{n} t_i)$, where h is defined in Eqn. 2. a) reports changes in the average heterogeneity of simulated colonies in the Legion population after each generation. b) reports changes in the number of behavioural classes.

in section 3. Fig. 3 reports data generated by a run of the food foraging task using this fitness function. Note the differences between Figs. 3 a) and 2 c), and between Figs. 3 b) and 2 d).

5 Conclusions

The results documented here support the claim that heterogeneity is a domain-specific property. Using an evolutionary algorithm applied to two task domains, selection pressure consistently evolved heterogeneous agent groups for the one task, and homogeneous groups for the other. Bloat [13] and random diffusion have been cited as two possible alternative explanations for the repeated appearance of multiple behavioural classes in the TMP, but these hypotheses are refuted by the repeated convergence to a single behavioural class in the food foraging problem (see Figs. 2 c) and d)).

The importance of heterogeneity (or the lack thereof) in agent groups is manifold. In the case of physical agents, homogeneous groups may suffer reduced robustness: a group of wheeled robots designed for smooth terrain will fail entirely in a rocky terrain; a mixed group of wheeled and legged robots may perform in both types of terrain. Conversely, morphological and behavioural redundancy may be addressed using the Legion system: for example, by automatically tuning the amount of heterogeneity in a robot group (similar to the technique used for generating the data reported in Fig. 3), one may be able to optimally tune the amount of sensor and effector overlap displayed among members of the group.

In addition to robustness and redundancy, division of labour is another concept intimately linked to heterogeneity. In some initial investigations, we have found that for agent groups with similar fitness values, heterogeneous groups tend to contain less s-expression nodes than homogeneous groups. This may suggest that agents within heterogeneous groups specialize to a specific set of sub-tasks within the main task, and thus exhibit reduced functionality in the form of smaller control architectures. It follows from this that the Legion system may be used to generate not only heterogeneous, but also specialized agent groups. We are currently pursuing this promising avenue of study.

Finally, it follows from the relationship between our heterogeneity measure and division of labour that collective tasks for which heterogeneous agent groups evolve may be *decomposable* tasks. Some tasks may be composed of a number of different subtasks; behavioural classes may then emerge and differentiate in agent groups to solve these subtasks. This was observed in the case of the TMP: behavioural classes emerged, each containing mailmen that serviced a subset of the streets in the city. Conversely, the homogeneity of evolved ant colonies may support the hypothesis that the simulated food foraging task is non-decomposable: all ants must be able to perform all basic behaviours to successfully achieve the collective task. The use of the Legion system for measuring the decomposability of collective tasks may be another interesting topic of future investigation.

In closing, we conclude that the Legion system, in conjunction with the domain-independent heterogeneity measure introduced here, is a powerful tool ideally suited for investigations of heterogeneity in agent-based systems and collective problem solving.

6   Acknowledgments

The author would kindly like to thank faculty and students of the School of Cognitive and Computing Sciences at the University of Sussex, without whom this work would not have been possible. Special thanks to Inman Harvey, for generous contributions of time, thought, criticism and encouragement.

References

1. Arkin, R. C. & J. D. Hobbs. Dimensions of Communication and Social Organization in Multi-agent Robotic Systems. In Meyer, J.-A., H. L. Roitblat & S. W. Wilson (eds.), *Procs. of the Second Intl. Conf. on Simulation of Adaptive Behavior*. MIT Press, pp. 486–493. (1992)
2. Arkin, R. C. & K. S. Ali. Integration of Reactive and Telerobotic Control in Multi-agent Robotic Systems. In Cliff, D., P. Husbands, J.-A. Meyer & S. W. Wilson (eds.), *Procs. of the Third Intl. Conf. on Simulation of Adaptive Behavior*. MIT Press, pp. 473–478. (1994)
3. Balch, T. Behavioral Diversity in Learning Robot Teams. PhD thesis, College of Computing, Georgia Institute of Technology. (1998)
4. T. Balch. Reward and Diversity in Multirobot Foraging. In *IJCAI-99 Workshop on Agents Learning About, From and With other Agents*. Sweden, July 31–August 6. (1999)
5. Balch, T. Hierarchic Social Entropy: An Information Theoretic Measure of Robot Group Diversity. *Autonomous Robots*, **8**:3, July, to appear. (2000)

6. Bennett, F. H. Automatic Creation of an Efficient Multi-Agent Architecture Using Genetic Programming with Architecture-Altering Operations. In Koza, J. R., D. E. Goldberg & D. B. Fogel (eds.), *Genetic Programming 1996 : Proceedings of the First Annual Conference*. MIT Press, pp. 30–38. (1996)

7. Bonabeau, E., A. Sobkowski, G. Theraulaz & J.-L. Deneubourg. Adaptive Task Allocation Inspired by a Model of Division of Labour in Social Insects. *Sante Fe Institute Tech. Rep. 98-01-004*. (1998)

8. Bull, L. & C. Fogarty. Evolutionary Computing in Multi-Agent Environments: Speciation and Symbiogenesis. In Voigt, H.-M., W. Ebeling & I. Rechenberg (eds.), *Parallel Problem Solving from Nature-PPSN IV*. Springer-Verlag, pp. 12–21. (1996)

9. Fahlman, S. & C. Lebiere. The Cascade-Correlation Learning Architecture. *Carnegie Mellon University Tech. Rep. CMU-CS-90-100*. (1990)

10. Fontan, M. S. & M. J. Mataric. A Study of Territoriality: The Role of Critical Mass in Adaptive Task Division. In Maes, P., M. Mataric, J.-A. Meyer, J. Pollack & S. W. Wilson (eds.), *Procs. of the Fourth Intl. Conf. on Simulation of Adaptive Behavior*. MIT Press, pp. 553–561. (1996)

11. Goldberg, D. & M. J. Mataric. Interference as a Tool for Designing and Evaluating Multi-Robot Controllers. In *AAAI-97: Procs. of the Fourteenth Natl. Conf. on Artificial Intelligence*. MIT Press, pp. 637–642. (1997)

12. Haynes, T. & S. Sen. Crossover Operators for Evolving a Team. In Koza, J. R., K. Deb, M. Dorigo, D. B. Fogel, M. Gazon, H. Iba & R. L. Riolo (eds.), *Genetic Programming 1997: Proceedings of the Second Annual Conference*. pp. 162–167, Morgan Kauffman. (1997)

13. Langdon, W. B. & R. Poli. "Fitness Causes Bloat". *Second On-Line World Conference on Soft Computing in Engineering Design and Manufacturing*. Springer-Verlag, London, pp. 13–22. (1997)

14. Luke, S. & L. Spector. Evolving Teamwork and Coordination with Genetic Programming. In Koza, J. R., D. E. Goldberg, D. B. Fogel & R. L. Riolo (eds.), *Genetic Programming 1996: Proceedings of the First Annual Conference*. MIT Press, pp. 141–149. (1996)

15. M. J. Mataric. Reinforcement Learning in the Multi-Robot Domain. In *Autonomous Robots*, 4(1):73–83. (1997)

16. M. J. Mataric. Designing and Understanding Adaptive Group Behavior. *Adaptive Behavior* 4(1):51–80. (1995)

17. McFarland, D. J. Animals as Cost-Based Robots. In Boden, M. (ed.), *The Philosophy of Artificial Life*. Oxford University Press, Oxford. (1996)

18. Ohno, S. *Evolution by Gene Duplication*. Springer-Verlag, New York. (1970)

19. Ohta, T. Multigene and Supergene Families. *Oxford Surv. Evol. Biol.*, 5:41–65. (1988)

20. Parker, L. Heterogeneous Multi-Robot Cooperation. PhD thesis, Massachussets Institute of Technology. (1994)

21. Potter, M. & K. De Jong. Evolving neural networks with collaborative species. In *Procs. of the 1995 Summer Computer Simulation Conference* Ottawa. (1995)

22. Sims, K. Evolving 3D Morphology and Behaviour by Competition. In Brooks, R. and P. Maes (eds.), *Artificial Life VI*. MIT Press, pp. 28–39. (1994)

23. Sneath, P. & R. Sokal. *Numerical Taxonomy* W. H. Freeman and Company, San Francisco. (1973)

24. Theraulaz, G., S. Goss, J. Gervet & J.-L. Deneubourg. Task Differentiation in Polistes Wasp Colonies: a Model for Self-organizing Groups of Robots. In Meyer, J. A. & S. W. Wilson (eds.), *Procs. of the First Intl. Conf. on the Simulation of Adaptive Behaviour. MIT Press, pp. 346–355. (1991)*

Metric Based Evolutionary Algorithms

Stefan Droste[1] and Dirk Wiesmann[2]

[1] FB Informatik, LS 2, Univ. Dortmund, 44221 Dortmund, Germany
droste@ls2.cs.uni-dortmund.de
[2] FB Informatik, LS 11, Univ. Dortmund, 44221 Dortmund, Germany
wiesmann@ls11.cs.uni-dortmund.de

Abstract. In this article a set of guidelines for the design of genetic operators and the representation of the phenotype space is proposed. These guidelines should help to systematize the design of problem-specific evolutionary algorithms. Hence, they should be particularly beneficial for the design of genetic programming systems.
The applicability of this concept is shown by the systematic design of a genetic programming system for finding Boolean functions. This system is the first GP-system, that reportedly found the 12 parity function.

1 Introduction

Ideally, the design of evolutionary algorithms (EA) should be based on sound theories about their working principles. Unfortunately, up to now there exist only few theoretical results, but many, even opposing, hypotheses on the working principles of EA.

Hence, the EA-designer has little more alternatives than to follow his or her favorite hypotheses, integrate some domain knowledge if available, and finally tune the algorithm through exhaustive testing. The integration of domain knowledge often improves the search process significantly, but in most cases this integration is not straightforward. In practice, this lack of theoretical knowledge makes the design process difficult and time consuming.

In this article we propose a set of guidelines for the design of genetic operators and the representation of the phenotype space. These guidelines should help to systematize the design of problem-specific EAs. Especially in genetic programming (GP) (see [6] for an introduction to GP), where the form of representation is often chosen dependent on the problem being attacked, this design-issue is of particular importance.

By defining two related distance measures within the geno- and phenotype spaces so that neighboring elements have similar fitness, problem-specific knowledge can be integrated. The guidelines for the genetic operators guarantee that the search process will then behave in a controlled fashion. As we assume that the choice of the distance measures has "smoothened" the search space, this strategy should increase the performance of the EA.

We do not claim that the set of rules defined in the following sections is sufficient for a successful EA-design. Nor do we say that this set of rules is different from traditional EA-design. Contrarily, we assume that they just reflect the common notion of EAs. But in the majority of published applications these points are rarely discussed explicitly. In this respect our proposal should be taken as a small step towards a systematic design of EAs.

In the next two sections the representation problem is formalized by giving an abstract definition of an EA and the proposed guidelines are presented. In Section 4 we show how the guidelines can be applied to design a genetic programming system for finding Boolean functions. The experiments and their results, showing a remarkable increase in efficiency of this system in comparison to other GP-systems, are presented in Section 5. The paper ends with a few comments about our choice of representation and a conclusion.

2 Representation and operators

To localize an optimal point of an objective function $f : \mathcal{P} \to W$ (W is at least partially ordered) an EA uses a population of elements of $\mathcal{P}$, which is modified by genetic operators, like mutation, recombination, and selection. The implementation of an EA requires that the abstract elements of $\mathcal{P}$ be represented by a data structure, i.e. elements of a space $\mathcal{G}$. The set $\mathcal{P}$ is called the *phenotype* space and $\mathcal{G}$ the *genotype* space. For instance, when using S-expressions in a GP-system, $\mathcal{G}$ consists of all S-expressions, while the phenotype space $\mathcal{P}$ consists of all functions $f : A \to B$ (where A and B are determined by the problem to be solved).

We assume that μ denotes the number of parents and λ the number of off-spring in one generation. In the context of EAs it is sufficient to store an individual in the form of its genotype, since the genotype-phenotype mapping $h : \mathcal{G}^\lambda \to \mathcal{P}^\lambda$ is deterministic and environmental influences are not taken into consideration. Hence, we can denote a population at generation t by $Pop(t) \in \mathcal{G}^\mu$.

The mapping $h : \mathcal{G}^\lambda \to \mathcal{P}^\lambda$ can be composed simply from λ reduced mappings $h' : \mathcal{G} \to \mathcal{P}$, which represent the geno-phenotype mapping for single individuals: $h(g) = (h'(g_1), \ldots, h'(g_\lambda))$, $g = (g_1, \ldots, g_\lambda) \in \mathcal{G}^\lambda$. The mapping $h' : \mathcal{G} \to \mathcal{P}$ determines the abstract element $h'(x)$ of the search space being represented by $x \in \mathcal{G}$. Thus, the mapping h' describes the relation between genotype and phenotype space.

The recombination operator $r : \mathcal{G}^\mu \times \Omega_r \to \mathcal{G}^\lambda$ generates λ offspring from the parent population by mixing the parental genetic information. The probabilistic influence during recombination is described by the probability space (Ω_r, P_r), i.e. the outcome of the recombination depends on the random choice of $\omega_r \in \Omega_r$ according to P_r.

The mutation $m : \mathcal{G}^\lambda \times \Omega_m \to \mathcal{G}^\lambda$ is working on the genotype space $\mathcal{G}$ only. Here (Ω_m, P_m) is the underlying probability space.

The new population $Pop(t+1) \in \mathcal{G}^\mu$ is selected from the set of offspring of $Pop(t)$, where the selection of an individual is based directly or indirectly on the objective function $f : \mathcal{P} \to W$. The objective function assesses only the phenotype.

This relationship is formalized by the selection operator $s : \mathcal{G}^\lambda \times \mathcal{P}^\lambda \times \Omega_s \to \mathcal{G}^\mu$ (with probability space (Ω_s, P_s)). With the auxiliary function $h^* : \mathcal{G}^\lambda \to \mathcal{G}^\lambda \times \mathcal{P}^\lambda, h^*(g) := (g, h(g))$ for $g \in \mathcal{G}^\lambda$ the equation

$$Pop(t+1) = s(h^*(m(r(Pop(t), \omega_r), \omega_m)), \omega_s)$$

holds, where $\omega_r \in \Omega_r$, $\omega_m \in \Omega_m$, and $\omega_s \in \Omega_s$ are chosen randomly according to P_r, P_m, and P_s.

3 Design Guidelines for Recombination and Mutation Operators: A Proposal

We assume that the objective function $f : \mathcal{P} \to \mathbb{R}$ has to be optimized and that the space $\mathcal{P}$ is defined in a way, such that f can be evaluated efficiently. For an EA as an optimization algorithm the following questions among others are of importance:

- Can each feasible solution be reached from an arbitrary initial population?
- Are the operators (mutation, recombination) generating an additional bias, which dominates the selection process?
- Is a gradual approximation of the optimum possible?

In the following we will present some design guidelines for genetic operators and the mapping h, as well. By presenting empirical results in Section 5 we show that these guidelines can be helpful in GP, although a theoretical proof is not given. The aim of our guidelines is to support a smooth motion of the search process through the fitness landscape, a property whose importance for EAs is for instance discussed in [8] and [10].

3.1 Metric

In the following we assume that we have certain information about the phenotype space concerning a specific problem domain. This domain-knowledge will be used to form a distance measure on the phenotype space. This distance measure shows the similarity between two phenotypes regarding a certain mark of quality, i.e. two elements of the phenotype space with small distance should have similar function values. Hence, the distance measure should be strongly connected to the objective function to be optimized. The available domain-knowledge should be used as well as possible by formulating the guidelines for the operators on the basis of the distance measure. The performance of the EAs then mainly depends on the suitability of the distance measure.

In order to fulfill the most basic properties, which are reasonable for distance measures, we only allow *metrics* $d_{\mathcal{P}} : \mathcal{P} \times \mathcal{P} \to \mathbb{N}_0^+$ to measure the distance between two points of the phenotype space (we restrict ourselves here to metrics having only values in $\mathbb{N}_0^+$).

We do not formalize the relationship between the metric $d_\mathcal{P}$ and the objective function f, but simply claim that there should be a strong correlation, so that elements of $\mathcal{P}$ with small distance should have similar function values. In section 4 we present an example, where such a strong relationship is given.

In the following both the genotype space and the phenotype space are assumed to be finite spaces.

3.2 Mapping

Assuming that the mapping h is injective, the metric $d_\mathcal{P}$ automatically induces a metric $d_\mathcal{G} : \mathcal{G} \times \mathcal{G} \to \mathbb{N}$ on the genotype space using the following definition:

$$\forall g_1, g_2 \in \mathcal{G} : d_\mathcal{G}(g_1, g_2) := d_\mathcal{P}(h(g_1), h(g_2)). \tag{1}$$

If a genetic operator causes a small change regarding $d_\mathcal{G}$, this will result in a small change in the phenotype space regarding $d_\mathcal{P}$ by making use of definition (1). The inverse statement holds, if h is bijective. If h is not injective (thus (1) cannot be used), the mapping should at least obey the following rule:

Guideline H 1 *For the mapping h and the metrics $d_\mathcal{G}$ and $d_\mathcal{P}$ the following relation should hold:*

$$\forall u, v, w \in \mathcal{G} :$$
$$(d_\mathcal{G}(u, v) \leq d_\mathcal{G}(u, w)) \Leftrightarrow (d_\mathcal{P}(h(u), h(v)) \leq d_\mathcal{P}(h(u), h(w))) .$$

This guarantees, that the distance relations between three points translate from the genotype to the phenotype space and back, although the exact distances do not have to be the same. If h is bijective, guideline H 1 is fulfilled by metrics generated through equation (1).

3.3 Recombination

Without loss of generality we look at the reduced recombination operator $r' : \mathcal{G}^2 \times \Omega_{r'} \to \mathcal{G}$ with a finite probability space $(\Omega_{r'}, P_{r'})$ instead of the general recombination operator r. Let $P_{r'}(r'(u, v) = w) := P_{r'}(\{\omega \in \Omega_{r'} \mid r'(u, v, \omega) = w\})$ be the probability, that two parents $u, v \in \mathcal{G}$ are recombined to an offspring $w \in \mathcal{G}$.

The first guideline on the recombination operator states that the genotype distance $d_\mathcal{G}$ between any of the two parents and the offspring should be smaller than the distance between the parents themselves. Hence, children will be similar to their parents, which should be beneficial in a smoothened fitness landscape:

Guideline R 1 *The recombination r' should fulfill:*

$$\forall u, v \in \mathcal{G}, \omega \in \Omega_{r'} \text{ with } z = r'(u, v, \omega) :$$
$$\max(d_\mathcal{G}(u, z), d_\mathcal{G}(v, z)) \leq d_\mathcal{G}(u, v)$$

Moreover, the recombination operator should not generate an additional bias independent of the parents, i. e. the offspring should not tend to one of its parents to avoid any bias in the search process:

Guideline R 2 *The recombination r' should fulfill:*

$$\forall u, v \in \mathcal{G}, \forall \alpha \geq 0:$$
$$P_{r'}(d_{\mathcal{G}}(u, r'(u, v)) = \alpha) = P_{r'}(d_{\mathcal{G}}(v, r'(u, v)) = \alpha)$$

In account of domain-knowledge, experiences or implementation details (see the example of use in the Section 4) the guidelines R 1 and R 2 can be extended or specialized.

3.4 Mutation

Without loss of generality we restrict our discussion to the reduced mutation operator $m' : \mathcal{G} \times \Omega_{m'} \to \mathcal{G}$ with the finite probability space $(\Omega_{m'}, P_{m'})$. With probability $P_{m'}(m'(u) = v) := P_{m'}(\{\omega \in \Omega_{m'} \mid m'(u, \omega) = v\})$ the mutation operator m' changes an element $u \in \mathcal{G}$ to a new element $v \in \mathcal{G}$.

The first rule claims that from each point $u \in \mathcal{G}$ any other point $v \in \mathcal{G}$ can be reached in one mutation step:

Guideline M 1 *The mutation m' should fulfill:*

$$\forall u, v \in \mathcal{G}: P_{m'}(m'(u) = v) > 0.$$

This guideline can be weakened by replacing the mutation with a finite sequence of mutations [9], although in this case it must be possible that elements with lower fitness can be preferred over elements with higher fitness under selection.

Moreover small mutations (with respect to $d_{\mathcal{G}}$) should occur more often than large mutations, such that the search process prefers small search steps. We assume this to be beneficial in a smoothened fitness landscape, as here good points will be at least partially surrounded by other good points:

Guideline M 2 *The mutation m' should fulfill:*

$$\forall u, v, w \in \mathcal{G}:$$
$$(d_{\mathcal{G}}(u, v) < d_{\mathcal{G}}(u, w)) \Rightarrow (P_{m'}(m'(u) = v) > P_{m'}(m'(u) = w))$$

For a given $u \in \mathcal{G}$ all genotypes $v \in \mathcal{G}$ which have the same distance to u should have the same probability to be produced by mutation of u:

Guideline M 3 *The mutation m' should fulfill:*

$$\forall u, v, w \in \mathcal{G}:$$
$$(d_{\mathcal{G}}(u, v) = d_{\mathcal{G}}(u, w)) \Rightarrow (P_{m'}(m'(u) = v) = P_{m'}(m'(u) = w)).$$

This guideline guarantees that the mutation operator does not induce a bias by itself.

3.5   Metric Based Evolutionary Algorithm (MBEA)

Since all guidelines are based on the definition of a suitable metric on the phenotype space, an EA which fulfills the guidelines H 1, R 1, R 2, M 1, M 2, and M 3 for a given metric and mapping is called *Metric Based Evolutionary Algorithm (MBEA)*.

4   Metric Based Genetic Programming

Taking into account all the guidelines that define an MBEA, it is not obvious that and how an MBEA can be built. Therefore, we will give an example of a *Metric Based Genetic Programming (MBGP)*-system in this section to show how crossover and mutation operators that conform with our guidelines can look like.

In GP we have the typical situation that in most existing systems the effect of crossover and mutation in the phenotype space is poorly understood. Therefore, most GP-systems do not fulfill the rules R 1 and H 1 as defined in Section 3. So, if our guidelines make sense, an MBGP-system should outperform other GP-systems for the chosen problem domain; otherwise, our approach would be challenged.

The task of our MBGP-system is to solve the following common benchmark-problem: find a representation of an unknown function, where it is only possible to compute the number of inputs where a given function and the unknown one have the same output (see for instance [2, 3, 5–7]).

In our case, the unknown function g is a Boolean function with n inputs and one output, i. e. $g : \{0,1\}^n \to \{0,1\}$. The training set T simply is the set $\{(x, g(x)) \mid x \in \{0,1\}^n\}$. Every element $f : \{0,1\}^n \to \{0,1\}$ of the phenotype space will be assigned the fitness $F(f) := |\{x \in \{0,1\}^n \mid f(x) = g(x)\}|$, i. e. the number of inputs, where f produces the same output as the unknown function g. Hence, the unknown function g is the only maximum of the fitness function F. The task of the MBGP-system is to find a representation of this maximum.

In our MBGP-system we will make use of a special data structure for Boolean functions, called *Ordered Binary Decision Diagrams (OBDD)* (see [1]) instead of S-expressions. OBDDs allow polynomially-sized representation in n of many practically important Boolean functions and many operations on OBDDs can be done in polynomially bounded time with regard to the size of the involved OBDDs. Some of these efficient operations are used in the implementation of our mutation and recombination operators. Before the MBGP-system is described, the next subsection gives a short introduction to OBDDs.

4.1   Ordered Binary Decision Diagrams

Given n Boolean input variables $x_1, \ldots, x_n$, let π be an ordering of these variables, i. e. a bijective function $\pi : \{x_1, \ldots, x_n\} \to \{1, \ldots, n\}$. An OBDD for π is an acyclic directed graph with one source, where every node is labeled either by

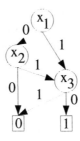

Fig. 1. OBDD representing the function $f(x_1, x_2, x_3) = x_1\overline{x}_3 \vee \overline{x}_1 x_2 \overline{x}_3$, where $\pi = (x_1, x_2, x_3)$.

one of the variables or one of the Boolean constants 0 or 1 (in the last two cases the node is called a *sink*).

Every sink node has no outgoing edges, while every non-sink node is labeled by a Boolean variable x_i, and has exactly two outgoing edges, one labeled by 0 and the other by 1 (whose endpoints are called *0-successor* and *1-successor*, resp.). Furthermore, if there is an edge from a node labeled x_i to a node labeled x_j, then $\pi(x_i)$ must be smaller than $\pi(x_j)$ (in the following it is always assumed that $\pi(x_i) = i$ for all $i \in \{1, \ldots n\}$).

To evaluate the function f_O represented by a given OBDD O for an input $(a_1, \ldots, a_n) \in \{0, 1\}^n$, one starts at the source. At a node labeled x_i one chooses the a_i-successor. The value $f_O(a)$ equals the label of the finally reached sink. The OBDD O in Figure 1 represents the Boolean function $f(x_1, x_2, x_3) = x_1\overline{x}_3 \vee \overline{x}_1 x_2 \overline{x}_3$.

OBDDs allow polynomially-sized representation in n of many practically important Boolean functions, and many operations on OBDDs can be done in polynomially bounded time with regard to the size of the involved OBDDs. Some of these efficient operations (e. g. synthesis) are used in the implementation of our mutation and recombination operators. Another big advantage of OBDDs is the existence of *reduced* OBDDs: for a given function f and variable ordering π the reduced OBDD representing f is the unique OBDD from the set of all OBDDs representing f having the minimal number of nodes (see [1]). A given OBDD D_1 can be reduced in time $O(|D_1|)$.

4.2 The MBGP-system

In the following we present a GP-system that fulfills the guidelines, thereby showing that they can be fulfilled. We start with the definition of a metric on the phenotype space, which is expected to lead to a smooth fitness landscape. Thus, the metric reflects our domain-knowledge. Then a rough outline of the GP-system is given. Due to space limitations we cannot present the operators in detail and we have to omit the rather easy, but lengthy proofs.

Metric We measure the fitness $F(f)$ of an individual $f : \{0,1\}^n \to \{0,1\}$ by counting the number of inputs, where f and the unknown function g agree. Hence, an appropriate metric $d_{\mathcal{P}}$ counts the number of inputs, where the two functions disagree:

$$d_{\mathcal{P}}(f, h) := |\{x \in \{0,1\}^n \mid f(x) \neq h(x)\}| \,.$$

$d_{\mathcal{P}}$ defines a metric, which can be easily seen, when realizing that it is equivalent to the Hamming distance on the space of all bitstrings of length 2^n. Because two functions f and h can differ in their fitness $F(f)$ and $F(h)$ by $d_{\mathcal{P}}(f, h)$ at most, the fitness landscape defined by $d_{\mathcal{P}}$ and F is smooth, i. e. neighboring functions have similar fitness. Hence, our metric $d_{\mathcal{P}}$ fulfills our informal guideline of smoothing the fitness landscape.

On the genotype space, the space of concrete representations, i. e. reduced OBDDs, we use the analogous function $d_{\mathcal{G}}$ defined by

$$d_{\mathcal{G}}(u, v) := d_{\mathcal{P}}(h(u), h(v)),$$

where $h : \mathcal{G} \to \mathcal{P}$ is the mapping between the genotype and phenotype space. As we use reduced OBDDs to represent our individuals, the mapping h is bijective. Hence, $d_{\mathcal{G}}$ is a metric, too.

A rough outline of the MBGP-system In the following, we assume that $MAXGEN \in \mathbb{N}$ (the maximum number of generations), $\mu \in \mathbb{N}$ (the size of a parental generation) and $\lambda \in \mathbb{N}$ (the number of children) are predefined by the user. Then, our MBGP-system looks as follows:

Algorithm A 1 The MBGP-system.

1. *Set $t := 0$.*
2. *Initialize P_0 by choosing μ Boolean functions with equal probability and including the reduced OBDDs representing these functions.*
3. *While $t \leq MAXGEN$:*

   (a) *For $i \in \{1, \ldots, \lambda\}$ do:*

      i. *Choose two reduced OBDDs D_1 and D_2 with equal probability from P_t.*
      ii. *Recombine D_1 and D_2 and call the resulting reduced OBDD D'.*
      iii. *Mutate D' and call the resulting reduced OBDD D^*.*
      iv. *Include D^* in P'_{t+1}.*

   (b) *Choose the μ OBDDs in P'_{t+1} with the highest fitness to form P_{t+1}.*

   (c) *Set $t := t + 1$.*

4. *Output the OBDD with the highest fitness that was found.*

Mutation Mutation of an OBDD D simply consists of changing each output bit of the 2^n possible inputs independently with probability $1/2^n$. Hence, the expected number of changed bits by a mutation is one. Because the probability of flipping a bit is less than $1/2$ (assuming $n \geq 2$), smaller mutations are more likely than larger ones (guideline M 2). Nevertheless, a mutation can lead to every bit string in $\{0,1\}^n$ regardless of the starting point (guideline M 1) and the probability of mutating from u to v depends on the Hamming distance between u and v only (guideline M 3).

Theorem 1. *The mutation operator fulfills the guidelines M 1, M 2, and M 3.*

The mutation operator does not directly refer to the OBDD structure, but OBDDs make it possible to implement it efficiently by doing a EXOR-synthesis of the parental OBDD and the reduced OBDD representing the flipping bits.

Recombination The recombination operator $r' : \mathcal{G}' \times \mathcal{G}' \times \Omega_{r'} \to \mathcal{G}$ creates a new reduced OBDD out of two parental OBDDs under random influence (represented by $\Omega_{r'}$).

The source of the resulting OBDD, when recombining two reduced OBDDs D_1 and D_2 results from calling a recursive algorithm, that gets the sources v_1 and v_2 of D_1 and D_2, resp., as arguments. Starting from the sources, the two OBDDs are traversed in parallel. If one node has a smaller label than the other node, the zero- and one-successor of this "smaller" node are visited before the successors of the other node. If both nodes have the same label, the successors of both nodes are visited. If sinks are reached in both OBDDs, one of the sinks is returned with equal probability.

The result of a recursive call for a pair (v_1, v_2) of nodes is not stored in order to guarantee that the random decisions are independent of each other. As a consequence, the runtime of this recombination operator cannot be upper bounded polynomially in the OBDD sizes. But all experiments indicate that the average running time is rather small, which is caused by the increasing similarity of the OBDDs in the course of evolution.

Without stating the rather lengthy and technical proof, we have:

Theorem 2. *The recombination fulfills the guidelines R 1 and R 2.*

Hence, we can conclude for the whole GP-system:

Corollary 1. *The GP-system described is an MBEA.*

5   Experimental results

We proposed guidelines for the design of problem-specific EAs. Thus, we have to check that the guidelines are making sense by examining the quality gain we obtain when using these guidelines. As an example we designed a MBGP-system for finding Boolean functions, which is presented in Section 4.

In the following we compare its results with GP- and EP-systems, which are not tailored for finding Boolean functions (except of the function set {AND,OR, NOR}). Furthermore, these systems are not MBEAs. Of course, if our MBGP-system does not perform much better than the general EP- and GP-systems, this would challenge the MBEA approach. We summarize the outcome of some empirical studies, which confirm the usefulness of the proposed guidelines at least for the problems presented here. We compare it to other systems trying to find representations of an unknown Boolean function, when its complete training set is given. Although this problem has no practical relevance, it is an often used benchmark problem. We take the *computational effort* (see [6] for a definition) to compare our results to those presented in [2, 3, 6, 7].

Furthermore, we were interested in the influence of using OBDDs on our MBGP-system. By using reduced OBDDs the size of the genotype space is reduced to $|\mathcal{G}| = |\mathcal{P}|$. Therefore, a GP-system that uses OBDDs as representation but makes use of traditional recombination and mutation operators may perform as well as the MBGP without the extra effort in designing the MBGP search operators. Hence, we also made experiments with a GP-system that uses OBDDs as representation but does not fulfill the MBEA-guidelines.

This OBDD-GP-system has mutation and recombination operators that work similarly to those of standard GP-systems using S-expressions as described in [6, 7]. Hence, the mutation operator replaces the subOBDD, starting at a randomly chosen node in the OBDD to be mutated, by a new random subOBDD. The recombination operator exchanges two randomly chosen subOBDDs in both parent OBDDs. Special measures are taken to guarantee that the resulting structures are still valid OBDDs that obey the same variable ordering as the parent OB-DDs. For a more detailed description of mutation and recombination see [4]. The rest of the OBDD-GP-system follows algorithm A 1, i.e. the OBDD-GP- and the MBGP-system differ only in their mutation and recombination operators.

We tested the MBGP- and the OBDD-GP-system on the 6- and 11-multiple-xer, the 3- to 12-parity functions, and on randomly chosen Boolean functions $rand_8$ with 8 inputs. The randomly chosen Boolean functions were chosen independently with equal probability for each of the 50 runs from the set of all Boolean functions with 8 inputs. By using this approach, we tested how well the MBGP-system can find representations of functions which have no structure at all. As the MBGP-system does not favor any Boolean function in particular, the computational effort to evolve them should be the same.

We chose $\mu = 15$ and $\lambda = 100$ for the MBGP-system. Because the OBDD-GP-system seemed to be more susceptible to premature stagnation in preliminary experiments, we used $\mu = 100$ and $\lambda = 700$ over a maximum number of 7000 generations for its runs.

In general one has to be careful when comparing systems that search for representations in different genotype-spaces (in our case, in the space of all reduced OBDDs or in the space of all S-expressions over a special terminal and function set), as the search process can have different complexities in different search spaces. But it is feasible to compare the MBGP-system to systems using

Function	GP	GP-ADFs	EP	EP-ADFs	OBDD-GP	MBGP
mux_6	160,000	-	93,000	-	9,200	1,615
mux_{11}	-	-	-	-	-	60,115
par_3	96,000	64,000	28,500	63,000	15,600	315
par_4	384,000	176,000	181,500	118,500	22,600	615
par_5	6,528,000	464,000	2,100,000	126,000	19,700	1,015
par_6	-	1,344,000	-	121,000	30,900	1,715
par_7	-	1,440,000	-	169,000	56,100	3,415
par_8	-	n.e.d.	-	321,000	183,600	7,415
par_9	-	n.e.d.	-	586,500	184,200	14,615
par_{10}	-	n.e.d.	-	-	315,100	31,415
par_{11}	-	n.e.d.	-	-	1,578,000	73,615
par_{12}	-	-	-	-	8,358,400	144,715
$rand_8$	-	-	-	-	-	7.615

Table 1. Comparison of the computational effort (with 99% probability of finding a solution)

S-expressions, since an OBDD D can be transformed in time $O(|D|)$ to an S-expression over the function set $F = \{AND, OR, NOT\}$, which is common for most systems.

Table 1 shows the results. Because in [7] only four runs were done for the 8-, 9-, 10-, and 11-parity function, the computational effort was not computed for these functions. In Table 1 this fact is noted as "n.e.d." (not enough data). A "-" in the columns for GP ([6,7]), GP with ADFs ([7]), EP ([2]) and EP with ADFs ([3]) indicates that no experiments were reported for this function type.

A "-" in the column for the OBDD-GP-system indicates that, although 50 runs with $\lambda = 100$ and $\mu = 700$ were done, no element with optimal fitness was found.

The comparison shows that the MBGP-system works well on the functions being tested, using only a small fraction of the number of individuals the GP-systems described in [6,7] or the EP-systems in [2,3] need. Furthermore, we see that even the 8-Random "function" (as the unknown function was chosen randomly for each of the 50 runs, this is not one single function) was found with a small computational effort. This indicates, that the MBGP-system works well on all Boolean functions, even on those without any "structure".

Although the GP-system using OBDDs, but without obeying the MBEA-guidelines, performs better than the GP-systems using S-expressions on these test functions with respect to the computational effort, it is worse than the MBGP-system. Furthermore, the robustness of the MBGP-system seems to be higher, because in contrast to the OBDD-GP-system it found the optimal function in every run, although the OBDD-GP-system was allowed to run over more generations and with a larger population. Hence, we can conclude that for the test functions presented the MBEA-concept seems to be useful.

5.1   The interplay of recombination and mutation

A GP-system based purely on a recombination operator fulfilling guideline R 1 is supposed to let the whole population converge to one single individual, as a child can only lie "between" the parents. Hence, over many generations the maximum distance between two individuals of the population will decrease. This makes sense, as long as we assume the fitness landscape to be smooth.

But a GP-system using only a recombination operator fulfilling R 1 and R 2 without any mutation can only lead to one of the best individuals that lie "between" the individuals of the initial generation. But in a search space large enough it is very unlikely that the optimum is located in the subspace covered in this way by the initial population. Thus, a recombination operator, that fulfills guidelines R 1 and R 2, alone will lead to stagnation in the search process. In other words, mutation is a prerequisite for reliable convergence to the optimum in MBEAs.

To empirically investigate these theories we modified the MBGP-system to two different versions:

- MBGP-system without mutation by eliminating step 3.(a).iii in Algorithm A 1 and including D' in P'_{t+1} in step 3.(a).iv.
- MBGP-system without recombination by eliminating step 3.(a).ii in Algorithm A 1 and mutating D_1 to D^* in step 3.(a).iii.

Both versions and the original MBGP-system were run 50-times for the 8-parity function. The average fitness of the best individual over the 125 generations is shown in Figure 2.

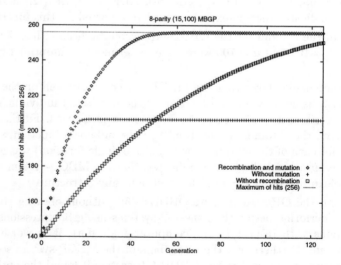

Fig. 2. Results for the 8-parity function over 50 runs with (1) mutation and recombination, (2) without mutation, and (3) without recombination.

The results confirmed that in our case recombination alone leads to stagnation, although its initial convergence speed is high. One can also observe that mutation alone is powerful enough to ensure convergence (the average number of generations to find the optimum was 154.44 with a maximum value of 186), but the process is much slower than that by recombination alone. On the other hand, by applying both operators in conjunction, the optimum was located after 53.27 generations on average (73 generations in the worst run). Thus, the MBGP-system with recombination and mutation shows both convergence velocity and reliability, when applied to the 8-parity problem.

6 MBEA and representation

In the preceding section we have described how a MBGP-system with OBDDs can be designed for finding Boolean functions. We feel appropriate to give some final comments on this choice of representation. In fact, one can use many other data structures than OBDDs to implement a MBGP-system. With a little effort one can construct mutation and recombination operators which use S-expressions and have the same semantics as the described mutation and recombination operators of the MBGP-system. Our MBGP-system with OBDDs and this new MBGP-system with S-expressions would have exactly the same computational effort.

But we know of no implementation of our genetic operators that work efficiently with S-expressions, i. e. in polynomial time with respect to the genotype. For instance, recombination by exchanging subtrees is simple to implement and fast to execute, but its consequences for the represented function are unclear. Hence, recombination implemented in this way would not meet the MBEA guidelines. However, this does not imply that MBGP-systems using S-expressions cannot be build for other problem domains in general. The following two points sum up the connection between operators and representation in MBEAs:

- The MBEA design guidelines affect the efficiency of the search process, i. e. its computational effort.
- The representation affects the time-efficiency of the operators.

Which representation is the most efficient one, depends on the problem at hand. For Boolean functions OBDDs have many advantages, which is also recognized in the CAD-community, where they are the state-of-the-art representation. Thus, the MBEA approach gives us a clear separation of representation and genetic operators. The representation of the phenotype space has to support an efficient calculation of the objective function, whereas the representation of the genotype space has to support an efficient implementation of the metric based genetic operators. Of course, both have to support an efficient implementation of a bijective mapping function h. Furthermore, mutation and recombination are per definition two distinct search operators, with different search characteristics (see section 5.1). Neither of them can be simulated by the other.

7 Conclusion

We proposed a set of guidelines for the design of EA and showed by experiments that these design guidelines can decrease the computational effort to solve a problem, assuming that knowledge about the problem domain exists. Furthermore, the choice of a proper representation (with respect to evaluation and genetic operators) is necessary in order to efficiently implement the EA. An algorithm can only be as good as its underlying data structure, i. e. its form of representation. The "algorithmic design approach" presented here provides a clear separation between operators and representation. The MBEA design guidelines affect only the efficiency of the search process, i. e., its computational effort, while the representation affects only the time-efficiency of the operators. Furthermore, experiments have shown that recombination and mutation have different search characteristics in the MBGP-system for the problems addressed.

We are aware that the proposed guidelines are not the only (and sometimes not the best) choice of incorporating domain-knowledge into the search algorithm. Additionally, it can be difficult to find a metric and genetic operators fulfilling the guidelines for a given problem. But with respect to the benchmark-problems presented here the MBEA concept proved as a suitable concept for the usage of domain knowledge in EAs. One has to be aware, that incorporating knowledge about the problem domain into the algorithm is necessary to have a chance of solving the problem with above-average quality.

Of course the MBEA concept has to be tested on other problem domains, where it is more difficult to find a representation, which allows a metric smoothing of the fitness landscape and to construct genetic operators following the guidelines. Additionally, one has to be aware that finding such a metric costs time and does not guarantee success. Hence, one has to compare the effort to find a better algorithm with the resulting performance gain. In order to circumvent this risk, a theoretical investigation of the properties of an MBEA could lead to more well-based guidelines, probably guaranteeing some properties like convergence speed or reliability.

Acknowledgements

Hereby we thank Thomas Bäck, Ulrich Hammel, Thomas Jansen, Günter Rudolph, Hans-Paul Schwefel, and Ingo Wegener for their help while preparing this paper. This research was supported by the Deutsche Forschungsgemeinschaft as part of the collaborative research center "Computational Intelligence" (531).

References

1. R.E. Bryant. Graph-based algorithms for Boolean function manipulation. *IEEE Transactions on Computers*, 35:677–691, 1986.
2. K. Chellapilla. Evolving computer programs without subtree crossover. *IEEE Transactions on Evolutionary Computation*, 1(3):209–216, 1997.

3. K. Chellapilla. A preliminary investigation into evolving modular programs without subtree crossover. In John R. Koza, W. Banzhaf, K. Chellapilla, K. Deb, M. Dorigo, D.B. Fogel, M.H. Garzon, D.E. Goldberg, H. Iba, and R.L. Riolo, editors, *Proc. Third Annual Genetic Programming Conference*, pages 23–31, Madison WS, July 22–25 1998. Morgan Kaufmann.

4. S. Droste. Efficient genetic programming for finding good generalizing Boolean functions. In J.R. Koza, K. Deb, M. Dorigo, D.B. Fogel, M.H. Garzon, H. Iba, and R.L. Riolo, editors, *Genetic Programming 1997: Proceedings of the Second Annual Conference*, pages 82–87, Stanford University, July 13–16 1997. Morgan Kaufmann.

5. C. Gathercole and P. Ross. Tackling the Boolean even n parity problem with genetic programming and limited-error fitness. In J.R. Koza, K. Deb, M. Dorigo, D.B. Fogel, M.H. Garzon, H. Iba, and R.L. Riolo, editors, *Genetic Programming 1997: Proceedings of the Second Annual Conference*, pages 119–127, Stanford University, July 13–16 1997. Morgan Kaufmann.

6. J.R. Koza. *Genetic Programming: On the Programming of Computers by Means of Natural Selection*. MIT Press, 1992.

7. J.R. Koza. *Genetic Programming II*. MIT Press, 1994.

8. I. Rechenberg. *Evolutionsstrategie '94*. Frommann-Holzboog, Stuttgart, 1994.

9. G. Rudolph. Finite Markov chain results in evolutionary computation: A tour d'horizon. *Fundamenta Informaticae*, 35(1-4):67–89, 1998.

10. H.-P. Schwefel. *Evolution and Optimum Seeking*. Sixth-Generation Computer Technology Series. Wiley, New York, 1995.

Register Based Genetic Programming on FPGA Computing Platforms

Heywood M.I.[1] Zincir-Heywood A.N.[2]

{[1]Dokuz Eylül University, [2]Ege University} Dept. Computer Engineering, Bornova, 35100
Izmir. Turkey
mheywood@cs.deu.edu.tr

Abstract. The use of FPGA based custom computing platforms is proposed for
implementing linearly structured Genetic Programs. Such a context enables
consideration of micro architectural and instruction design issues not normally
possible when using classical Von Neumann machines. More importantly, the
desirability of minimising memory management overheads results in the
imposition of additional constraints to the crossover operator. Specifically,
individuals are described in terms of the number of pages and page length,
where the page length is common across individuals of the population. Pairwise
crossover therefore results in the swapping of equal length pages, hence
minimising memory overheads. Simulation of the approach demonstrates that
the method warrants further study.

1 Introduction

Register based machines represent a very efficient platform for implementing Genetic
Programs (GP) which are organised as a linear structure. That is to say the GP does
not manipulate a tree but a register machine program. By the term 'register machine'
it is implied that the operation of the host CPU is expressed in terms of operations on
sets of registers, where some registers are associated with specialist hardware
elements (e.g. as in the 'Accumulator' register and the Algorithmic Logic Unit). The
principle motivation for such an approach is to both significantly speed up the
operation of the GP itself through direct hardware implementation and to minimise
the source code footprint of the GP kernel. This in turn may lead to the use of GPs in
applications such as embedded controllers, portable hand held devices and
autonomous robots. Several authors have assessed various aspects of such an
approach, early examples being [1], [2] and [3]. The work of Nordin *et al.*, however,
represents by far the most extensive work in the field with applications demonstrated
in pattern recognition, robotics and speech recognition to name but three [4], [5].
Common to Nordin's work however is the use of standard Von Neumann CPUs as the
target computing platform; an approach which has resulted in both RISC [4] and
CISC [6] versions of their AIM-GP system. In all cases a 3-address 32-bit format is
used for the register-machine, where this decision is dictated by the architecture of the
host CPU. In the case of this work, the target host computational platform is that of an
FPGA-based custom computing machine. Such a choice means that we are free to
make micro-architecture decisions such as the addressing format, instructions and

word-lengths, the degree of parallelism and support for special purpose hardware (e.g. fast multipliers). Moreover, the implementation of bit-wise operations, typical to GP crossover and mutation operators, is very efficiently supported on FPGA architectures. However, it should be emphasised that the work proposed here is distinct from the concept of Evolvable Hardware for which FPGA based custom computing platforms have received a lot of interest [7]. In particular the concept of Evolvable Hardware implies that the hardware itself begins in some initial (random) state and then physically evolves to produce the solution through direct manipulation of hardware. In the past the technique has been limited by the analogue nature of the solutions found [8]. Recent advances facilitate evolution of digital circuits with repeatable performance [9]. In contrast, the purpose of this work is to provide succinct computing cores of a register machine nature, thus the FPGA is used to produce a GP computing machine from the outset as oppose to mapping the requirements of GP to a general-purpose machine.

In the following text, section 2 defines the concept of register machines and introduces the instruction formats used later for 0-, 1-, 2- and 3-addressing modes. Limitations and constraints on the register machine, as imposed by an FPGA custom computing platform, are discussed in section 3. This sets the scene for the redefinition of the crossover operator and the introduction of a second mutation operator. Section 4 summarises performance of each approach on a benchmark problem. Finally the results are discussed and future directions indicated in section 5.

2 Address Register Instruction formats

As indicated above a linear GP structure is employed, hence individuals take the form of register-level transfer language instructions. The functional set is now in the form of opcodes, and the terminal set becomes the operands (e.g. permitted inputs, outputs and internal registers or general address space). Before detailing the specific format used here to implement register machines, it is first useful to summarise what a register machine is. As indicated in the introduction the modern CPU, as defined by Von Neumann, is a computing machine composed from a set of registers, where the set of registers is a function of the operations, hence application of the CPU [10, 11]. The basic mode of operation therefore takes the form of (1) fetching and (2) decoding of instructions (where this involves special purpose resisters such as the program counter, instruction and address registers) and then (3) manipulating the contents of various registers associated with implementing specific instructions. The simplest Von Neumann machines therefore have a very limited set of registers in which most operations are performed using, say, a single register. This means, for example, that the ability to store sub-results is a function of execution sequence, as in the case of a stack-based evaluation of arithmetic and logical operators. If the instruction sequence is so much as a single instruction wrong, then the entire calculation is likely to be corrupted. In this case additional registers are provided to enable the popping/ pushing of data from/ to the stack before the program completes (at which point the top of the stack is (repeatedly) popped to produce the overall answer(s)). However, as the number of internal registers capable of performing a calculation increases, then our ability to store sub-results increases. Within the context of a GP these provide for a divide and conquer approach to code development. Such a capability is particularly important in the case of linear GP structures as no structured organisation of instruction sequences is assumed. Identification of useful instruction 'building blocks' is a function of the generative activity of GP. Moreover, such a property is a function of the flexibility of the addressing format associated with the register language (c.f. 0-, 1-, 2- and 3- register addressing). Table 1 provides an example of the significance of the register addressing employed when evaluating the function $x^4 + x^3 + x^2 + x$. It is immediately apparent that the 2- and 3- register addressing modes provide a much more compact notation than 0- and 1- register addressing modes. However, this does not necessarily imply that the more concise register addressing modes are easier to evolve, although it may well be expected given that the three-address format roughly corresponds to a functional node in the classical tree-based structure. In short, the 0-address case requires a stack to perform any manipulation of data – arithmetic or register transfer. The 1-address format relaxes this constraint by enabling direct inclusion of register values and input ports within arithmetic operations. The target register is always the accumulator however, and one of the operands (terminals) is always the present value of the accumulator. A 2-address format replaces the accumulator with any designated internal register, whereas 3-address register machines permit the use of internal registers as both the operands (terminals) and result registers as well as arbitrary target and source registers.

Table 1. Evaluating $x^4 + x^3 + x^2 + x$ using different register addressing modes.

0-address	1-address	2-address	3-address
TOS ← P0	AC ← P0	R1 ← P0	R1 ← P0 * P0
TOS ← P0	AC ← AC * P0	R1 ← R1 * P0	R2 ← R2 + R1
TOS ← TOS_i * TOS_{i-1}	R1 ← AC	R2 ← R1	R1 ← R2 * R1
R1 ← TOS	AC ← AC * P0	R2 ← R2 * R1	R1 ← R1 + P0
TOS ← R1	AC ← AC + P0	R2 ← R2 + R1	R1 ← R1 + R2
TOS ← P0	R2 ← AC	R1 ← R1 * R1	
TOS ← TOS_i * TOS_{i-1}	AC ← R1	R1 ← R1 + P0	
TOS ← P0	AC ← AC * R1	R2 ← R2 + R1	
TOS ← TOS_i + TOS_{i-1}	AC ← AC + R2		
TOS ← R1			
TOS ← R1			
TOS ← TOS_i * TOS_{i-1}			
TOS ← TOS_i + TOS_{i-1}			
TOS ← R1			
TOS ← TOS_i + TOS_{i-1}			

KEY: TOS – top of stack; R× - internal register ×; P× - external input on port ×; AC – accumulator register; {*, +}- arithmetic operators; ← - assignment operator.

Where smaller addressing modes benefit however, is in terms of the number of bits necessary to represent an instruction. That is to say if the word-length of a shorter address instruction is half or a quarter that of a 3-address instruction, then the smaller footprint of the associated register machine, may well enable several register machines to exist concurrently on a single FPGA. Hence, a GP defined in terms of a smaller word-length can perform the evaluation of multiple individuals in parallel.

In summary it is expected that the smaller register address formats will produce a less efficient coding, but provide a smaller hardware footprint. Hence although more generations may be necessary to achieve convergence, multiple evaluations are performed at each iteration. However, the assumption of a smaller algorithm footprint for the shorter address cases does not always hold true. In particular, for the special case of a 0-address register machine, a stack is required to perform the calculations. This will therefore consume a significant amount of hardware real estate. The method is retained in the following study, however, given the historical interest in the method [3].

The details for the address formats employed here are summarised in table 2. In each case we design to provide: up to 8 internal registers; up to 7 opcodes (the eighth is retained for a reserved word denoting end of program); an eight bit integer constant field; and up to 8 input ports. The output is assumed to be taken from the stack in the case of the 0-address format, the accumulator in the case of the 1-address format and the internal register appearing as the target most frequently in the case of the 2- and 3-address formats (implies the use of a multiplexed counter).

When initializing linear GPs it is not merely sufficient to randomly select instructions until the maximum code length (for the individual) is reached (and then append the program stop code). This results in a predominance of the instruction occupying most range in the instruction format. In this case instructions defining a constant. Instead a two-stage procedure, as summarised by [4], is necessary. That is, firstly the instruction 'type' is selected and then the instance (uniform random distribution in each case). Here 3 instruction types are identified: those defining an eight-bit integer constant; those involving input ports (any register or port); and those involving internal ports (any opcode)[1].

Table 2. Formats of 0-, 1-, 2- and 3- register address instructions.

3-register address instruction format		
Field	Bits/ field	Description
Mode	2	<0 0> two internal register sources; <0 1> internal and external register sources; <1 0> two external register sources; <1 1> 8 bit const. to target reg.
Opcode	3	Only 6 defined <+, -, %, *, NOP, EOP>
Target register	3	Internal register identifier
Source 1 register	3	Register identifier c.f. mode bits
Source 2 register	3	Register identifier c.f. mode bits
2-register address instruction format		
Mode	2	<0 0> internal register source; <0 1> external register source; <1 x> 8 bit const. to target reg.
Opcode	3	Only 6 defined <+, -, %, *, NOP, EOP>
Target/ source 1 register	3	Internal register identifier

---

[1] The parameterized cases refer to 0- and 1-address instruction formats, the case of 3- and 4- addressing remain as stated.

Source 2 register	3	Register identifier c.f. mode bits
1-register address instruction format		
Mode	2	<0 0> internal register source; <0 1> external register source; <1 x> 8 bit const. to accumulator.
Opcode	3	Only 6 defined <+, -, %, *, NOP, EOP>
Source register	4	Register identifier c.f. mode bits (bit 3 not used)
0-register address instruction format		
Mode	3	<1 x x> push 8 bit constant; <0 0 0> opcode; <0 0 1> pop from stack; <0 1 0> push to stack;
Source bit	2	<x 0> use external input in push/ pop; <x 1> use internal register in push/ pop;
Register ID	3	Either indicate internal register or external source c.f. mode bit.

EOP denotes end-of-program and is protected from manipulation from genetic operators.

3 Genetic Operators and Memory Management

The principle operations of the GP based register machine are (1) the fetch decode cycle (as in a general purpose computer); (2) evaluation of a suitable cost function; (3) application of suitable crossover and mutation operators; (4) memory management in support of the latter two activities; and (5) generation of random bit sequences (stochastic selection). It is assumed that the responsibility for forming an initial population of (linearly structured) individuals is performed off-line by the host (most custom computing platforms take the form of PCI cards [12]). Point 5 implies that either a hardware random number generator is employed based on the classical tapped shift register approach or that random numbers are created off-line[2] using a more statistically robust algorithm and read by the custom computing machine as and when required. Point 2 is application specific, but for the case of the results discussed in section 4, a scalar square error cost function is assumed. However, such an evaluation represents the inner loop of the GP, for which the principle method of accelerated evaluation remains paralleled register machines. An alternative approach would be to provide an explicit hardware embodiment of the GP individual. In this case the concept of real-time reconfigurable custom computing platforms have a unique contribution to make, as illustrated by recent results in the efficient evaluation of sorting networks [13, 14], image processing tasks [15, 16] and optimisation [16]. However, to do so would require the efficient automatic identification of the spatial implementation of an unconstrained program, something that remains an open problem. This study does not dwell further on this issue as the principle interest remains in the development of software individuals and their efficient execution where this is still significantly faster than lisp or interpreted C code [5]. This means that the platform remains application independent, but relies on the provision of

---

[2] By using a reconfigurable computing platform based on FPGA and DSP we are able to generate random number sequences using the DSP whilst dedicating the FPGA to evaluation of the GP.

parallel register machines to quickly evaluate the fitness of individuals. Points 1, 3 and 4 are inter-related and therefore the focus of the following discussion. In particular we note that the most expensive operation is that of a memory access. Moreover, the crossover operator is responsible for creating significant overheads in terms of memory management and memory accesses. Classically the central bottleneck of any computing system is the memory–processor interface. Application specific designs may alleviate this to a certain extent but the constraint still exists in terms of the number of physical device I/O pins. It is therefore desirable to minimise I/O as much as possible, a design philosophy which has resulted in most Von Neumann CPUs dedicating half of the available silicon real estate to memory. Such a solution is not applicable to an FPGA based custom computing machine due to the low (silicon) efficiency in realising memory using random logic elements.

In the particular case of the crossover operator, as classically applied, the main memory management problem comes from the requirement to swap code fragments of differing lengths between pairs of individuals. This means that for example, enough memory space needs reserving for each individual up to the maximum program length (irrespective of the individual's actual program length), and entire blocks of code physically shuffled; figure 1. Alternatively, jump instructions are used to manipulate program flow, thus supporting the swapping of code blocks, with periodic re-writes of the entire population when program flow of individuals becomes overtly fractured. The first case results in most clock cycles being lost to memory manipulation, whereas the latter requirement results a non-sequential manipulation of the program counter (of the register machine) as well as periodic halts in the computation (memory defragmentation). Moreover, both instances have to be

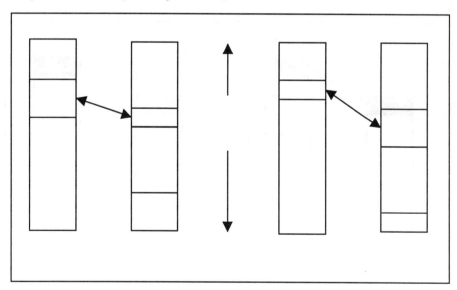

Fig. 1 Operation of classically defined pairwise crossover operator. From a computational perspective this implies that the most time consuming activity in applying the crossover operator is memory management.

implemented directly in hardware c.f. the spatial implementations of any algorithm on a custom computing platform, thus resulting in a lot of 'messy' control circuitry.

One approach to removing the undesirable effects of crossover is to drop the crossover operator completely. This has indeed been shown to produce very effective GPs in the case of tree based structures [17]. However, this approach requires the definition of a more varied set of mutation operators, some of which effectively replace a terminal with a randomly grown sub-tree, hence another memory management problem and a requirement for intervention from the host, or random instruction generator, in order to generate sub-trees. The approach chosen here therefore is to define the initial individuals in terms of the number of program pages and the program page size. Pages are composed of a fixed number of instructions and the crossover operator is constrained to selecting which pages are swapped between two parents. Hence we do not allow more than one page to be swapped at a time; figure 2. This means that following the initial definition of the population of programs – the number of pages each individual may contain (uniformly randomly selected over the interval [min program length, max program length]) – the program length of an individual never changes. The memory management now simplifies to reading programs and copying the contents of parents to their children; figure 2. Moreover, we also assume a steady state as opposed to a generational manipulation of the pool of individuals. That is to say, a tournament is conducted between 'N' individuals, where N << M, the number of individuals in the entire population. This results in N/2 parents, which are retained in their original form in the initial population. These parents are also copied to the children, and manipulated stochastically by way of the crossover and mutation operators. The children than replace the N/2 individuals

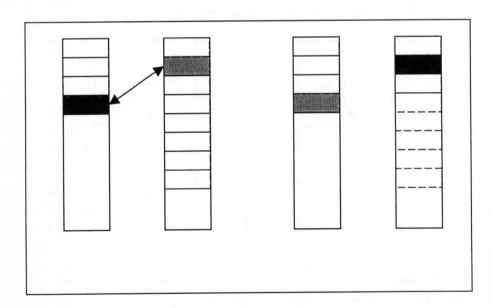

judged not to have performed well. This method has been shown to be as effective as the classical generational approach, but is employed in this case due to,

1. The ranking process necessary to identify parents and children is independent of the size of the population. This provides an explicit constraint on the hardware necessary to support the process;
2. The potential for parallel hardware implementation.

That is to say, depending on the word-lengths, hence instruction set supported, operand requirements (number of registers and I/O), and available hardware resources, it is now possible to design a process directly implementing the GP. For such a processor to be useful however, the specific word-length, instruction set and operands need to be application specific or sufficiently general to enable application to a wide cross-section of applications. By using an FPGA platform we are able to provide application specific GPs without the losses in efficiency associated with a Von Neumann system (i.e. word lengths fit those of the GP exactly and specialised functional units are tailored for the opcodes defined). From a design context however, support for terminal and functional sets does not require a unique design process, just the alteration of the parameter definition in the original hardware description language.

In the case of the mutation operator the standard approach of randomly selecting instructions and then performing a Ex-OR with a second random integer. Although fast, it is acknowledged that this approach is probably lacking in finesse, future work will identify how true this is in practice. In addition a second mutation operator is introduced. In this case an arbitrary pairwise swap is performed between two instructions in the *same* individual. The motivation here is that the sequence in which instructions are executed within a program has a significant effect on the solution. Thus a program may have the correct composition of instructions but specified in the wrong order. This is particularly apparent in the case of the more constrained addressing formats of 0- and 1- address register machines, where the result always appears in the same register (stack and accumulator respectively).

4 Evaluation

The modifications proposed above, to the crossover operator and use of a second mutation operator, represent significant departures from the normal linear GP. Moreover, the purpose of the following study is to demonstrate the significance of decisions made at the micro-architectural level. To this end a software implementation[3] is therefore used to assess the effects of these re-defined operators using a well-known function approximation (symbolic regression) benchmark. In terms of specific tests to the micro-architecture, interest lies in identifying the degree of sensitivity to the population size over the various register formats, c.f. table 1. Also investigated is the significance of different register provisions on convergence ratios as population size varies. The following is a preliminary summary of some of these findings. The symbolic regression benchmark is summarised in table 3. The functional set takes the form of five operators: $F = \{+, -, \times, \%, NOP, RND\}$, where %

---

[3] Watcom C++, version 11.0, Windows 95 target, Pentium II 266Mhz, 64Mb RAM.

is the protected division operator returning a value of unity when division by zero is attempted and NOP implies no operation. Given that a linearly structured GP is employed, there is no need to define a set of pre-defined integer constants as part of the terminal set, table 3. Instead the constants appear in the function (instruction) set, c.f. RND, and are subject to mutation as any other instruction. A sum square error stop criteria of 0.01 or smaller is employed, the tournament is held between four individuals selected randomly from the population and a maximum of 30,000 generations (tournaments) are performed. Time constraints dictate that data is only collected for 20 initialisations of the population in each case. Section 2.1 describes the experiments in detail and summarises results in terms of average performance figures whereas section 2.2 assesses performance using Koza's metric for computational efficiency [18].

Table 3. Benchmark function approximation problem.

Problem	Relation	Input range	Terminal set
Symbolic regression	$x^4 + x^3 + x^2 + x$	[-1,1]	{x}

4.1 Symbolic Regression

In the first case we are interested in identifying the significance of the second mutation operator, hereafter referred as the swap operator. Table 4 rows 1 and 2 summarise the parameters characterising the first experiment. The only deviation from normal practice is to bias the composition of the initial population away from including constants with an equal probability (c.f. the last three columns of table 4). From table 5 it is immediately apparent that the swap operator both increases the number of converging generations and reduces the iterations necessary to reach convergence. A second series of tests is performed to evaluate the effect of increasing the degree of exploration performed by individuals, table 4, row 3. To do so the mutation operator is increased to 0.5. Test 2 in table 6 summarises the effect of such a modification. With respect to the more traditional approach of minimal mutation (0.05) in experiment 2, table 5, a further incremental improvement is observed, although this may well be problem specific. The remainder of table 6 (experiments 4 and 5) illustrate the effect of decreasing the size of the initial population. It is apparent that the longer address format programs appear to benefit from a smaller initial population than their short address format co-patriots. Moreover, the worst case results from the 2-address linear GP are better than the best case 0-address linear GP results.

Table 4. Parameters of GP register machine tests.

Num. Pages	Instr. Per pg	P(cross over)	P(mutate)	P (swap)	Num. Regis.	P (Type 1)	P (Type 2)	P (Type 3)
32	4	0.9	0.05	0.0	8	0.5	1	1
32	4	0.9	0.05	0.5	8	0.5	1	1
32	4	0.9	0.5	0.9	8	0.5	1	1
32	4	0.9	0.5	0.9	8	0.5	1	1

Table 5. Test 1 – Swap and no swap operator (Table 4, row 1 vs row 2).

Experiment 1 – no swap operator (population of 500 individuals)		
Register address format	Average SSE	Comment
0	13.38	No cases converge
1	1.8156	4 cases converge in an average of 10,909 generations.
2	0.14035	15 cases converge in average of 4,987 generations.
3	0.432	12 cases converge in average of 3,034 generations.
Experiment 2 – swap operator included (population of 500 individuals)		
0	6.801	10 cases converge in average of 24,145 generations.
1	0.2279	18 cases converge in average of 3,744 generations.
2	0.0687	18 cases converge in average of 2,774 generations.
3	0.0921	17 cases converge in average of 9,186 generations.

Table 6. Test 2 – Differing mutation and population size (Table 4, row 3).

Experiment 3 – high rate of mutation (population of 500 individuals)		
Register address format	Average SSE	Comment
0	0.984	11 cases converge in average of 24,145 generations.
1	0.05741	18 cases converge in average of 5,452 generations.
2	0.0	20 cases converge in average of 5,607 generations.
3	0.216	17 cases converge in average of 6,154 generations.
Experiment 4 – high rate of mutation (population of 250 individuals)		
0	1.8753	10 cases converge in average of 16,177 generations.
1	0.0618	18 cases converge in average of 5,495 generations.
2	0.0	20 cases converge in average of 1,771 generations.
3	0.054	18 cases converge in average of 4,873 generations.
Experiment 5 – high rate of mutation (population of 125 individuals)		
0	8.521	10 cases converge in average of 23,725 generations.
1	0.6546	16 cases converge in average of 9,398 generations.
2	0.0618	18 cases converge in average of 1,662 generations.
3	0.035	18 cases converge in average of 6,228 generations.

In the final experiment the effect of differing internal register provisions is investigated. That is to say, the all register machines considered so far contain 8 internal registers. A resister machine with 4 internal registers is considered in test 3, table 7. Here it appears that the performance of the 0-address register machine actually improves with a more limited set of registers, whereas the performance of the 2-address formatted instructions decreases. Again, it is acknowledged that these results require verification across a larger set of problems before a general trend is claimed. These results do however, illustrate the significance that design issues at the micro operation level have on the performance of the GP.

Table 7. Test 3 – Register machines with 4 internal registers (Table 4, row 4).

Experiment 6 – population of 500 individuals		
Register address format	Average SSE	Comment
0	0.54215	11 cases converge in an average of 18,275 generations.
1	0.0618	18 cases converge in an average of 6,260 generations.
2	0.115	16 cases converge in an average of 4,077 generations.
3	0.1854	16 cases converge in an average of 7,747 generations.
Experiment 7 – population of 250 individuals		
0	0.502	10 cases converge in an average of 18,722 generations.
1	0.0	20 cases converge in an average of 8,710 generations.
2	0.1454	17 cases converge in an average of 3,539 generations.
3	0.1834	16 cases converge in an average of 7,599 generations.
Experiment 5 – population of 125 individuals		
0	5.9246	11 cases converge in an average of 14,874 generations.
1	0.0978	17 cases converge in an average of 5,196 generations.
2	0.1301	16 cases converge in an average of 4,905 generations.
3	0.0303	19 cases converge in an average of 4,933 generations.

4.2 Individuals Processed

A useful method for expressing the computational effort of the algorithm (as opposed to the computational effort of the processor) was identified by Koza [18]. This defines a probabilistic relationship between the generation and number of individuals processed, and takes the following form,

$$E = T \times i \times \frac{\log(1-z)}{\log(1-C(T,i))}$$

where T is the tournament size; i is the generation at which convergence of an individual occurred; z (= 0.99) is the probability of success; and $C(t, i)$ is the cumulative probability of seeing a converging individual in the experiment.

In the first case interest lies in the effectiveness of the different search operators for a population of 500 individuals; rows 1, 2 and 3 in table 3. Figure 3 summarises this for the case of test 2 (probabilities of 0.9, 0.05 and 0.9 for Crossover, Mutation and Swap respectively) on address formats of 3-, 2- and 1). The knee of the curve is lowest for the case of a 2-address format. However, this result may well be a function of the program length. What is interesting however is that the 1-address format remains competitive, at least problems with single input and output and no constants. Results for test 3, table 3, produce a similar set of curves, however, the 0-address case is several orders of magnitude worse; summarised independently on figure 4.

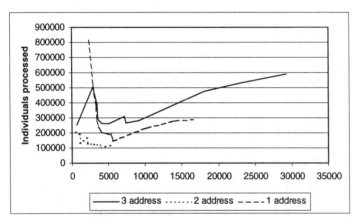

Fig. 3. Performance of 3-, 2- and 1-address linear GPs on a population of 500 and search operator probabilities of 0.9, 0.05, 0.9 (Crossover, Mutation, Swap).

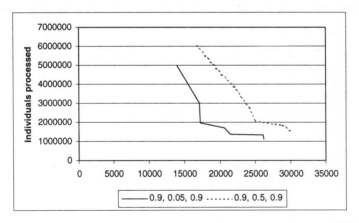

Fig. 4. Performance of 3-, 2- and 1-address linear GPs on a population of 500 and search operator probabilities of 0.9, 0.05, 0.9 (Crossover, Mutation, Swap).

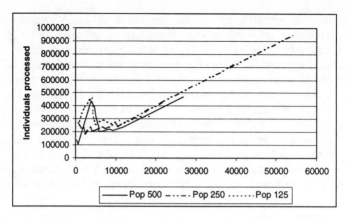

Fig. 5. 3-address for varying population size and search operator probabilities of 0.9, 0.5, 0.9.

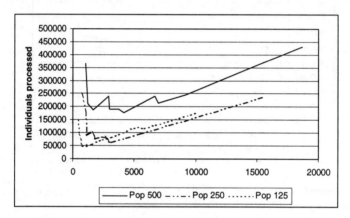

Fig. 6. 2-address for varying population size and search operator probabilities of 0.9, 0.5, 0.9.

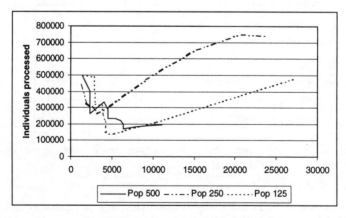

Fig. 7. 1-address for varying population size and search operator probabilities of 0.9, 0.5, 0.9.

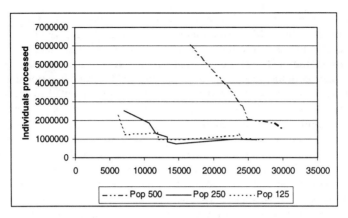

Fig. 8. 0-address for varying population size and search operator probabilities of 0.9, 0.5, 0.9.

The next series of tests illustrate the effect of varying the population size on the different address formats using a fixed set of search operators (0.9, 0.5, 0.9). This is summarised by figures 5 to 8. The robustness of the higher addressing formats is again readily apparent, even on the simple problem context investigated here.

5. Discussion and Conclusion

A case is made for using FPGA custom computing machines to implement linearly structured GPs as register machines. Such a situation enables the inclusion of considerations at the micro architectural and instruction levels of the register machine not possible within the context of classical Von Neumann machines [11]. However, this also implies that if efficient utilisation of the available (FPGA) computing resource is to be made, then memory management overheads require minimisation, where memory management activities are responsible for consuming most *computing* clock cycles. To do so individuals are described in terms of a common page size (number of instructions per page) but arbitrary page count (up to a predefined limit). Crossover is then constrained to appear at page boundaries. Moreover, this also means that once initially defined the program size remains fixed. This significantly simplifies the memory management activity to a level suitable for efficient direct hardware implementation. Such a definition also raises several interesting questions regarding the nature of the code developed, and the significance of the NOP instruction for aligning code fragments with page boundaries. In terms of related work, Nordin *et al.* have proposed a further constraint, that of crossover which is limited to appear in the same location for both individuals of the pairwise exchange [6] – homologous crossover. At the time of writing however, results where not available on the effect of such a constraint or whether such an operator is able to function alone or in conjunction with classically defined crossover operators.

The above modifications to the GP algorithm effectively minimise the number of I/O intensive operations, thus regularising the algorithm. This means that a dataflow design is now appropriate, where once an individual is selected it remains in the 'processor' until it is finally written back to the population (case of a parent) or over written (case of a child). Moreover, dataflow designs have received a lot of interest in the custom computing community [19, 20], and therefore represent a comparatively mature methodology [21, 22]. For example, high levels of hardware reuse are readily achieved when combined with the multiplexing of different functional units, where the significance of this approach has already been demonstrated in the case of GA fitness evaluation [16]. With respect to specifying a specific custom computing platform the authors would tentatively recommend the use of Xilinx® Virtex™ technology on account of the large configurable logic block counts available and support for multiple modes of device reconfiguration (global, local and run-time reconfiguration) [23]. However, recent results in the field of custom computing machines indicate that this is by no means a pre-requisite [9]. Support for DSP may also prove to be an advantage, particularly in the case of support activities such as random number generation and the evaluation of complex functions (trigonometric and logarithms)[4].

As the result section suggests, work completed to date is still in the initial stages. Hence, in addition to evaluating the approach on a wider application base, specific emphasis will be given to the efficient manipulation of constants, where this has long been recognised as a significant impediment to the efficient operation of GPs [24, 25]. Moreover, future work will address the spatial implementation of competitive cost functions and an evaluation of different techniques for generating random numbers. Finally, the constraints imposed by fixed length individuals are acknowledged. In this case we are particularly interested in the hierarchical evolution of populations, with the results of a preceding population forming the Automatically Defined Functions in the next.

References

1. Cramer N.L.: A Representation for Adaptive Generation of Simple Sequential Programs. Proc. 1st Int. Conf. on Genetic Algorithms and their Applications. (1985) 183-187.
2. Nordin J.P.: A Compiling Genetic Programming system that directly manipulates the machine code. In Kinnear K.E. Jr (ed): Advances in Genetic Programming. Vol 1. Chapter 14. MIT Press (1994) 311-331.
3. Perkis T.: Stack Based Genetic Programming. Proc IEEE Congress on Computational Intelligence. IEEE Press (1994).
4. Nordin J.P., Banzhaf W.: Evolving Turning Complete Programs for a Register Machine with Self-Modifying code. In Proc. 6th Int. Conf. on Genetic Programming. Morgan Kaufmann (1995) 318-325.
5. Nordin J.P.: Evolutionary Program Induction of Binary Machine Code and its Applications. Corrected Edition. Krehl Verlag. IBSN 3-931546-07-1 (1999).
6. Nordin J.P., Banzhaf W., Francone F.D.: Efficient Evolution of Machine Code for CISC Architectures. In Spector L., Langdon W.B., O'Reilly U.-M., Angeline P.J. (eds): Advances in Genetic Programming. Vol 3. Chapter 12. MIT Press (1999) 275-299.

---

[4] Nordin shows that support for such functions is not a pre-requisite for solving a cross-section of real world problems (e.g. phoneme recognition, robot navigation, compression of image and sound data) [5].

7. Yao X., Higuchi T.: Promises and Challenges of Evolvable Hardware. IEEE Trans. on Systems Man and Cybernetics – Part C: Applications and reviews. 29(1) (1999) 87-97.

8. Thompson A.: Hardware Evolution: Automatic Design of Electronic Circuits in Reconfigurable Hardware by Artificial Evolution. Springer-Verlag. ISBN 3-540-76253-1 (1998).

9. Levi D., Guccione S.A.: GeneticFPGA: Evolving Stable Circuits on Mainstream FPGA Devices. Proceedings of the 1st NASA/ DoD Workshop on Evolvable Hardware. (1999) 12-17.

10. Mano M.M.: Computer System Architecture. 3rd Edition. Prentice-Hall. IBSN 0-13-175563-3 (1993).

11. Heywood M.I., Zincir-Heywood N.A.: Reconfigurable Computing – Facilitating Micro-operation Design? In Bilişim'99. Istanbul (1999) 77-82.

12. http://www.io.com/~guccione/HW_list.html

13. Koza J.R., et al.: Evolving Computer Programs using Reconfigurable FPGAs and Genetic Programming. ACM Symposium on Field Programmable Gate Arrays. (1997) 209-219.

14. Dandalis A., Mei A., Prasanna V.K.: Domain Specific Mapping for Solving Graph Problems on Reconfigurable Devices. Proceedings of the 9th International Workshop on FPGAs and Applications. (1999).

15. Chaudhuari A.S., Cheung P.Y.K., Luk W.: A Reconfigurable Data-Localised Array for Morphological Operations. In Luk W., Cheung P.Y.K., Glesner M. (eds.): 7th International Workshop on Field Programmable Logic and Applications. Lecture Notes in Computer Science, Vol. 1304. Springer-Verlag. (1997) 344-353.

16. Porter R., McCabe K., Bergmann N.: An Application Approach to Evolvable Hardware. In Proceedings of the 1st NASA/ DoD Workshop on Evolvable Hardware. (1999) 170-174.

17. Chellapilla K.: Evolving Computer Programs without subtree Crossover. IEEE Trans. on Evolutionary Computation. 1(3) (1997) 209-216.

18. Koza J.R.: Genetic Programming: Automatic Discovery of Reusable Programmes. The MIT Press (1994).

19. Duell D.A., et al.: Splash 2: FPGAs in a Custom Computing Machine. IEEE Computer Society Press, ISBN 0-8186-7413-X (1996).

20. Vuillemin J.E., et al.: Programmable Active Memories: Reconfigurable Systems Come of Age. IEEE Transactions on Very Large Scale Integration Systems. 4(1) (1996) 56-69.

21. Kamal A.K., Singh H., Agrawal D.: A Generalised Pipeline Array. IEEE Transactions on Computers. 23 (1974) 533-536.

22. Kung T.H.: Why Systolic Architectures. IEEE Computer. 15(1) (1982) 37-46.

23. Kelem S.: Virtex™ Configuration Architecture Advanced User's Guide. Xilinx® Application note XAPP151. Version 1.2 (1999).

24. Koza J.R.: Genetic Programming: On the Programming of Computers by Means of Natural Selection. The MIT Press (1992).

25. Daida J.M., Bertram R.R., Polito J.A., Stenhope S.A.: Analysis of Single-Node (Building) Blocks in Genetic Programming. In Spector L., Langdon W.B., O'Reilly U.-M., Angeline P.J. (eds): Advances in Genetic Programming. Vol 3. Chapter 10. MIT Press (1999) 216-241.

An Extrinsic Function-Level Evolvable Hardware Approach

Tatiana Kalganova

School of Computing, Napier University
219 Colinton Road, Edinburgh, EH14 1DJ, UK
t.kalganova@dcs.napier.ac.uk,
WWW home page: http://www.dcs.napier.ac.uk/~tatiana/

Abstract. The function level evolvable hardware approach to synthesize the combinational multiple-valued and binary logic functions is proposed in first time. The new representation of logic gate in extrinsic EHW allows us to describe behaviour of any multi-input multi-output logic function. The circuit is represented in the form of connections and functionalities of a rectangular array of building blocks. Each building block can implement primitive logic function or any multi-input multi-output logic function defined in advance. The method has been tested on evolving logic circuits using half adder, full adder and multiplier. The effectiveness of this approach is investigated for multiple-valued and binary arithmetical functions. For these functions either method appears to be much more efficient than similar approach with two-input one-output cell representation.

1 Introduction

Evolvable Hardware (EHW) is technique to synthesize electronic circuits using genetic algorithms. The search for an electronic circuits realization of a desired transfer characteristic can be made in software as in *extrinsic* evolution, or in hardware as in *intrinsic* evolution. In extrinsic evolution the entire evolution process is implemented in the software simulator based on a model of the implementation technology. In intrinsic evolution the hardware actively participate in evolution process [1].

In the context of electronic synthesis the configuration of evolved circuit as well as connecting elements inside circuit is represented by chromosome. Any genetic operators are applied to the chromosome and obtained new circuits are compared with target logic function. The process in usually ended after a given number of generation or when the closeness to the target response has been reached. If connecting element is represented by primitive logic function, a *gate-level EHW approach* is applied. In the *function-level EHW approach*, high level hardware functions such as adders, multipliers, etc. rather than simple logic functions are used as primitive functions in evolution [3], [4]. Therefore the building block implements the multi-input one-output or multi-input multi-output logic function.

A variety of extrinsic EHW methods have been used to synthesise digital circuits (Table 1). In most cases primitive functions or multiplexers have been considered as connecting elements. In our work we use multi-input multi-output logic functions to define the behaviour of connecting elements.

Table 1. Summary of extrinsic EHW approaches for digital circuit design

Author	Building Block	Fitness function	Evolutionary approach	Application
Kitano H., 1996 [5]	$f(2,1,2)$	-	Evolutionary programming	Digital design.
Zebulum R.S. et al, 1996 [6]	$f(2,1,2)$	F1	GA	Binary logic design.
Murakawa M. et al, 1996 [7]	$f(n,1,2)$	F1	VGA	Adaptive equalization.
Miller J. et al, 1997 [8]	$f(2,1,2)$ $f(n,m,2)$	F1	Cartesian GP	Arithmetic binary circuit design.
Kalganova T. et al, 1998 [9]	$f(2,1,r)$ $f_m(r+1,1,r)$	F1	Cartesian GP	Multi-valued circuit design.
Damiani E. et al, 1999 [10]	$f(4,1,2)$	F1	Conventional GA	Hash function design.
Aguirre A.H. et al, 1999 [11]	$f_m(3,1,2)$	F1, F2	GP	Binary circuit design.
Kalganova T. et al, 1999 [12]	$f(2,1,2)$ $f_m(3,1,2)$	F1, F2	Cartesian GP	Binary circuit design.
Masher J. et al, 1999 [13]	$f(2,1,2)$	F1, F3	Conventional GA	Sorting network design.
Miller J., 1999 [14]	$f(2,1,2)$ $f(3,1,2)$	F4	Cartesian GP	Low pass filter design.
Proposed method	$f(2,1,r)$ $f(n,m,r)$	F1, F2	Cartesian GP	Multi-valued logic design.

VGA - an variable-length chromosome GA; GP - Genetic Programming $f(n,m,r)$ is an n-input m-output r-valued logic function; $f_m(3,1,2)$ is an logic function described the behaviour of multiplexer; F1 defines the correctness of outputs of logic circuit evolved; F2 is the minimal number of logic cells used; F3 is the correctness of input combinations; F4 is an error based fitness.

The proposed method is an extension of EHW approach applied for binary circuit design [8] and multi-valued logic (MVL) functions [9]. Some aspects of this approach have been investigated in the past. Thus, it has been found that functional set of logic gates [15] as well as circuit layout and connectivity restrictions [9] influence on the GA performance. Some attempts to evolve circuit layout together with circuit functionality have been reported in [12]. A dynamic fitness function has been proposed in [12], that allow us to evolve functional complete circuit with minimal number of logic gates employed.

In this paper we proposed to use multi-input multi-output logic functions as logic cell (so called building block) in an extrinsic evolvable hardware approach. We limit our focus to binary and multi-valued combinational logic circuit design problems. We introduce the new chromosome representation, that allows us to evolve logic circuits using high-level logic functions. A number of binary and multi-valued circuit structures evolved are discussed. The experimental results show us that the proposed method performs better than one earlier reported in case if the suitable functional set of logic gates is chosen.

2 A Problem Statement

Any design problem can be represented as follows. Let $\mathbb{C}_1$ and $\mathbb{C}_2$ be the sets of values and let $X = \{x_0, x_1, \cdots, x_{n-1}\}$ be an input vector of n variables, where x_k takes on values from $\mathbb{C}_1$. Let $Y = \{y_0, y_1, \cdots, y_{m-1}\}$ be an output vector of m variables, where $y_k \in \mathbb{C}_2$. Let $\mathbb{G}$ be the set of primitive operations over values from sets $\mathbb{C}_1$ and $\mathbb{C}_2$. The cardinality of X defines the number of members of X and denoted as $|X|$. Then, $|X| = n$. Let f be the function such that $\mathcal{F}: \mathbb{C}_1^n \longrightarrow \mathbb{C}_2^m$. Let function $\mathcal{F}$ be represented as a matrix mapping denoted as $X \to Y$, where X is a $(k \times n)$ matrix of all the given inputs and k is the number of input combinations, and Y is a $(k \times m)$ matrix of the corresponding m outputs. Then the synthesis of function can be stated as follows. Design a sequence of operations that accomplishes the mapping $X \to Y$. This mapping is achieved by applying a sequence of primitive or complex operations. This statement of design problem can be applied to any type of functions. We will consider the logic design of binary and multi-valued logic functions.

Binary circuit design: Let $\mathbb{B} = \{0, 1\}$ be the set of binary logic values. Then the design of binary circuits can be represented by the equations given above, if $\mathbb{C}_1 = \mathbb{B}$, $\mathbb{C}_2 = \mathbb{B}$.

Multi-valued circuit design: Let $\mathbb{R} = \{0, 1, \cdots, r-1\}$ be the set of r-valued logic values. Then the multi-valued circuit design can be described by the equations given above, if $\mathbb{C}_1 = \mathbb{R}$ and $\mathbb{C}_2 = \mathbb{R}$. The primitive two-input multi-valued logic operators are defined as follows:

$$
\begin{aligned}
NOT &: \quad !x_1 = \overline{x_1} = (r-1) - x_1 \\
SUCCESSOR &: \quad ?x_2 = x_2 + 1 \pmod{r} \\
MIN &: \quad x_1 \cdot x_2 = min(x_1, x_2) \\
MAX &: \quad x_1 \vee x_2 = max(x_1, x_2) \\
MODSUM &: \quad x_1 \oplus x_2 = x_1 + x_2 \pmod{r} \\
MODPRODUCT &: \quad x_1 \otimes x_2 = x_1 * x_2 \pmod{r} \\
TSUM &: \quad x_1 \Diamond x_2 = min(x_1 + x_2, r-1) \\
TPRODUCT &: \quad x_1 \star x_2 = min(x_1 + x_2 - (r-1), 0).
\end{aligned}
$$

where "+" is an ordinary addition. The TPRODUCT operator has been first introduced in [16], [17]. The symbolic representation of MVL operators mentioned above are shown in Fig. 1. More information about multiple-valued logic can be found in [18].

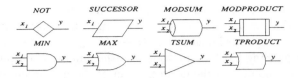

Fig. 1. Symbols and analytic representation of two-input r-valued logic gates

3 Function-Level Extrinsic Evolvable Hardware Approach

There are several issues of interest concerning the use of cartesian generic programming [19] in this domain, such as the encoding of building blocks implemented multi-input multi-output logic functions, mutation operator, repair procedures. The representation chosen for our work allows us to synthesise the logic circuits using multi-input multi-output logic sub-functions defined in advance.

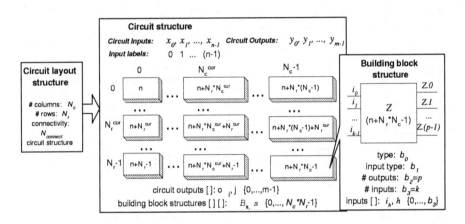

Fig. 2. Schematic of chromosome structure: N_c, N_r are the number of columns and rows respectively; N_c^{cur}, N_r^{cur} are current column and row respectively.

3.1 Encoding

A combinational logic circuit is represented as a rectangular array of building blocks (Fig. 2). Each building block is uncommitted and can be removed from the actual circuit design if they prove to be redundant. The building block can implement any primary logic operation or multi-input multi-output logic function defined in advance. The inputs to any building block in the combinational network may be the primary inputs or any outputs of building blocks which are in columns to the left of the building block in question. The circuit inputs $x_0, x_1, \cdots, x_{n-1}$ are numbered $0, 1, \cdots, (n-1)$ respectively. The building blocks which from the array are numbered column-wise from n to $(n+N_c*N_r-1)$, where N_c and N_r are the number of columns and rows in rectangular array. The outputs of building block $\mathcal{B}(z)$ are labeled from $(z + 0/D_{order})$ to $(z + (p-1)/D_{order})$, where z is the building block in question, p is the number of outputs in building block $\mathcal{B}_z$ and D_{order} defines the decimal order of the maximum number of outputs allowed to be used in building block. For example, if the maximum number of output in building block can not exceed 99, then $D_{order} = 100$. If 9 outputs are allowed to be used in building block, then $D_{order} = 10$. Thus, each output of building block is defined by real number. In the work reported we consider evolving binary and MVL functions. Table 2 shows the set of both binary and MVL functions employed in evolution.

Table 2. Gate functionality according to the $b_0(z)$ gene in chromosome; "-" denotes that the logic function is not assigned for encoding value.

Gene functionality, $b_0(z)$	Gate logic function	
	binary	multi-valued
0	Logic constant	Logic constant
1	-	$SUCCESSOR : ?x_0$
2	$NOT : \overline{x_0}$	$NOT : !x_0$
3	-	Literal: $L(x_0, c)$
4	-	Clockwise Operator: $c \leftarrow x_0$
5	-	Counter Clockwise Operator: $c \to x_0$
6	Wire: x_0	Wire: x_0
7	$AND : x_0 \cdot x_1$	$MIN : x_0 \cdot x_1$
8	$OR : x_0 \vee x_1$	$MAX : x_0 \vee x_1$
9	$EXOR : x_0 \oplus x_1$	$MODSUM : x_0 \oplus x_1$
10	-	$MODPRODUCT : x_0 \otimes x_1$
11	-	$TSUM : x_0 \Diamond x_1$
12	-	$TPRODUCT : x_0 * x_1$
13	-	$x_0 + x_1$ (over $GF(r)$)
14	-	$x_0 * x_1$ (over $GF(r)$)
15	Multiplexer	T-gate
16	-	1-digit multiplier
17	1-digit full adder	1-digit full adder
18	2-digit multiplier	2-digit multiplier
19	2-digit full adder	2-digit full adder
20	3-digit multiplier	3-digit multiplier
21	3-digit full adder	3-digit full adder
22	Half adder	Half adder

The chromosome is represented by a 3-level structure: 1) Circuit layout structure; 2) Circuit structure; 3) Building block structure.

Circuit layout structure: At the first level the global characteristics of the circuit are defined. These are connectivity parameter $N_{connect}$, the number of rows N_r and columns N_c. Note that these parameters are allowed to be variable, but in this paper we consider that the circuit layout is defined in advance and it is not allowed to be changed during evolution process.

Circuit structure: At the second level the array of building blocks $\mathcal{B}_i$ is created and the circuit outputs $\mathcal{O} = \{o_0, o_1, \cdots, o_m\}$ are determined.

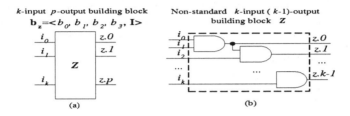

Fig. 3. Schematic of building block structure.

Building block structure: Finally, the third level represents the structure of each building block in the network $\mathcal{N}$. This data consists of the functional gene b_0, the type of inputs b_1, the number of outputs b_2 and inputs b_3 and the input connections i_h, (Fig. 2, Fig. 3(a)) . In this work the gene b_1 defining the type of inputs in building block has not been taken into account. The number of inputs and outputs in building block depend on the functional type of building block b_0 and are defined if the value of functional gene is known. Thus, if functional gene defines the primitive logic function, then variable number of inputs in building block can be used. In this case the primitive logic gates are connected as shown in Fig. 3(b). For example, let " $\vee$ " defines the 2-input 1-output logic primitive function and $b_3 = 4$. The number of outputs in building block is 3. These outputs can be analytically represented as follows:

$$o_0(z) = i_0 \vee i_1;$$
$$o_1(z) = (i_0 \vee i_1) \vee i_2;$$
$$o_2(z) = ((i_0 \vee i_1) \vee i_2) \vee i_3.$$

If multi-input multi-output logic functions defined in advance is employed, then the number of inputs and outputs in building block are fixed and can not be changed. For example, the binary two-bit full adder has 5 inputs and 3 outputs. So, if the functional gene of building block is 19 ($b_0 = 19$), then $b_2 = 3$ and $b_3 = 5$.

3.2 An example

An example of chromosome representation with the actual circuit structure is given in Fig. 4. This circuit represents the full 2-bit adder evolved using AND, OR, EXOR and half adder binary logic functions. This function has 5 inputs and 3 outputs and is implemented on a combinational network with 5x2 circuit

66

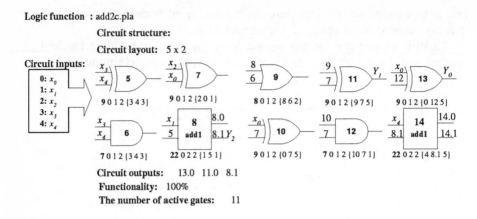

Fig. 4. An example of the phenotype and corresponding genotype of a chromosome with 5x2 circuit layout

layout ($N_c \times N_r$). The labels of circuit inputs 0, 1, 2, 3, 4 correspond to the input variables x_0, x_1, x_2, x_3 and x_4 respectively. The type of building block is defined by functional gene shown in bold. The encoding table of functional gene is given in Table 2.

Each cell is assigned an individual address. Thus the building block located in 0th column and 0th row is labeled as 5. The building block located in 4th column 1st row is labeled as 14. Each output of building block is labeled with real number. The main part of this number defines the code of building block and the fractional part determines the position of output in building block. For example, the 8th building block B_8 located in 1st row and 1st column has 2 outputs. The first output is numbered as 8.0 and the second one as 8.1. The number of circuit outputs is defined by the number of outputs in the logic function implemented. Let us examine the encoding of the 14th building block represented as $< 22\ 0\ 2\ 2\ \{4\ 8.1\ 5\} >$. We refer to this representation of the building block as *building block genotype*. The functional gene defining the type of this gate is 22. This corresponds to the half adder in encoding table (Table 2). The examined cell has two inputs and two outputs. The first input is connected to the input x_4 and second to the second output of building block 8, labeled as 8.1. The inputs of building block 8 are connected to input variable x_1 and output of building block 5. The logic function of building block 5 depends on the input variables x_3 and x_4. Therefore the logic function implemented in building block 14 depends on three input variables: x_1, x_3 and x_4. The outputs of circuit are connected to the outputs of building blocks 13, 11 and 8.1.

3.3 Fitness Function

Our goal is to produce a fully functional design and minimize the number of building blocks actually used in circuit. A *fully functional design* produces the expected behavior stated by its truth table. Therefore, we decided to use two-stage fitness strategy [12]. At the beginning of search, only compliance with the truth table is taken into account. Once the first functional solution appears, we switch to a new fitness function in which fully functional circuits with less cost are rewarded. The *cost* or size of the fully functional circuit $\mathcal{N}$ is defined as

$$cost(\mathcal{N}) = \sum_{j=0}^{j<N_c*N_r-1} cost(\mathcal{B}_j) \qquad (1)$$

and the cost of building block $\mathcal{B}_j$ is calculated as

$$cost(\mathcal{B}_j) = \begin{cases} N_j^p, & \mathcal{B}_j \quad \text{is committed building block} \\ 0, & \mathcal{B}_j \quad \text{is uncommitted building block.} \end{cases} \qquad (2)$$

where N_j^p is the minimum number of primitive logic cells required to implement the logic function described behaviour of building block $\mathcal{B}_j$.

We consider the building block $\mathcal{B}_j$ as a sub-circuit with the structure that is not allowed to be changed. So, the cost of building block does not take into account whatever all outputs of building block has been involved or not. For example, let the two-bit multiplier be represented as building block $\mathcal{B}_j$ and the first digit of this two-bit multiplier be *only* involved in circuit $\mathcal{N}$. The cost of two-bit multiplier is 7 [8]. Despite the first digit of two-bit multiplier is implemented using only one primitive logic gate, the cost of building block $\mathcal{B}_j$ is 7. Members of the population with changed genotype have their fitness calculated.

3.4 Evolutionary algorithm

The circuit evolution has been performed using a rudimentary $(1 + \lambda)$ evolutionary strategy (ES) with uniform mutation [20]. In this case a population of random chromosomes is generated and the fittest chromosome is selected. The new population is then filled with mutated versions of this.

Initialisation: The initial population is generated randomly. During initialisation of cell inputs and circuit outputs is performed in accordance with the levels-back constraint and the type of variables which are able to be present throughout all circuit. Thus if the logic constants are allowed as input connections throughout the circuit, then during initialisation procedure the inputs of gates can be chosen from the set of inputs constrained by levels-back or from the set of logical constants. The same procedure is true for the primary and inverted primary inputs.

Mutation: The circuit mutation allows us to change the following three features of the circuit: (1) Cell input; (2) Cell type; (4) Circuit output. Each of these parameters is considered as an elementary unit of the genotype. Cell type is defined by the functional gene and the functionality gene. The mutation rate defines how many genes in the population are involved in mutation. The chromosome contains 4 different types of genes, whose number is :

$$N_{genes} = \sum_{i=1}^{\lambda}(4 \cdot N_{gates}^3 + N_{outputs})$$ (3)

where $N_{outputs}$ is the number of outputs in the circuit, N_{gates}^i is the number of gates in the i-th chromosome, λ is the population size.

4 Experimental Results

In this section we will consider some experimental results obtained for function- and gate-level EHW. Two applications to EHW approach has been examined: 1) Binary arithmetic circuit design; 2) Multi-valued arithmetic circuit design. For the purposes of this paper, 5 examples were chosen to illustrate our approach The performance of function-and gate- level EHW approaches has been compared. The initial data for the experiments is given in Table 3.

Table 3. Initial data

| | Logic functions | | | | |
| | binary | | | multi-valued | |
Circuit	mult2	mult3	add2c	add3_3c	mult3_3
Circuit layout	10x1	30x1	15x1	10x1	10x1
Connectivity parameter	10	30	15	10	10
Population size	5	5	5	5	5
Number of generations	5 000	100 000	15 000	15 000	25 000
Number of GA runs	100	100	100	100	100
Mutation rate	5%	5%	5%	5%	5%

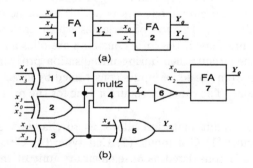

Fig. 5. Evolved 2-bit full adder; FA is one-bit full adder, mult2 is two-bit multiplier.

Two-bit full adder, add2c.pla. One of the principles used by human to construct larger adders is known as the *ripple-carry* principle. The block diagram for a two bit adder is shown in Fig. 5(a). Each of the blocks in Fig. 5(a) are identical to one-bit full adder. The two bit full adder evolved using one-bit full adder is produced by connecting the two smaller adders in a configuration identical to that shown in Fig. 5(a). This structure has been appeared in the most circuit designs (approximately 70 % of all fully functional designs) evolved using one-bit full adder as a building block. This demonstrates that the ES finds principles of the ripple-carry adder.

The circuit shown in Fig. 5 (b) has been evolved using one-bit full adder and two-bit multiplier as building blocks. This circuit requires 18 logic cells. Note that only third output of two bit multiplier is used. This shows that there is not necessary that all outputs of multi-input multi-output building blocks can be exploited in the circuit.

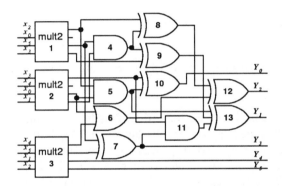

Fig. 6. Evolved three-bit multiplier (most efficient - 26 gates); mult2 is two-bit multiplier.

Three-bit multiplier, mult3.pla. The 3-bit multiplier can calculate the product of two integers a and b in range 0-7. The conventional 3-bit multiplier designed at gate-level requires 30 primitive logic gates [21]. The most efficient 3-bit multiplier evolved at gate-level requires only 26 gates [21]. Note that this structure contains 3 logic cells implemented AND logic function with one inverted input. In our approach we count only the number of NOT, AND, EXOR and OR gates. Therefore in our calculations, AND with one inverted input requires 2 primitive logic gates: AND and NOT. From this point of view the circuit structure reported in [21] contains 29 primitive logic gates, such as AND, OR, EXOR and NOT. It is more important to point out that this solution has been evolved after 3,000,000 generations, whereas in case of using the function-level EHW the fully functional solution has been evolved after 100,000 generations. In this particular example the evolution process has been improved in 30 times. It shows that using function-level EHW allows us to improve the ES performance.

The most efficient evolved 3-bit multiplier at function level is shown in Fig. 6. This circuit requires 32 gates. The cost of this circuit has been calculated regardless the outputs used in two-bit multiplier building blocks. Note that the second output of multipliers 1 and 2 is not used. This output requires 4 logic gates to be implemented. Only one logic gate is used to implement another output of two-bit multiplier ([22]). This means that 3 logic gates are employed to implement the second output and are not used in implementation of other circuit outputs. Therefore, the circuit shown in Fig. 6 requires 26 primitive logic gates (32-2*3=26), such as AND, OR and NOT. Therefore we can conclude that this is the most efficient circuit structure evolved using both gate and function level EHW.

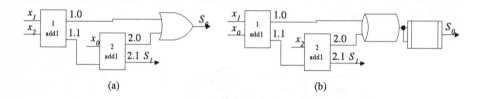

(a)   (b)

Fig. 7. Evolved 3-valued 1-digit full adders; add1 is 3-valued half adder.

One-digit 3-valued full adder, add3_3c.pla. Evolving a fully functional 3-valued one-digit adder using a circuit layout of 10 columns and 1 row with connectivity parameter equaled 10 proved to be relatively easy and the designs shown in Fig. 7 were obtained. The circuit shown in Fig. 7 (a) contains only 3 building blocks. Note that the optimal implementation of half adder contains 4 primitive logic cells [9]. Therefore the circuit in question requires 9 primitive logic cells. It is interesting to note that in this structure all outputs of half adder have been used actively. In case when the MAX gate is not allowed to be involved in evolution, the circuit structure shown in Fig. 7 (b) has been evolved. This structure contains 4 building blocks and requires 10 primitive logic cells. Note that both structures mentioned above use all outputs of half adder.

The 1.5-digit Multiplier, mult3_3.pla. An 1.5-digit multiplier multiplies the r-valued numbers $(A_1 A_0)$ by B_0 to produce the two-digit r-valued number $(P_1 P_0)$, where A_1, B_0 and P_1 are the most significant digits. Thus this is a circuit with 3 inputs and 2 outputs and it requires 27 input and output conditions for full specification in case of 3-valued logic. An example of circuit structures evolved using proposed method are shown in Fig. 8. It is interesting to note that in case of circuit shown in Fig. 8(a) ES uses the outputs of half adder and one-digit multiplier as sub-functions and some of its outputs are not in use. Thus the first output of 3-valued 1-digit multiplier labeled as 6 does not employed. At the same time all outputs of half adder are used. The circuit contains 7 building blocks and involves 12 primitive logic cells. Note that the most efficient 1-digit multiplier evolved requires 3 primitive logic cells. Fig. 8(b) shows the circuit evolved using only half adder. This circuit requires 6 building blocks and 12 primitive logic cells. Note that the implementation of digit P_1 is the same for both cases. It is interesting to note that the circuit evolved with half adder and one-digit multiplier requires less number of primitive logic cells then the alternative logic function evolved using only half adder. The circuits shown in Fig. 8 can not be obtained using the rules of standard multiplication process.

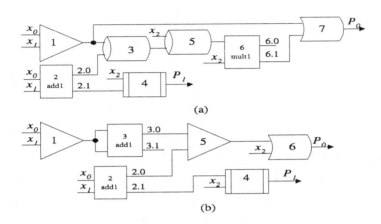

(a)

(b)

Fig. 8. Evolved 3-valued 1.5-digit multiplier. add1 is a 3-valued half adder; mult1 is 3-valued 1-digit multiplier.

Binary circuit design In this section we will discuss some experimental results obtained for two-bit full adder (add2c.pla), two- (mult2.pla) and three- bit multipliers mult3.pla evolved at gate and function level EHW. The ES performs the fixed number of generations for both approaches. The functional set of logic gates for gate-level EHW is a subset of {AND, OR, EXOR, NOT}. Note that using functional set of NOT, AND, OR, EXOR(i.e. 2-7-8-9 according to encoding table (Table 2)) in evolution corresponds performing EHW at gate-level. The number of inputs in building block for gate-level EHW can not be more then 2. Half bit adder, one-bit full multiplier, two-bit multiplier together with primitive logic gates have been used at function-level EHW. The experimental results obtained using initial data shown in Table 3 are summarized in Table 4. The number of fully functional solutions evolved at function-level EHW is shown in bold (4).

Let us consider how ES performs at gate- and function-level EHW during evolution of two-bit multiplier (4). Analysing these data we can conclude that in terms of the number of active primitive logic gates used in circuit, the gate- and function-level EHW perform better. Thus, the average number of active logic gates (av.100F2) in fully functional circuits evolved at gate-level is 7.14 and at function-level is 8 or more. In terms of the number of fully functional circuits evolved during 100 ES runs, both methods perform nearly the same: 31-

Table 4. Comparison function- and gate-levels EHW. **av.F1** is the mean functionality fitness function of the best chromosomes obtained 100 runs; **av.F2** is the mean number of active logic gates in the best chromosomes obtained after 100 runs; **av.100F2** is the mean number of active logic gates in the fully functional designs evolved; **#100% cases** is the number of fully functional circuits evolved.

Circuit	n	m	Functional set	av.F1	av.F2	av.100F2	#100% cases
mult2.pla	4	4	2-7-8-9	98.2968	7.22	7.14	36
			2-7-8-9-22	98.2031	11.56	8	37
			2-7-8-9-17	97.7344	15.51	8.75	16
			2-7-8-9-17-22	98.3438	15.83	10.9355	31
mult3.pla	6	6	2-7-8-9	95.5686	32.2667	0	0
			2-7-8-9-18	97.1807	59.6087	33.3	3
			2-7-8-9-18-22	98.0547	58.8	42	5
			2-7-8-9-17-18-19-22	99.5734	95.1525	65.33	6
add2c.pla	5	3	2-7-8-9	93.75	10.25	11.1429	14
			2-7-8-9-22	93.9167	12.62	12.5833	24
			2-7-8-9-17-22	99.6042	10.69	10.4375	96
			2-7-8-9-17-18-22	99.7708	11.74	11.2128	94
add3_3c	3	2	2-7-8-11-12	63.5185	8.44	0	0
			2-7-8-11-12-22	98.7778	9.12	8.86	77
			2-7-8-9-10-11-12	98.7593	6.97	7	72
			2-7-8-9-10-11-12-22	99.7593	6.77	7	96
			2-9-10-11-12	86.1852	5.31	8	1
			2-9-10-11-12-16-22	99.2222	7.87	8.069	86
mult3_3	3	2	2-7-8-9-10-11-12	76.5556	6.64	0	0
			2-7-8-9-10-11-12-22	98.3889	8.4	9.39	61
			2-7-8-9-10-11-12-16-22	97.5741	8.69	9.56	41
			1-7-8-11-12	59.2592	8.4	0	0
			1-7-8-11-12-16-22	92.1851	13.89	12.4	10
			2-9-10-11-12	77.9444	8.3	0	0
			2-9-10-11-12-22	93.9999	7.94	9.8	26
			2-9-10-11-12-16-22	96.5741	8.13	10.11	17

37 fully functional designs have been evolved during 100 ES runs. Considering ES performance for three-bit multiplier carried out at gate- and function-level we can conclude that the function-level EHW executes much better. Thus during evolution at gate level the average functionality of evolved circuits is 95.5686 and at function level it has been significantly increased to 99.5734. No fully functional solutions have been evolved for three-bit multiplier at gate-level EHW. Some functional solutions has been found using function-level EHW. Comparing function- and gate-level EHW used to evolve two-bit full adder, we find that the number of fully functional designs obtained at function-level EHW has been significantly improved in comparison with the similar EHW performed at gate-level. The average functionality fitness function is higher for function-level EHW rather then for gate-level EHW.

Note that in all cases mentioned above the average best functionality fitness functions (av.F1) for tested logic functions we can conclude that the best functionality fitness function is lower in case when the gate-level EHW has been applied. Analysing the average number of active primitive logic gates in best fully functional chromosomes (av.100F2), we find that there is no significant difference between function- and gate-level EHW in case when three-bit multiplier and two-bit full adder have bit evolved. Note that in case of evolving the two-bit multiplier, the ES at gate-level performs better in terms of the number of active primitive logic gates in circuit. Thus, we can conclude that function-level EHW performs better then gate-level EHW in terms of the number of fully functional binary circuits evolved when the suitable functional set of logic gates has been chosen.

Multi-valued circuit design. The similar experimental results have been carried out for 3-valued one-digit full adder (add3_3.pla) and 3-valued 1.5 digit multiplier (mult3_3.pla).

Let us consider 3-valued one-digit full adder. Three different functional sets of primitive logic gates have been employed during gate-level EHW execution. At function-level EHW, the 3-valued half adder has been added to these functional sets. In all cases the function-level EHW method performs better. Thus, when we use functional set of {COMPLEMENT, MIN, MAX, TSUM, TPRODUCT} no fully functional solutions have been evolved. Adding only half digit adder in functional set improves the GA performance in 77 times. The same conclusion we can made about other functional sets of logic gates used. Three different functional sets of primitive logic gates have been used at gate-level EHW. Half adder and one-digit multiplier have been added to functional set of logic gates at function-level EHW. Analysis of 1.5 digit multiplier allows us to make the conclusion that the function-level EHW performs better in term of the number of fully functional circuits evolved. Note that no fully functional designs have been evolved at gate-level. This number has been significantly increased when function-level evolution has been applied.

Thus, we can conclude that the function-level EHW approach applied to multi-valued logic design performs better then the similar approach implemented at gate-level if the suitable functional set of logic cells has been chosen.

5 Conclusion

We have introduced a new representation of logic cell in extrinsic EHW. This representation allows us to evolve circuits at function-level evolvable hardware. WE showed that this approach can be applied to synthesize both multiple-valued and binary logic functions. The advantage of proposed method is that it does not restricted by the radix of logic or the set of logic cells (functions) chosen to describe the behaviour of building blocks.

We have compared the ES performance for function- and gate-level EHW methods applied to design binary and multi-valued logic circuits. The obtained results show that the function-level EHW performs better in terms of the number of fully functional solutions evolved. Also it has been shown that the most efficient 3-bit multiplier evolved at function level evolution contains 26 primitive logic gates. This design less the number of active primitive logic gates then the similar conventional one. Also it has been shown that the evolved design contains less the number of primitive logic gates such as NOT,AND, EXOR and OR then the similar most efficient design evolved at gate level [21]. The experimental results show that the proposed chromosome representation allows us easier to evolve arithmetic MVL functions if the appropriate set of standard logic functions has been chosen.

The future work can be focused on the using multi-input multi-output automatically defined logic functions as a building blocks in proposed method as well as attempts to evolve logic functions of large number of variables.

References

1. Stoica A., Keymeulen D., Tawel R., Salazar-Lazaro C., and Li W. Evolutionary experiments with a fine-grained reconfigurable architecture for analog and digital cmos circuits. In Stoica A., Keymeulen D., and Lohn J., editors, *Proc. of the First NASA/DoD Workshop on Evolvable Hardware*, pages 76–84. IEEE Computer Society, July 1999.
2. Murakawa M., Yoshizawa S., Kajitani I., Furuya T., Iwata M., and Higuchi T. Hardware evolution at function level. In *Proc. of the Fifth International Conference on Parallel Problem Solving from Nature (PPSNIV)*, Lecture Notes in Computer Science. Springer-Verlag, Heidelberg, 1996.
3. Higuchi T., Murakawa M., Iwata M., Kajitani I., Liu W., and Salami M. Evolvable hardware at function level. In *Proc. of IEEE 4th Int. Conference on Evolutionary Computation, CEC'97*. IEEE Press, NJ, 1997.
4. Kitano H. Morphogenesis for evolvable systems. In Sanchez E. and Tomassini M., editors, *Towards Evolvable Hardware. The Evolutionary Engineering Approach.*, volume 1062 of *Lecture Notes in Computer Science*, pages 99–117. Springer-Verlag, 1996.
5. Zebulum R., M. Vellasco, and M. Pacheco. Evolvable hardware systems: Taxonomy, survey and applications. In *Proc. Of the 1st Int. Conf. on Evolvable Systems: From Biology to Hardware (ICES'96)*, volume 1259 of *Lecture Notes in Computer Science*, pages 344–358, Tsukuba, Japan, 1996. Springer-Verlag, Heidelberg.
6. Murakawa M., Yoshizawa S., and Higuchi T. Adaptive equalization of digital communication channels using evolvable hardware. In *Proc. Of the 1st Int. Conf. on Evolvable Systems: From Biology to Hardware (ICES'96)*, volume 1259 of *Lecture Notes in Computer Science*, pages 379–389, Tsukuba, Japan, 1996. Springer-Verlag, Heidelberg.
7. Miller J. F., Thomson P., and Fogarty T. C. Genetic algorithms and evolution strategies. In Quagliarella D., Periaux J., Poloni C., and Winter G., editors, *Engineering and Computer Science: Recent Advancements and Industrial Applications.* Wiley, 1997.
8. Kalganova T., Miller J., and Fogarty T. Some aspects of an evolvable hardware approach for multiple-valued combinational circuit design. In Sipper M., Mange D., and Perez-Uribe A., editors, *Proc. Of the 2nd Int. Conf. on Evolvable Systems: From Biology to Hardware (ICES'98)*, volume 1478 of *Lecture Notes in Computer Science*, pages 78–89, Lausanne, Switzerland, 1998. Springer-Verlag, Heidelberg.

9. Damiani E., Tettamanzi A.G.B., and Liberali V. On-line evolution of fpga-based circuits: A case study on hash functions. In Stoica A., Keymeulen D., and Lohn J., editors, *Proc. of the First NASA/DoD Workshop on Evolvable Hardware*, pages 26–33. IEEE Computer Society, July 1999.

10. Aguirre A.H., Coello Coello C.A., and Buckles B.P. A genetic programming approach to logic function sunthesis by means of multiplexer. In Stoica A., Keymeulen D., and Lohn J., editors, *Proc. of the First NASA/DoD Workshop on Evolvable Hardware*, pages 46–53. IEEE Computer Society, July 1999.

11. Kalganova T. and Miller J. Evolving more efficient digital circuits by allowing circuit layout evolution and multi-objective fitness. In Stoica A., Keymeulen D., and Lohn J., editors, *Proc. of the First NASA/DoD Workshop on Evolvable Hardware*, pages 54–63. IEEE Computer Society, July 1999.

12. Masher J., Cavalieri J., Frenzel J., and Foster J.A. Representation and robustness for evolved sorting networks. In Stoica A., Keymeulen D., and Lohn J., editors, *Proc. of the First NASA/DoD Workshop on Evolvable Hardware*, pages 255–261. IEEE Computer Society, July 1999.

13. Miller J. On the filtering properties of evolved gate arrays. In Stoica A., Keymeulen D., and Lohn J., editors, *Proc. of the First NASA/DoD Workshop on Evolvable Hardware*, pages 2–11. IEEE Computer Society, July 1999.

14. Kalganova T., Miller J., and Lipnitskaya N. Multiple-valued combinational circuits synthesized using evolvable hardware approach. In *Proc. of the 7th Workshop on Post-Binary Ultra Large Scale Integration Systems (ULSI'98) in association with ISMVL'98*, Fukuoka, Japan, May 1998. IEEE Press.

15. Utsumi T., Kamiura N., Nata Y., and Yamato K. Multiple-valued programmable logic arrays with universal literals. In *Proc. of the 27th Int. Symposium on Miltiple-Valued Logic (ISMVL'97)*, pages 169–174, Nova Scotia, Canada, May 1997. IEEE-CS-Press.

16. Hata Y. and K. Yamato. Multiple-valued logic functions represented by tsum, tproduct, not and variables. In *Proc. of the 23th Int. Symposium on Miltiple-Valued Logic (ISMVL'93)*, pages 222–227. IEEE-CS-Press, May 1993.

17. D. C. Rine. *An Introduction to Multiple-Valued Logic.* North-Holland, Amsterdam, 1977.

18. Miller J. An empirical study of the efficiency of learning boolean functions using a cartesian genetic programming approach. In *Proc. of the Genetic and Evolutionary Computation Conference (GECCO'99)*, volume 1 of *ISBN 1-55860-611-4*, pages 1135–1142, Orlando, USA, July 1999. Morgan Kaufmann, San Francisco, CA.

19. Miller J. Digital filter design at gate-level using evolutionary algorithms. In *Proc. of the Genetic and Evolutionary Computation Conference (GECCO'99)*, volume 1 of *ISBN 1-55860-611-4*, pages 1127–1143, Orlando, USA, July 1999. Morgan Kaufmann, San Francisco, CA.

20. Miller J.F., Kalganova T., Lipnitskaya N., and Job D. The genetic algorithm as a discovery engine: Strange circuits and new principles. In *Proc. of the AISB'99 Symposium on Creative Evolutionary Systems, CES'99*, ISBN 1-902956-03-6, pages 65–74. Edinburgh, UK, The Society for the Study of Arificial Intelligence and Simulation of Behaviour, April 1999.

21. Coello C. A., Christiansen A. D., and Hernndez A. A. Use of evolutionary techniques to automate the design of combinational circuits. *International Journal of Smart Engineering System Design*, 1999.

Genetic Programming, Ensemble Methods and the Bias/Variance Tradeoff – Introductory Investigations

Maarten Keijzer and Vladan Babovic

Danish Hydraulic Institute,
Agern Alle 5,
Hørsholm Denmark
mak@dhi.dk vmb@dhi.dk

Abstract. The decomposition of regression error into bias and variance terms provides insight into the generalization capability of modeling methods. The paper offers an introduction to bias/variance decomposition of mean squared error, as well as a presentation of experimental results of the application of genetic programming. Finally ensemble methods such as bagging and boosting are discussed that can reduce the generalization error in genetic programming.

1 Introduction

Genetic programming [9] is a general purpose search technique that can be applied to both regression and classification problems. In contrast with linear and non-linear regression, the technique does not presuppose a functional form as it generally proceeds from a definition of lower level building blocks (the function set). Therefore Koza [9] speaks of symbolic regression, as the object of search is both the functional form and the optimal coefficients.

Although a setup in symbolic regression is more general than in standard regression, the question may be posed whether this technique leads to programs with better generalization properties. In order to address this issue, the present paper describes a decomposition of error into its bias and variance constituents. The decomposition is subsequently used to investigate whether the method of symbolic regression is (a) unbiased and (b) reliable.

The bias component reflects the generic ability of a method to fit a particular data set, while the variance error reflects the intrinsic variability of the method in arriving at a solution. Section 2 provides a theoretical introduction into the decomposition of error in bias and variance terms.

Section 3 presents some methods of dealing with the bias/variance tradeoff with a particular emphasis on ensemble methods. In ensemble modeling the

* This work was in part funded by the Danish Technical Research Council (STVF) under the Talent Project N 9800463 entitled "Data to Knowledge – D2K". More information on the project can be obtained through http://www.d2k.dk

results of many runs are combined to form a single, larger model. The expected generalization error for such an ensemble model is reduced from the total error to the error due to the bias error alone. This because the variance component is effectively reduced to zero. The implication for symbolic regression and genetic programming generally is that the generalization error of an ensemble can be dramatically lower than the error for the single best model.

Finally, Section 4 describes some experiments investigating the inherent bias and variance error associated with symbolic regression.

2 Background: Bias, Variance and Estimation Error

As the technique of error decomposition in bias and variance terms is not widely known in the field of genetic programming, this section presents the basic background for the decomposition of the squared error measure. Most of the material in the section follows the notation and derivations presented by Bishop [1] (chapter 9, pp. 333-336). The particular decomposition performed here is based on the sum-of-squares error, but decompositions are possible for other cost functions as well.

Consider a modeling method that produces a model $y(\mathbf{x})$ when it is confronted with some target function $\langle t|\mathbf{x}\rangle$. Usually, the estimate $y(\mathbf{x})$ will depend on the training data D. When a different training set is used, a different estimate $y(\mathbf{x})$ will be produced.

Consider such a model trained using a sum-of-squares error measure. When applying a modeling method that produces estimates $y(\mathbf{x})$ given some data D, a measure of how close the function $y(\mathbf{x})$ is to the desired response $\langle t|\mathbf{x}\rangle$ is the squared error measure:

$$\{y(\mathbf{x}) - \langle t|\mathbf{x}\rangle\}^2 \tag{1}$$

This error depends on the particular data set D that is used for training the model $y(\mathbf{x})$ and on the particular data point $\mathbf{x}$. Integrating this quantity over $\mathbf{x}$ will give the usual sum-of-squares measure. However, a better estimate of the performance of a *method* is to eliminate the dependence on the particular data set D. This is done by considering the average over all possible data sets drawn from D, written as:

$$E_D[\{y(\mathbf{x}) - \langle t|\mathbf{x}\rangle\}^2] \tag{2}$$

Where $E_D[\cdot]$ denotes the expectation operator. The value of this quantity is the error the models $y(\mathbf{x})$ will make when they are trained using (a sample of) data set D and applied to point $\mathbf{x}$. If the modeling method was perfect, then this error would be zero regardless of $\mathbf{x}$. A non-zero error can occur for two distinct reasons. It can happen that the model is on average different from $\langle t|\mathbf{x}\rangle$. This is referred to as *bias*. On the other hand, it may be that the method is very sensitive to the particular data set used in training, so that, at a given $\mathbf{x}$, it is larger or smaller than the target value, this depending on the data set used for training. This is called *variance*. The decomposition of equation (2) into bias and variance terms proceeds by applying some algebraic manipulations. First expand the term inside the expectation operator in equation (2):

$$
\begin{aligned}
\{y(\mathbf{x}) - \langle t|\mathbf{x}\rangle\}^2 &= \{y(\mathbf{x}) - E_D[y(\mathbf{x})] + E_D[y(\mathbf{x})] + E_D[y(\mathbf{x})] + \langle t|\mathbf{x}\rangle\}^2 \\
&= \{y(\mathbf{x}) - E_D[y(\mathbf{x})]\}^2 + \{E_D[y(\mathbf{x})] - \langle t|\mathbf{x}\rangle\}^2 \\
&\quad + 2\{y(\mathbf{x}) - E_D[y(\mathbf{x})]\}\{E_D[y(\mathbf{x})] - \langle t|\mathbf{x}\rangle\}
\end{aligned}
\tag{3}
$$

The third term on the right hand side of equation (3) vanishes since:

$$
E_D[y(\mathbf{x})E_D[y(\mathbf{x})] - E_D[y(\mathbf{x})]^2] = 0
$$
$$
E_D[y(\mathbf{x})\langle t|\mathbf{x}\rangle - E_D[y(\mathbf{x})]\langle t|\mathbf{x}\rangle] = 0
$$
$$
\tag{4}
$$

Which leads to the definition of bias and variance for the sum-of-squares error at point $\mathbf{x}$:

$$
\begin{aligned}
&E_D[\{y(\mathbf{x}) - \langle t|\mathbf{x}\rangle\}^2] \\
&= \underbrace{\{E_D[y(\mathbf{x})] - \langle t|\mathbf{x}\rangle\}^2}_{(bias)^2} + \underbrace{E_D[\{y(\mathbf{x}) - E_D[y(\mathbf{x})]\}^2]}_{variance}
\end{aligned}
\tag{5}
$$

$$
\tag{6}
$$

The bias term measures the extent to which the average model (calculated over all data sets), differs from the desired function $\langle t|\mathbf{x}\rangle$. The variance term measures the extent to which the model is sensitive to the particular data set used in training. Note that this notion of bias and variance critically relies on the notion of the *average model* $E_D[y(\mathbf{x})]$ which is in effect the average of all predictions at point $\mathbf{x}$.

It is instructive to consider two limits for the choice of the functional form for $y(\mathbf{x})$. Suppose that the target data is generated from a smooth function $h(x)$ to which zero mean noise ϵ is added, so that:

$$
t = h(\mathbf{x}) + \epsilon
\tag{7}
$$

One modeling method could be to pick a fixed function $g(\mathbf{x})$ independent of the data set D. The variance term from equation (5) would then vanish as all generated models would be equivalent. However, without any prior knowledge used in selecting $g(\mathbf{x})$, the bias term can be very high since no attention at all was given to the data.

The opposite extreme is to take a function that fits the training data perfectly, such as an exact linear interpolation. In this case the bias term vanishes at the data points themselves, since:

$$E_D[y(\mathbf{x})] = E_D[h(\mathbf{x}) + \epsilon] = h(\mathbf{x}) = \langle t|\mathbf{x} \rangle \qquad (8)$$

The bias will be typically small in the neighbourhood of the data points because $h(\mathbf{x})$ is smooth. The variance will however be significant as:

$$E_D[\{y(\mathbf{x}) - E_D[y(\mathbf{x})]\}^2] = E_D[\{y(\mathbf{x}) - h(\mathbf{x})\}^2] = E_D[\epsilon^2] \qquad (9)$$

The overall error the linear interpolation makes on the data set D is then proportional to the square of the noise. The error will typically be much larger for points not drawn from D as here the bias term will be non-zero.

Equation (5) clearly illustrates that it is desirable to have both low bias and low variance since both terms contribute to the squared estimation error to the same extent. However, these two objectives are contradicting each other, hence the bias/variance trade-off. A function that is closely fitted to the data will tend to have a large variance and subsequently a large expected error. This variance can be decreased by smoothing the function, but smoothing too much will then increase the bias term and again the expected error. Balancing the bias and variance terms by controlling a smoothing parameter is referred to as *regularization*.

3 Dealing with the Bias/Variance tradeoff

3.1 Declarative Bias

In statistical learning theory the bias/variance trade-off is often referred to as a dilemma. Geman *et al.* [5] go so far as to state that the dilemma can be circumvented only if one is willing to sacrifice generality, that is, *purposefully* introduce bias, in order to reduce the variance. When this bias is appropriate in the problem domain, such a method will reliably give good results.

In genetic programming several methods have been investigated for introducing such bias. This bias then takes the form of *declarative* bias. Few examples are: context free grammars [17, 16], strong typing [13, 3], attribute grammars [11] and physical units of measurement [8]. Background knowledge about the problem domain is introduced to increase the level of bias and — due to the tradeoff — decrease the magnitude of variance. When the *declarative bias* so introduced does not increase the *bias error*, the method will produce better solutions.

3.2 Ensemble Methods

There are however methods that reduce the overall error by reducing the error due to variance, without increasing the bias.

One such method for obtaining better performance makes explicit use of the fact that the error due to variance is caused by deviation from the *average* model $E_D[y(\mathbf{x})]$. As this *average* model can be easily calculated, it can be used instead of a single best model. By using this *average* model as the *resulting model of many runs*, the error due to variance is effectively eliminated. The remaining error contribution is due to the bias alone[1]—which represents the generic ability of the method to deal with the problem.

The technique of combining multiple models into a single one is referred to as *ensemble* modeling [6]. When the process of averaging is used in conjunction with bootstrapping to obtain the training sets, the technique is referred to as bootstrap aggregating (bagging) [2]. Other familiar ensemble methods are boosting [4] and stacking [18]. All of these methods reduce the error due to variance while leaving the bias constituent unaltered. Ensemble methods hold great promise for unbiased, high variance methods. Much experimental evidence has been gathered in the fields of decision tree induction and neural networks where these methods offer considerable improvement in estimating and reducing generalization error. Even more interestingly, empirical investigations show that combining relatively simple, diverse predictors seems to be a very powerful approach.

Iba [7] applied the ensemble methods of boosting and bagging within the framework of genetic programming and obtained encouraging results. Sections 2 and 4 offer some arguments why this is to be expected. The paper also proposes a simple averaging method of obtaining more reliable approximations using basically unaltered symbolic regression.

The average model as introduced above is calculated without using testing data. Therefore this offers a simple and feasible approach to reducing the generalization error in symbolic regression.

4   Experimental Setup

The theoretical underpinnings of bias/variance decomposition given in Section 2 are given a concrete experimental form by considering the following setup. Given a data set D, firstly a disjoint training D_{train} and testing set D_{test} are created. D_{train} is further subdivided by randomly drawing cases to form M independent subsets. These M sets are then used as the training sets for M independent runs of the genetic programming algorithm. Given these M models $y_1(\mathbf{x}), \ldots, y_M(\mathbf{x})$ and testing data $D_{test} = (\mathbf{x}_1, t_1), \ldots, (\mathbf{x}_N, t_N)$, the *generalization* error is defined as

$$\frac{1}{NM} \sum_{i=1}^{N} \sum_{j=1}^{M} (t_i - y_j(\mathbf{x}_i))^2 \qquad (10)$$

---

[1] There is no free lunch here: using the average model instead of a single model leads to a new bias/variance tradeoff. Similar experiments can be performed for this new modeling technique. The difference is that a new decomposition over different runs producing different average models can be performed. The expected improvement originates in the fact that the associated variance of these average models is lower than the variance of the original setup.

By introducing the *average* model $\bar{y}(\mathbf{x}_i) = \frac{1}{M} \sum_{j=1}^{M} y_j(\mathbf{x}_i)$ the error can be decomposed into bias or variance by a straightforward rearrangement of terms.

$$(bias)^2 = \frac{1}{N} \sum_{i=1}^{N} (t_i - \bar{y}(\mathbf{x}_i))^2 \tag{11}$$

$$variance = \frac{1}{NM} \sum_{i=1}^{N} \sum_{j=1}^{M} (\bar{y}(\mathbf{x}_i) - y_j(\mathbf{x}_i))^2 \tag{12}$$

Again the bias term depends on the target distribution $\langle t|\mathbf{x}\rangle$, while the variance term does not.

4.1 Stability of the Estimates and Trimming

Genetic programming often produces models with extremely poor generalization properties. In order to address this issue, the actual calculation methods used in this study are slightly different from the ones presented in equations (11) and (12). When the resulting formula of a single run is applied to unseen data, it often happens that it is undefined at certain points or has extreme values (see for example Figures 5(a) and (b)). By including such predictions in the *average model*, the extreme value would dominate the outcome of the average model at these points, leading to an unstable model and subsequently unstable bias and variance estimates [2]

In order to prevent this, the predictions participating in the *average model* $\bar{y}(\mathbf{x})$ are *trimmed*. This trimming is done by excluding the highest and lowest predictions with a pre-specified percentage, usually of 10%. This trimming is performed before the bias and variance errors are calculated. Practically, this means that when 50 predictions are made to predict a single point in the validation set, only the middle 40 predictions are used for calculating the average. In most cases this stabilizes the averaging process.

4.2 Regression of $t = x^3 - x^2 - x - 1 + \epsilon$

To investigate the relationship between bias, variance and total error in genetic programming, a regression run was set up to estimate the equation $t = x^3 - x^2 - x - 1 + \epsilon$. In this investigation, a flexibility parameter is used to test its effect on the bias and the variance. Such a flexibility parameter is supposed to affect the sensitivity to the training data: it is supposed to govern the fitting capability. Examples of such parameters are: the number of neighbours in nearest neighbour methods, the number of hidden nodes in neural networks, the depth of

---

[2] Variance, being a second-order moment statistic becomes unstable when the tails of the distribution are heavier than those of a Gaussian distribution. Bias is calculated from an averaging process. An average is a first-order moment statistic and becomes unstable when the tails of the distribution are heavier than those of a Cauchy distribution [14].

a decision tree, the learning time of a neural network. For genetic programming obvious choices are: the maximum size of the trees, duration of the run, number of generations and population size... The flexibility parameter investigated here is the maximum size of individual trees. Chosen values were: 10, 20, 30, ..., 80, 100, 150, 200.

The function set consisted of addition, multiplication and subtraction. For the experiments 500 data points were uniformly generated between -1 and 1. White noise of standard deviation 0.1 was added to the outcomes. From this set, 250 points were kept aside for testing – all graphs and results are presented for this data set. From the remaining data, for every run, 10 points were drawn randomly without replacement to form the training set. Ten points are chosen as this does not give a dense coverage of the input space — a situation that is common in multidimensional real world problems. For each of the 11 parameter settings, 50 runs were performed. Each run lasted 100 generations with a population size of 1000. The best individual in the last generation was taken to be the result of the run.

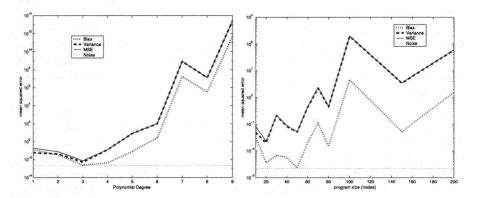

Fig. 1. (a) Bias and variance of (a) polynomial regression and (b) genetic programming on $x^3 - x^2 - x - 1 + \epsilon$ (not trimmed).The error for both figures is plotted on a logarithmic scale

Figure 1 illustrates the difficulty in inferring a cubic polynomial based on 10 noisy data points. Figure 1(a) shows the bias/variance decomposition for a polynomial regression while varying the degree of the polynomial. Using the same setup, Figure 1(b) shows the decomposition of error for symbolic regression. For both methods the variance is so large that the *average* model is unstable, leading to a high bias. Note that the error made by genetic programming does not increase nearly as rapidly as the error by polynomial regression.

Figure 2(a) shows the errors attributed to bias and variance for this setup together with the total error and the artificially introduced noise. The trimming percentage used was 10%. Even with the trimmed predictions the variance varies greatly, whereas the bias term remains quite constant and stable. Bias drops to the level of the inherent noise (the optimal value) at program size 20, and remains close to it for larger program sizes. The error due to variance increases steadily over increased program size. Setting the maximum program size to high values leads to a higher number of overfitted programs.

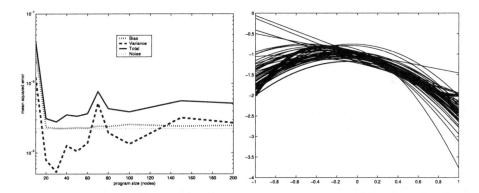

Fig. 2. (a) 10% trimmed bias/variance decomposition for the regression of $x^3 - x^2 - x - 1 + \epsilon$)(b) The 50 equations produced by GP for a maximum program size of 10 nodes. The target function is plotted bold.

Figure 2(a) shows that with small size programs the error due to bias is large. This is not surprising as the space to represent these programs is limited. What is surprising is that in these runs with small program sizes, the error due to variance is relatively large. Figure 2(b) shows the trimmed predictions for programs of size 10. The target function is shown in bold. It is clear that the best genetic programming could do is to fit a parabolic function. The large error due to variance seems to be caused by the inability of the method to determine a set of coefficients for the polynomial.

Figure 3(a) shows the trimmed predictions for the best setting in this experiment, a maximum program size of 30 nodes. In this setting there was enough space to represent an approximate solution, yet not enough to overfit on a regular basis. Figure 3(b) shows a similar plot for a maximum of 70 nodes. The high variance for this setting (see Figure 2(a)) can be attributed to one *rogue* solution in $-1.0 < x < -0.8$, a similar outlier in the range $0.8 < x < 1.0$, as well as a generally large variance in the range $0.8 < x < 1.0$.

Figure 2 shows that for this problem genetic programming quickly reaches the optimal bias level. In this range variance is low as well. This indicates that genetic programming finds good approximations reliably.

With increasing program size the bias remains relatively unchanged, but the variance increases. This indicates that genetic programming starts to produce overfitted solutions. This in turn implies that the maximum program size parameter is indeed an effective flexibility parameter, governing the sensitivity of the method to the data.

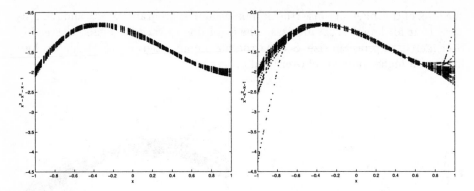

Fig. 3. The 40 middle predictions for a maximum program size of (a) 30 and (b) 70 nodes. The functions are trained on 10 randomly chosen points of $x^3 - x^2 - x - 1 + \epsilon$.

In contrast with the hypothesis that increased program size (bloat) is semantically neutral and impedes search [15, 12, 10] genetic programming does seem to utilize the increased number of degrees of freedom to fit the training data. As the target function in the present experiment was perfectly attainable by genetic programming, the optimal bias was achieved with small program sizes. The next section addresses a setup where the optimal solution can not be represented using the provided function set.

4.3 Regression of $t = x + 0.3\sin(2\pi x) + \epsilon$

To investigate the effect of both the program size and the number of cases used for training, a new experiment was performed. Here the function to approximate is a symbolic equivalent of the function $t = x + 0.3\sin(2\pi x) + \epsilon$, where ϵ represents uniform noise drawn from the range -0.1 – +0.1. The range from which the x variable was drawn was 0 – 1. In addition to multiplication, addition and subtraction used in the previous experiment, the function set included division. This function set is uncapable of perfectly representing the target function as it does not contain the trigonometric functions.

Training data were selected according to the following scheme. Firstly training sets of varying sizes were created: 10, 20, 30, 40, 50, 75, 100, ..., 250. Subsequently, for every independent run of the genetic programming system 75% of these cases were used for training, rounding down when fractions occurred. This setup resembles the real world situation with sparse data coverage. Again each parameter setting was tested using 50 independent runs.

Figure 4 shows the corresponding bias and variance plot. Both the bias and variance calculations were additionally treated using a trimming percentage of 10%. Trimming was even more necessary here than in the previous experiment due to the influence of the division operator. Figure 5 clearly shows two examples of genetic programming using division to introduce an asymptote at a convenient location.

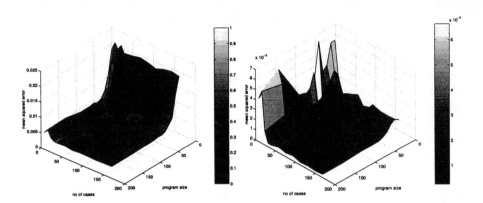

Fig. 4. (a) Bias and (b) variance plot for symbolic regression of $x + 0.3sin(2\pi x) + \epsilon$. The trimming percentage was set to 10%

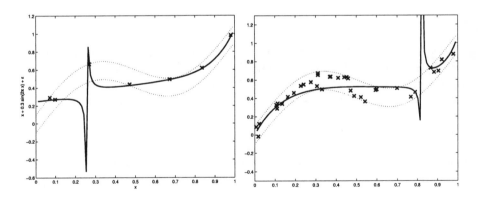

Fig. 5. Two examples of the use of asymptotes to overfit. (a) a situation with sparse data and (b) a situation with a more dense covering.

Figure 4 shows the regions were genetic programming produces models with unsafe behaviour. When program sizes are small, both bias and variance errors are high, indicating inadequacy of the method. With few data points, the error due to variance oscillates wildly, larger programs sizes having the higher variance. Due to trimming, the bias error remains relatively stable.

When data is abundant, best results are achieved with a maximum program size greater than 60 nodes. Although 60 nodes provides more degrees of freedom than necessary to produce a good approximation of the data (for example: 60 nodes can represent a polynomial of degree 14), genetic programming seems to need this much space to reduce the bias for this problem.

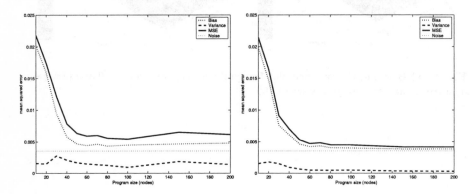

Fig. 6. Bias/variance decomposition for inferring $y = x + 0.3sin(2\pi x) + \epsilon$: (a) 30 cases used in training, (b) 250 cases

Figure 6(a and b) show two slices of Figure 4 for different numbers of cases used for training. With 30 data points, the optimal bias level is not attained. At a program size of 60 the best bias level is reached, and larger program sizes do not further decrease this bias. On the contrary, there is a slight increase in bias error. Figure 6(b) shows decomposition for a training set consisting of 250 data points. The bias is asymptotically approaching the optimal level (i.e. the level of noise) with increasing program size. The error due to variance decreases as well, indicating that given enough data genetic programming is perfectly capable of obtaining a good fit. Therefore, increasing the maximum size helps in attaining this performance level.

Here the performance of the average model is tested as well. Figure 7(b) shows the predictions of the average model (trimmed by 10%) on this problem, whereas Figure 7(a) shows all 50 model predictions that form the basis for the calculation of an average model. It is clear that the figures that the 10% trimmed average model produces a reasonable fit on the data, while its 50 constituents are not nearly as reliable.

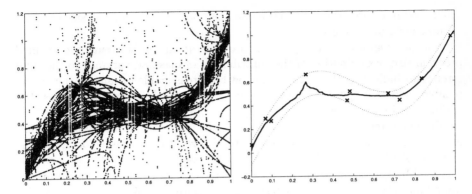

Fig. 7. (a) The predictions of all 50 models on the test set. (b) The 10% trimmed average model (maximum program size 200), the bounds of the target function and the 10 data points used to arrive at this prediction.

5 Discussion

The bias/variance decomposition of error offers a fresh perspective on the performance of genetic programming. In model induction both the bias and the variance of a method should be minimized as they both contribute to the total error. These two terms contradict each other to a certain extent, implying a seemingly unavoidable tradeoff. The experiments indicate that genetic programming applied to symbolic regression is a low bias, but high variance induction technique. The variance in the results is often so high that the trimming needs to be employed to obtain a robust estimate of the average model. This is a clear indication that simple symbolic regression as applied here misses a smoothing or regularization component.

In the limit of data abundance, genetic programming seems to be capable to approximate the target function reliably, provided that the size of the programs is unconstrained. At the same time small programs appear to be strongly biased (see the ridge in Figure 4(a) where bias remains constant regardless of the number of data points available). Small models also exhibit high variance, leading to the straightforward conclusion that genetic programming needs more space than 15 nodes even when applied to relatively simple problems.

The situation of data abundance is however not always a realistic scenario, as data coverage in multi-dimensional spaces is often poor. Furthermore many techniques exist that can approximate a target function to any degree of accuracy with much less effort than genetic programming.

When data is sparse, the situation is a little more complicated. Small programs are again strongly biased and highly variant. Large programs exhibit high variance as well, even to an extent that this destabilizes bias (see Figure 1(b)). Here the trade-off takes effect: larger models lead to a higher variance, yet the bias error decreases. If the total error is the only quantity investigated, the premature conclusion would be that large program size hinders the search, yet when used in combination with ensemble methods, larger program size can lead to more accurate models.

The bias/variance decomposition provides an aid in selecting the optimal program size. Examination of the contribution of bias and variance to the total error helps indicating the location of a low bias, reasonable variance region where reliable models might be found (this in contrast with a reasonable bias, low variance region which is of no interest).

Ensemble modeling provides an alternative approach to obtaining a single model. Since this method reduces the total error to the bias error alone, the question of destabilized bias needs to be addressed with care. Destabilized bias is caused by genetic programming's ability to produce extremely poor solutions (an infinite overfit) on a regular basis. Appropriate trimming can eliminate this effect to some extent. An optimal trimming scenario should stabilize bias while leaving most of the variance intact, as a high degree of variance is instrumental in establishing an ensemble model. As the calculation of variance is independent of the target function, trimming scenarios can be designed after the programs have been created, even when the *ensemble* model is applied. In the authors' experience, it is rather straightforward to establish a reasonable trimming setup: one should remove those predictions that have an dominant effect in the calculation of the average model.

Conclusions

This paper presented an introduction to the bias/variance tradeoff studied intensively in the fields of neural networks and machine learning. It provides some empirical evidence that genetic programming can benefit greatly from this work in those fields. Considerations of the contribution of bias and variance to the total error, together with *ensemble* methods to reduce the error due to variance, lead to the following recipe for applying genetic programming to regression problems:

- Perform many runs using some sampling method to create unique datasets for different runs.
- Do not constrain the maximum size of the programs much. Values between 100 nodes and 200 nodes seem to be reasonable.
- Combine the results of all the runs in a single *average* model. Use this model together with a trimming scenario in order to eliminate predictions that destabilize this averaging process.

This recipe follows both from theoretical considerations and experimental evidence. Larger size programs effectively handle bias error. Ensemble methods, eliminate the contribution of variance to the total error. The total error of the *average* model is then reduced to the bias term.

Simple symbolic regression as studied here suffers from extremely poor solutions that show up regularly. It was deemed necessary to provide a trimming scheme to stabilize the average model. How to trim optimally is as of yet unresolved.

The computational burden of this procedure does not exceed that of a cross-validation approach as an equal number of runs need to be performed. Furthermore, due to the high variance of the results between runs of genetic programming, relying on results produced by single runs should be strongly discouraged.

References

1. Christopher M. Bishop. *Neural Networks for Pattern Recognition*. Clarandon Press, Oxford, 1995.
2. Leo Breiman. Bagging predictors. *Machine Learning*, 26:123–140, 1996.
3. Chris Clack and Tina Yu. Performance enhanced genetic programming. In Peter J. Angeline, Robert G. Reynolds, John R. McDonnell, and Russ Eberhart, editors, *Proceedings of the Sixth Conference on Evolutionary Programming*, volume 1213 of *Lecture Notes in Computer Science*, Indianapolis, Indiana, USA, 1997. Springer-Verlag.
4. Y. Freund and R. Schapire. Experiments with a new boosting algorithm. In *Proceedings of the International Conference on Machine Learning*, pages 148–156, 1996.
5. Stuart Geman, Elie Bienenstock, and René Doursat. Neural networks and the bias/variance dilemma. *Neural Computation*, 4:1–58, 1992.
6. Lars Kai Hansen and P. Salamon. Neural network ensembles. *IEEE Transactions on pattern analysis and machine intelligence*, 12:993–1001, 1990.
7. Hitoshi Iba. Bagging, boosting, and bloating in genetic programming. In Wolfgang Banzhaf, Jason Daida, Agoston E. Eiben, Max H. Garzon, Vasant Honavar, Mark Jakiela, and Robert E. Smith, editors, *Proceedings of the Genetic and Evolutionary Computation Conference*, volume 2, pages 1053–1060, Orlando, Florida, USA, 13-17 July 1999. Morgan Kaufmann.
8. Maarten Keijzer and Vladan Babovic. Dimensionally aware genetic programming. In Wolfgang Banzhaf, Jason Daida, Agoston E. Eiben, Max H. Garzon, Vasant Honavar, Mark Jakiela, and Robert E. Smith, editors, *Proceedings of the Genetic and Evolutionary Computation Conference*, volume 2, pages 1069–1076, Orlando, Florida, USA, 13-17 July 1999. Morgan Kaufmann.
9. John R. Koza. *Genetic Programming: On the Programming of Computers by Means of Natural Selection*. MIT Press, Cambridge, MA, USA, 1992.
10. William B. Langdon, Terry Soule, Riccardo Poli, and James A. Foster. The evolution of size and shape. In Lee Spector, William B. Langdon, Una-May O'Reilly, and Peter J. Angeline, editors, *Advances in Genetic Programming 3*, chapter 8, pages 163–190. MIT Press, Cambridge, MA, USA, June 1999.
11. Lionel Martin, Frédéric Moal, and Cristel Vrain. Declarative expression of biases in genetic programming. In Wolfgang Banzhaf, Jason Daida, Agoston E. Eiben, Max H. Garzon, Vasant Honavar, Mark Jakiela, and Robert E. Smith, editors, *Proceedings of the Genetic and Evolutionary Computation Conference*, volume 1, pages 401–408, Orlando, Florida, USA, 13-17 July 1999. Morgan Kaufmann.
12. Nicholas Freitag McPhee and Justin Darwin Miller. Accurate replication in genetic programming. In L. Eshelman, editor, *Genetic Algorithms: Proceedings of the Sixth International Conference (ICGA95)*, pages 303–309, Pittsburgh, PA, USA, 15-19 July 1995. Morgan Kaufmann.
13. David J. Montana. Strongly typed genetic programming. *Evolutionary Computation*, 3(2):199–230, 1995.

14. Chrysostomos L. Nikias and Min Shao. *Signal Processing with Alpha-Stable Distributions and Applications.* John Wiley and Sons, 1995.
15. Peter Nordin and Wolfgang Banzhaf. Complexity compression and evolution. In L. Eshelman, editor, *Genetic Algorithms: Proceedings of the Sixth International Conference (ICGA95)*, pages 310–317, Pittsburgh, PA, USA, 15-19 July 1995. Morgan Kaufmann.
16. Conor Ryan, J. J. Collins, and Michael O Neill. Grammatical evolution: Evolving programs for an arbitrary language. In Wolfgang Banzhaf, Riccardo Poli, Marc Schoenauer, and Terence C. Fogarty, editors, *Proceedings of the First European Workshop on Genetic Programming*, volume 1391 of *LNCS*, pages 83–95, Paris, 14-15 April 1998. Springer-Verlag.
17. Peter Alexander Whigham. *Grammatical Bias for Evolutionary Learning.* PhD thesis, School of Computer Science, University College, University of New South Wales, Australian Defence Force Academy, 14 October 1996.
18. David Wolpert. Stacked generalization. *Neural Networks*, 5:241–259, 1992.

Evolution of a Controller with a Free Variable Using Genetic Programming

John R. Koza
Stanford University, Stanford, California
koza@stanford.edu

Jessen Yu
Genetic Programming Inc., Los Altos, California
jyu@cs.stanford.edu

Martin A. Keane
Econometrics Inc., Chicago, Illinois
makeane@ix.netcom.com

William Mydlowec
Genetic Programming Inc., Los Altos, California
myd@cs.stanford.edu

ABSTRACT

A mathematical formula containing one or more free variables is "general" in the sense that it provides a solution to an entire category of problems. For example, the familiar formula for solving a quadratic equation contains free variables representing the equation's coefficients. Previous work has demonstrated that genetic programming can automatically synthesize the design for a controller consisting of a topological arrangement of signal processing blocks (such as integrators, differentiators, leads, lags, gains, adders, inverters, and multipliers), where each block is further specified ("tuned") by a numerical component value, and where the evolved controller satisfies user-specified requirements. The question arises as to whether it is possible to use genetic programming to automatically create a "generalized" controller for an entire category of such controller design problems — instead of a single instance of the problem. This paper shows, for an illustrative problem, how genetic programming can be used to create the design for both the topology and tuning of controller, where the controller contains a free variable.

1   Introduction

Automatic controllers are ubiquitous in the real world (Astrom and Hagglund 1995; Boyd and Barratt 1991; Dorf and Bishop 1998). The output a controller is fed into the to-be-controlled system (conventionally called the *plant*). The purpose of a controller is to force, in a meritorious way, the plant's response to match a desired response (called the *reference signal* or *setpoint*). For example, the cruise control device in an car continuously adjusts the engine (the plant) based on the difference between the driver-specified speed (reference signal) and the car's actual speed (plant response).

The measure of merit for controllers typically incorporate a variety of interacting and conflicting considerations. Examples are the time required to force the plant response to reach a desired value, the degree of success in avoiding significant overshoot beyond the desired value, the time required for the plant to settle, the ability of the controller to operate robustly in the face of minor variations in the actual characteristics of the plant, the ability of the controller to operate robustly in the face of disturbances that may be introduced between the controller and the plant's input, the ability of the controller to operate robustly in the face of sensor noise that may be added to the plant output, the controller's compliance with various frequency domain constraints (e.g., bandwidth), and the controller's compliance with constraints on the magnitude of the control variable or the plant's internal state variables.

Controllers are often represented by block diagrams. A block diagram is a graphical structure containing (1) directed lines representing the (one-directional) flow of time-domain signals within the controller, (2) time-domain signal processing blocks (e.g., integrator, differentiator, lead, lag, gain, adder, inverter, and multiplier), (3) external input points from which the controller receives signals (e.g., the reference signal and the plant output that is externally fed back from the plant to the controller), and (4) output point(s) for the controller, conventionally called the *control variable*). Some signal processing blocks have multiple inputs (e.g., an adder), but they all have exactly one output. Because of this restriction, block diagrams for controllers usually contain *takeoff points* that enable the output of a block to be disseminated to more than one other point in the block diagram. Many blocks used in controllers (e.g., gain, lead, lag, delay) possess numerical parameters. The determination of these values is called *tuning*. Block diagrams sometimes also contain internal feedback (internal loops in the controller) or feedback of the controller output directly into the controller.

The design (synthesis) of a controller entails specification of both the topology and parameter values (tuning) for the block diagram of the controller such that the controller satisfies user-specified high-level design requirements. Specifically, the design process for a controller entails making decisions concerning the total number of signal processing blocks to be employed in the controller, the type of each block (e.g., integrator, differentiator, lead, lag, gain, adder, inverter, and multiplier), the interconnections between the inputs and outputs of each signal processing block and between the controller's external input and external output points, and the values of all required numerical parameters for the signal processing blocks.

Genetic programming has recently been used to create a block diagram for satisfactory controller for a particular two-lag plant and a particular three-lag plant (Koza, Keane, Yu, Bennett, and Mydlowec 2000). Both of these genetically evolved controllers outperformed the controllers designed by experts in the field of control using the criteria originally specified by the experts. However, both of these controllers were for plants having lags with a particular value of the time constant, τ.

A mathematical formula containing one or more free variables is "general" in the sense that it provides a solution to an entire category of problems. For example, the familiar formula for solving a quadratic equation contains free variables representing the coefficients of the equation. The question arises as to whether it is possible to evolve a controller for an *entire category* of plants — instead of one particular plant. In other words, is it possible to design a "generalized" controller containing one or more free variables. The fact that genetic programming permits the evolution of mathematical expressions containing free variables suggests that this might be possible. This paper shows, for an illustrative problem, that such a "generalized" controller can indeed be automatically created by means of genetic programming, that the genetically evolved controller has reasonable and interpretable characteristics, and that the genetically evolved controller performs better than the textbook human-designed controller for the illustrative problem.

Section 2 shows how genetic programming can automatically synthesize the design of a controller. Section 3 itemizes the preparatory steps for applying genetic programming to an illustrative problem with a free variable. Section 4 presents results.

2 Genetic Programming and Control

Genetic programming is commonly used in several different ways. As one example, genetic programming is often used as an automatic method for creating a program tree to solve a problem (Koza 1992). The individual programs that are evolved by genetic programming are typically multi-branch programs consisting of result-producing branches, automatically defined functions (subroutines), and other types of branches (e.g., iterations, loops, recursions). In this approach, the program tree is simply executed. The result of the execution may be a set of returned values, a set of side effects on some other entity (e.g., an external entity such as a robot or an internal entity such as computer memory), or a combination of returned values and side effects. In this approach, the functions in the program are sequentially executed, in time, in accordance with a specified order of evaluation such that the result of executing one function is available at the time when the next function is to be executed. Early work on the problem of automatically creating controllers used this conventional approach to genetic programming. This work includes, but is not limited to, Andersson, Svensson, Nordin, and Nordin, and Mats 1999; Banzhaf, Nordin, Keller, and Olmer 1997; Whitley, Gruau, and Preatt 1995; and Koza 1992.

As a second example, genetic programming is often also used to automatically create program trees which can be used in conjunction with a developmental process to design complex structures, such as neural networks (Gruau 1992) and analog electrical circuits (Koza, Bennett, Andre, and Keane 1996; Koza, Bennett, Andre, and Keane 1999; Koza, Bennett, Andre, Keane, and Brave 1999). In this approach, the program tree is interpreted as a set of instructions for constructing the desired structure. The construction process is implemented by applying the functions of a program tree to an embryonic structure so as to develop the embryo into a fully developed structure. As in the first approach, the functions in the program are executed separately, in time, in accordance with a specified order of evaluation.

In a closed-loop continuous-time feedback system consisting of a plant and its controller, the output of the controller is input to the plant and the output of the plant is, in turn, input to the controller. Because of these continuous-time interactions, the sequential order of evaluation inherent in the two above approaches to genetic programming not a suitable. In the past, both genetic algorithms and genetic programming have been used for synthesizing controllers having mutually interacting continuous-time variables and continuous-time signal processing blocks (Man, Tang, Kwong, and Halang; 1997, 1999; Crawford, Cheng, and Menon 1999; Dewell and Menon 1999; Menon, Yousefpor; Lam, and Steinberg 1995; Sweriduk, Menon, and Steinberg 1998, 1999). In addition, the PADO system and neural programming (Teller 1996a, 1996b; Teller and Veloso 1995a, 1995b, 1995c, 1995d, 1995e, 1996) provide a graphical program structure representation for such multiple independent processes. Also, Marenbach, Bettenhausen, and Freyer (1996) used automatically defined functions to represent internal feedback in a system (used for a system identification problem) where the overall multi-branch program represented a continuous-time system with continuously interacting processes. The essential features of their approach has been subsequently referred to as "multiple interacting programs" (Angeline 1997, 1998a, 1998b; Angeline and Fogel 1997).

In this paper, a program tree will constitute a description of the block diagram of a controller. The block diagram consists of time-domain signal processing functions linked by directed lines representing the flow of information. There is no order of evaluation of the functions and terminals of the program tree. Instead, the signal processing blocks of the controller and the plant interact with one another other as part of a closed system in the manner specified by the topology of the block diagram.

Figure 1 is a block diagram for an illustrative system containing a controller and a plant. The directed lines in a block diagram represent continuous-time signals while the blocks represent signal processing functions that operate in the time domain. The output of the controller 500 is a control variable 590 which is, in turn, the input to the plant 592. The plant has one output (plant response) 594. The plant response is fed back (externally as signal 596) and becomes one of the controller's two inputs. The controller's second input is the reference signal 508. The fed-back plant response 596 and the externally supplied reference signal 508 are compared (by subtraction here). Notice that the takeoff point 520 of figure 1 provides a way to disseminate the a particular result (the result of the subtraction 510) to three places in the block diagram. The system in this figure is called a "closed loop" system because there is external feedback of the plant output back to the controller.

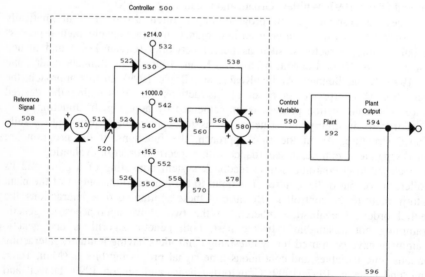

Figure 1 Block diagram of a plant and a PID controller.

Since this controller's output (control variable 590) is the sum of a proportional (P) term (gain block 530, with an amplification factor of 214.0), an integrating (I) term (integrator 560 preceded by the gain block 540, with an amplification factor of 1,000.0), and a differentiating (D) term (derivative block 570 preceded by the gain block 550, with an amplification factor of 15.5), this controller is called a PID controller.

Figure 2 presents the block diagram for the PID controller of figure 1 as a program tree. The internal points of this tree represent the signal processing blocks contained in the block diagram of figure 1 (i.e., derivative, integrator, gain, subtraction, addition) and other functions. The external points represent numerical constants and time-domain signals, such as the reference signal and plant output (and other terminals). Notice that automatically defined function ADF0 (left branch) produces a time-domain signal that equals the result of subtracting the plant output from the reference signal. The three references to ADF0 in the result-producing (right) branch disseminate the

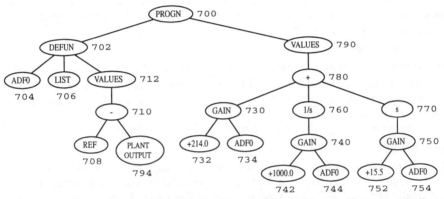

Figure 2 Program tree representation of the PID controller of figure 1. The automatically defined function ADF0 (left) subtracts the plant output from the reference signal and makes this difference available at three points in the result-producing branch (right).

result of this subtraction and correspond to the takeoff point 520 of figure 1.

In the style of ordinary computer programming, a reference to an automatically defined function (subroutine) such as ADF0 from inside the function definition for ADF0 would be considered to be a recursive reference. However, in the context of control systems, an automatically defined function that references itself represents a loop in the controller's block diagram (i.e., internal feedback inside the controller).

3   Illustrative Problem

The problem is to create both the topology and parameter values for a robust "generalized" controller for a three-lag plant, where the controller contains a free variable representing the plant's time constant, τ. This free variable, τ, is allowed to range over two orders of magnitude. The transfer function of the three-lag plant is

$$G(s) = \frac{K}{(1+\tau s)^3}$$

The controller is to be designed so that plant output reaches the level of the reference signal so as to minimize the integral of the time-weighted error (ITAE), such that the overshoot in response to a step input of the reference signal (setpoint) is less than 2%, and such that the controller is robust in the face of variation in the plant's internal gain, K (tested by values of 1.0 and 2.0). The input to the plant is limited to the range between -40 and +40 volts.

The techniques in Astrom and Hagglund 1995 yield a credible controller for this problem.

3.1   Program Architecture

Since the to-be-synthesized controller has one output (control variable), each program tree in the population has one result-producing branch. Each program tree in the initial random population (generation 0) has no automatically defined functions. However, after generation 0, the architecture-altering operations may insert (and delete) automatically defined functions. Automatically defined functions may be used for takeoff points, internal feedback within the controller, and reuse of portions of the block diagram. The permitted maximum of five automatically defined functions is more than sufficient for this problem.

3.2   Terminal Set

An arithmetic-performing subtree containing perturbable numerical terminals, arithmetic operations, and a free variable representing the plant's time constant, τ, is used to establish the numerical parameter value for each signal processing block possessing a parameter.

The terminal set, T_{aps}, for such arithmetic-performing subtrees is

$$T_{aps} = \{\Re, \tau\},$$

where $\Re$ denotes perturbable numerical values in the range from -3.0 and +3.0. In the initial random generation of a run, each perturbable numerical value is set, individually and separately, to a random value in this range. In later generations, this perturbable numerical value is perturbed by a Gaussian probability distribution (with standard deviation of 1.0). A constrained syntactic structure maintains a function and terminal set for the arithmetic-performing subtrees and a different function and terminal set (below) for all other parts of the program tree.

The remaining terminals are time-domain signals. The terminal set, T, for the result-producing branch and any automatically defined functions (except the arithmetic-performing subtrees described above) is

T = {REFERENCE_SIGNAL, CONTROLLER_OUTPUT, PLANT_OUTPUT, CONSTANT_0}.

Space does not permit a detailed description of the various terminals used herein (although the meaning of the above terminals should be clear from their names). See Koza, Keane, Yu, Bennett, and Mydlowec 2000 for details.

3.3   Function Set

The function set, F_{aps}, for the arithmetic-performing subtrees is

F_{aps} = {ADD_NUMERIC, SUB_NUMERIC, MUL_NUMERIC, DIV_NUMERIC}.

The two-argument DIV_NUMERIC function divides the first argument by the second argument, except that the quotient is never allowed to exceed 10^5.

The function set, F, for the result-producing branch and any automatically defined functions (except the arithmetic-performing subtrees described above) consists of continuous-time signal processing functions and automatically defined functions.

F = {GAIN, INVERTER, LEAD, LAG, LAG2,
    DIFFERENTIAL_INPUT_INTEGRATOR, DIFFERENTIATOR,
    ADD_SIGNAL, SUB_SIGNAL, ADD_3_SIGNAL, ADF0, ADF1, ADF2,
    ADF3, ADF4}.

The definition of the above signal processing functions is suggested by their names. See Koza, Keane, Yu, Bennett, and Mydlowec 2000 for details. ADF0, ..., ADF4 denote automatically defined functions added during the run by the architecture-altering operations.

3.4   Fitness Measure

Genetic programming conducts a probabilistic search algorithm through the space of compositions of the available functions and terminals. The search is guided by a fitness measure. The fitness measure is a mathematical implementation of the high-level requirements of the problem and is couched in terms of "what needs to be done" — not "how to do it." Construction of the fitness measure requires translating the high-level requirements of the problem into a mathematically precise computation. The fitness measure may incorporate any measurable, observable, or calculable behavior or characteristic or combination of behaviors or characteristics. The fitness measure for most problems of controller design is multi-objective in the sense that there are several different (generally conflicting and interacting) requirements.

The fitness of each individual is determined by using the program tree (i.e., the result-producing branch and any automatically defined functions that may have been created during the run by the architecture-altering operations) to generate an interconnected sequence of signal processing blocks — that is, a block diagram for the controller. We use our modified version of the SPICE simulator (Quarles, Newton, Pederson, and Sangiovanni-Vincentelli 1994) from the University of California at Berkeley to simulate the continuous-time behavior of each controller. The various signal processing functions in the controller are each represented by means of electrical subcircuits (or mathematical expressions that can be handled by SPICE). A SPICE netlist (itemizing the subcircuits and the topological connections between them) is generated from the block diagram. The netlist is wrapped inside an appropriate set of SPICE commands and the overall behavior of the controller is then simulated. See Koza, Keane, Yu, Bennett, and Mydlowec 2000 and Koza, Bennett, Andre, and Keane 1999 for details.

The fitness of each controller in the population is measured by means of 42 separate invocations of the SPICE simulator. This 42-part fitness measure consists of

(1) 40 time-domain-based elements based on a modified integral of time-weighted absolute error (ITAE) measuring how quickly the controller causes the plant to reach the reference signal, the robustness of the controller in face of significant variations in the plant's internal gain, K, the success of the controller in avoiding overshoot, and the effect of the externally supplied value of the time constant of τ,

(2) one frequency-domain-based element measuring bandwidth, and

(3) one frequency-domain-based element measuring the effect of sensor noise.

The fitness of an individual controller is the sum (i.e., linear combination) of the detrimental contributions of these 42 elements of the fitness measure. The smaller the sum, the better.

The first 40 elements of the 42-part fitness measure together represent

- five different externally supplied values for the plant's time constant τ,
- in conjunction with two choices of values for the plant's internal gain, K,
- in conjunction with two choices of values for the height of the reference signal, and
- in conjunction with two choices of values of the variation in time constant τ, around its externally supplied value.

The five values of the externally supplied values for the plant's time constant τ, span a range of two orders of magnitude. They are 0.1, 0.3, 1.0, 3.0, and 10.0. The five externally supplied values of the plant's time constant τ, force generalization in the sense that the to-be-evolved controller becomes a function of the externally supplied value of the plant's time constant τ (the free variable).

The two values of the plant's internal gain, K, are 1.0 and 2.0. These two values of the plant's internal gain, K, are used in order to ascertain robustness in the face of variation in K.

The reference signal is step function that rises from 0 to a specified height at $t = 100$ milliseconds. The two values for the height are 1 volt and 1 microvolts. The two values for the height of the step function are used to deal with the non-linearity caused by the limiter.

For each of these first 40 elements of the 42-part fitness measure, a transient analysis is performed in the time domain using the SPICE simulator. The contribution to fitness for each of these 40 elements of the fitness measure is based on the integral of time-weighted absolute error

$$\frac{\int_{t=0}^{10\tau} t|e(t)|A(e(t))Bdt}{\tau^2}.$$

Here $e(t)$ is the difference (error) at time t between the plant output and the reference signal. The integral is approximated from $t = 0$ to $t = 10\tau$, where τ is the externally supplied value of the time constant, τ. We multiply each value of $e(t)$ in the body of the integral by the reciprocal of the amplitude of the reference signals (so that both reference signals are equally influential). Specifically, B multiplies the difference $e(t)$ associated with the 1-volt step function by 1 and multiplies the difference $e(t)$ associated with the 1-microvolt step function by 10^6. In addition, we multiply each value of $e(t)$ in the body of the integral by an additional weight, A, that varies depending on $e(t)$ and that heavily penalizes non-compliant amounts of overshoot. The function A weights all variations below the reference signal and all variations up to 2% above the reference signal by a factor of 1.0, but A penalizes overshoots above 2% by a factor 10.0. Finally, we divide each value of the integral by τ^2 so as to equalize the influence of each of the five externally supplied values of τ.

The 41st element of the 42-part fitness measure is designed to constrain the frequency of the control variable so as to avoid extreme high frequencies in the demands placed upon the plant. This term reflects an unspoken constraint that is typically observed in real-world systems in order to prevent damage to delicate components of plants. This element of the fitness measure is based on 121 fitness cases representing 121 frequencies. Specifically, SPICE is instructed to perform an AC sweep of the reference signal over 20 sampled frequencies (equally spaced on a logarithmic scale) in each of six decades of frequency between 0.01 Hz and 10,000 Hz. A gain of less than 6 dB is acceptable for frequencies between 0.01 Hz and 60 Hz and a gain of less than -80 dB is acceptable for frequencies between 60 Hz and 10,000 Hz. For each of the 121 frequencies, the contribution to fitness is 0 if the gain is acceptable, but 1,000/121 otherwise.

The 42nd element of the fitness measure is based on 121 fitness cases representing 121 frequencies. Specifically, SPICE is instructed to perform an AC sweep of the signal resulting by adding the plant response to a noise source ranging over 20 sampled frequencies (equally spaced on a logarithmic scale) in each of six decades of frequency between 0.01 Hz and 10,000 Hz. A gain of less than 6 dB is acceptable for frequencies between 0.01 Hz and 60 Hz and a gain of less than -80 dB is acceptable for frequencies between 60 Hz and 10,000 Hz. For each of the 121 frequencies, the contribution to fitness is 0 if the gain is acceptable but 1,000/121 otherwise.

A controller that cannot be simulated by SPICE is assigned a fitness of 10^8.

3.5   Control Parameters

The population size, M, was 500,000. A (generous) maximum size of 150 points (for functions and terminals) was established for each result-producing branch and a (generous) maximum size of 100 points was established for each automatically defined function. The percentages for the genetic operations were the same as in Koza, Keane, Yu, Bennett, and Mydlowec 2000. The other parameters are the same default values that we have used previously on a broad range of problems (Koza, Bennett, Andre, Keane 1999).

3.6   Termination

The run was manually monitored and manually terminated when the fitness of the best-of-generation individuals appeared to have reached a plateau. The single best-so-far individual is harvested and designated as the result of the run.

3.7   Parallel Implementation

This problem was run on a home-built Beowulf-style (Sterling, Salmon, Becker, and Savarese 1999; Bennett, Koza, Shipman, and Stiffelman 1999) parallel cluster computer system consisting of 1,000 350 MHz Pentium II processors (each with 64 megabytes of RAM). The system has a 350 MHz Pentium II computer as host. The processing nodes are connected with a 100 megabit-per-second Ethernet. The processing nodes and the host use the Linux operating system. The distributed genetic algorithm with unsynchronized generations and semi-isolated subpopulations was used with a subpopulation size of $Q = 500$ at each of $D = 1,000$ demes. Two processors are housed in each of the 500 physical boxes of the system. As each processor (asynchronously) completes a generation, four groups of emigrants from each subpopulation (selected probabilistically based on fitness) are dispatched to each of four toroidally adjacent processors. The migration rate is 2% (but 10% if the toroidally adjacent node is in the same physical box).

4 Results

The best-of-generation circuit from generation 0 has a fitness of 3,448.8. The best-of-run controller (figure 3) appears in generation 42. It has an overall fitness of 38.72. Notice that the free variable, τ, appears in five blocks of this controller.

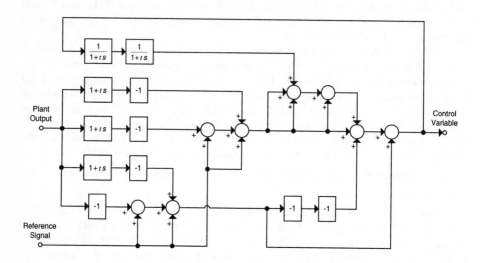

Figure 3   Block diagram of best-of-run genetically evolved controller from generation 42.

After simplification, it can be seen that the best-of-run controller from generation 42 has recognizable characteristics that are interpretable in the context of previously patented inventions or previously known control engineering techniques. The genetically evolved controller has the topology of a "PID-D2 controller" (i.e., proportional integrative, derivative, and second derivative), followed by a first-order lag block.. The PID controller was patented in 1939 by Albert Callender and Allan Stevenson of Imperial Chemical Limited of Northwich, England (Callender and Stevenson 1939) and the use of the second derivative was patented in 1942 by Harry Jones of the Brown Instrument Company of Philadelphia (Jones 1942). Also, the genetically evolved controller employs the technique called "setpoint weighting" in which the reference signal is weighted before the plant output is subtracted from it.

The control signal, $U(s)$, for this controller is given by the transfer function

$$U(s) = \frac{K_p(bR(s) - Y(s)) + K_i(R(s) - Y(s))/s + K_d(cR(s) - Y(s))s + K_{d2}(eR(s) - Y(s))s^2}{1 + ns}$$

Here the four PID-D2 coefficients in the numerator are $K_p = 17.0$ for the proportional (P) term, $K_i = 6/\tau$ for the integrative (I) term (with the $1/s$ in the numerator), $K_d = 16\tau$ for the derivative (D) term (with the s), and $K_{d2} = 5\tau^2$ for the second derivative (D2) term (with the s^2). The coefficient for the setpoint weighting of the proportional term is $b = 0.706$. That is, the plant output, $Y(s)$, is subtracted from b multiplied by the reference signal, $R(s)$. Similarly, the coefficient for the setpoint weighting of the derivative term is $c = 0.375$. The coefficient for the setpoint weighting of the second derivative term is $e = 0.0$. The time-constant of the first-order lag (shown here in the denominator of the transfer function) is $n = 0.5\tau$.

Figure 4 compares the time-domain response of the best-of-run genetically evolved controller (triangles) from generation 42 and the Astrom and Hagglund controller (squares) with $K = 1$ and $\tau = 1$. This (and other comparisons below) are substantially similar for other values of K and τ. Figure 5 compares the time-domain response of the best-of-run controller (triangles) from generation 42 and the Astrom and Hagglund controller (squares) to a 1-volt disturbance signal with $K = 1$ and $\tau = 1$.

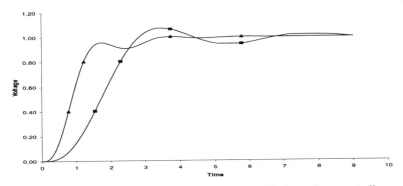

Figure 4 Comparison of the time-domain response of the best-of-run controller (triangles) from generation 42 and the Astrom and Hagglund controller (squares) with $K = 1$ and $\tau = 1$.

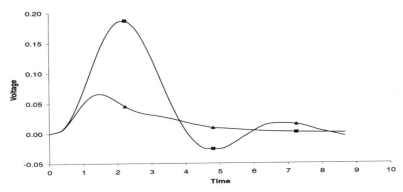

Figure 5 Comparison of the time-domain response of the genetically evolved controller (triangles) from generation 42 and the Astrom and Hagglund controller (squares) to a 1-volt disturbance signal with $K = 1$ and $\tau = 1$.

Table 1 compares the characteristics of the best-of-run genetically evolved controller (triangles) from generation 42 with those of the Astrom and Hagglund 1995 controller for $K = 1.0$, $\tau = 1.0$, and step size of 1.0.

Table 1 Comparison of characteristics for $K = 1.0$, $\tau = 1.0$, and step size of 1.0.

	Units	Genetically evolved controller	Astrom and Hagglund controller
ITAE for unit-step reference	volt sec^2	0.77	2.84
Rise time	seconds	1.43	2.51
Settling time	seconds	4.66	8.50
Disturbance sensitivity	volts/volts	0.07	0.19
IATE for unit-step disturbance	volt sec^2	0.42	1.35
Noise Attenuation Corner Frequency	Hz	0.72	0.34
Noise Attenuation Roll Off	dB/Decade	40	40
Maximum Sensitivity	volts/volts	2.09	2.11

The controller created by genetic programming is better than 3.69 times as effective as the Astrom and Hagglund 1995 controller as measured by the integral of the time-weighted absolute error, has only 57% of the rise time in response to the reference input, and has only 55% of the settling time.

4.1   Interpretation of the Results

Suppose $G(s)$ is the transfer function of a plant and we wanted to change all the time constants of this plant by the same amount, say τ. To do this, we would produce a plant with transfer function, $G_\tau(s) = G(\tau s)$. In this new transfer function $G_\tau(s)$, each coefficient of s would be multiplied by τ; each coefficient of s^2 would be multiplied by τ^2; and each coefficient of $1/s$ would be multiplied by $1/\tau$.

A human designer might approach this problem as follows: Call the plant with $\tau = 1$ the reference plant. Design a controller for this reference plant satisfying the design requirements and call the new controller, H, the reference controller. Call the controlled plant, J, the reference controlled plant. Let $H(s)$ be the transfer function of controller $H(s)$, and let $J(s)$ be the transfer function of the controlled plant. Now, to handle the time scaling, let $H_\tau(s) = H(\tau s)$ be the controller design for the time-scaled plant. This will produce an entire controlled plant, $J_\tau(s)$, scaled by the same factor τ; $J_\tau(s) = J(\tau*s)$. Effectively all the behavior of the time-scaled controlled plant is that of the reference controlled plant with the time scale changed.

In fact, we did exactly this in order to produce a PID controller with set point weighting for comparison purposes above (using the design rules using $M_s = 2$ on page 225 of Astrom and Hagglund 1995).

Referring now to genetically evolved controller shown above, notice that we could have written the following four coefficients in the control signal, $U(s)$, as follows:

$K_i/s = 6/\tau s$

$K_d s = 16\tau s^2$

$K_{d2} s^2 = 5\tau^2 s^2$

$ns = 0.5\tau s$

This is, in fact, the very same change in time scale that a human designed would have used.

Notice that we did not provide genetic programming with any of the above knowledge and insight about how a human designer might have proceeded on this problem. We did not bias genetic programming toward multiplying s by τ; toward multiplying s^2 by τ^2; and toward multiplying $1/s$ by $1/\tau$. We simply provided the free variable, τ, as a terminal that genetic programming could use as it saw fit.

4.2   Computer Time

Most of the computer time was consumed by the fitness evaluation of candidate individuals in the population. The fitness evaluation (involving 40 time-consuming time-domain SPICE simulations and two relatively fast frequency-domain SPICE simulations) averaged about 3.92 seconds per individual (using a 350 MHz Pentium II processor). The best-of-run individual from generation 42 was produced after evaluating 2.15×10^7 individuals (500,000 times 43). This required 23.43 hours (84,360 seconds) on our 1,000-node parallel computer system — that is, the expenditure of 2.95×10^{16} computer cycles (about 30 peta-cycles of computer time).

5   Conclusion

This paper demonstrated, for an illustrative problem involving a three-lag plant, that genetic programming can be used to create the design for both the topology and parameter values (tuning) of signal processing functions for a controller, where the controller contains a free variable. The genetically evolved controller outperformed the textbook human-designed controller.

References

Andersson, Bjorn, Svensson, Per, Nordin, Peter, and Nordahl, Mats. 1999. Reactive and memory-based genetic programming for robot control. In Poli, Riccardo, Nordin, Peter, Langdon, William B., and Fogarty, Terence C. 1999. *Genetic Programming: Second European Workshop. EuroGP'99. Proceedings.* Lecture Notes in Computer Science. Volume 1598. Berlin, Germany: Springer-Verlag. Pages 161 - 172.

Angeline, Peter J. 1997. An alternative to indexed memory for evolving programs with explicit state representations. In Koza, John R., Deb, Kalyanmoy, Dorigo, Marco, Fogel, David B., Garzon, Max, Iba, Hitoshi, and Riolo, Rick L. (editors). *Genetic Programming 1997: Proceedings of the Second Annual Conference, July 13–16, 1997, Stanford University.* San Francisco, CA: Morgan Kaufmann. Pages 423 - 430.

Angeline, Peter J. 1998. Multiple interacting programs: A representation for evolving complex behaviors. *Cybernetics and Systems.* 29 (8) 779 - 806.

Angeline, Peter J. 1998. Evolving predictors for chaotic time series. In Rogers, S., Fogel, D., Bezdek, J., and Bosacchi, B. (editors). *Proceedings of SPIE (Volume 3390): Application and Science of Computational Intelligence,* Bellingham, WA: SPIE - The International Society for Optical Engineering. Pages 170-180.

Angeline, Peter J. and Fogel, David B. 1997. An evolutionary program for the identification of dynamical systems. In Rogers, S. (editor). *Proceedings of SPIE (Volume 3077): Application and Science of Artificial Neural Networks III.* Bellingham, WA: SPIE - The International Society for Optical Engineering. Pages 409-417.

Astrom, Karl J. and Hagglund, Tore. 1995. *PID Controllers: Theory, Design, and Tuning.* Second Edition. Research Triangle Park, NC: Instrument Society of America.

Banzhaf, Wolfgang, Nordin, Peter, Keller, Richard, and Olmer, Markus. 1997. Generating adaptive behavior for a real robot using function regression with genetic programming. In Koza, John R., Deb, Kalyanmoy, Dorigo, Marco, Fogel, David B., Garzon, Max, Iba, Hitoshi, and Riolo, Rick L. (editors). *Genetic Programming 1997: Proceedings of the*

Second Annual Conference, July 13–16, 1997, Stanford University. San Francisco, CA: Morgan Kaufmann. Pages 35 - 43.

Bennett, Forrest H III, Koza, John R., Shipman, James, and Stiffelman, Oscar. 1999. Building a parallel computer system for $18,000 that performs a half peta-flop per day. In Banzhaf, Wolfgang, Daida, Jason, Eiben, A. E., Garzon, Max H., Honavar, Vasant, Jakiela, Mark, and Smith, Robert E. (editors). 1999. *GECCO-99: Proceedings of the Genetic and Evolutionary Computation Conference, July 13-17, 1999, Orlando, Florida USA.* San Francisco, CA: Morgan Kaufmann. Pages 1484 - 1490.

Boyd, S. P. and Barratt, C. H. 1991. *Linear Controller Design: Limits of Performance.* Englewood Cliffs, NJ: Prentice Hall.

Callender, Albert and Stevenson, Allan Brown. 1939. *Automatic Control of Variable Physical Characteristics.* United States Patent 2,175,985. Filed February 17, 1936 in United States. Filed February 13, 1935 in Great Britain. Issued October 10, 1939 in United States.

Crawford, L. S., Cheng, V. H. L., and Menon, P. K. 1999. Synthesis of flight vehicle guidance and control laws using genetic search methods. *Proceedings of 1999 Conference on Guidance, Navigation, and Control.* Reston, VA: American Institute of Aeronautics and Astronautics. Paper AIAA-99-4153.

Dewell, Larry D. and Menon, P. K. 1999. Low-thrust orbit transfer optimization using genetic search. *Proceedings of 1999 Conference on Guidance, Navigation, and Control.* Reston, VA: American Institute of Aeronautics and Astronautics. Paper AIAA-99-4151.

Dorf, Richard C. and Bishop, Robert H. 1998. *Modern Control Systems.* Eighth edition. Menlo Park, CA: Addison-Wesley.

Gruau, Frederic. 1992. Genetic synthesis of Boolean neural networks with a cell rewriting developmental process. In Schaffer, J. D. and Whitley, Darrell (editors). *Proceedings of the Workshop on Combinations of Genetic Algorithms and Neural Networks 1992.* Los Alamitos, CA: The IEEE Computer Society Press.

Holland, John H. 1975. *Adaptation in Natural and Artificial Systems.* Ann Arbor, MI: University of Michigan Press.

Jones, Harry S. 1942. *Control Apparatus.* United States Patent 2,282,726. Filed October 25, 1939. Issued May 12, 1942.

Kinnear, Kenneth E. Jr. (editor). 1994. *Advances in Genetic Programming.* Cambridge, MA: The MIT Press.

Koza, John R. 1992. *Genetic Programming: On the Programming of Computers by Means of Natural Selection.* Cambridge, MA: MIT Press.

Koza, John R., Bennett III, Forrest H, Andre, David, and Keane, Martin A. 1996. Automated design of both the topology and sizing of analog electrical circuits using genetic programming. In Gero, John S. and Sudweeks, Fay (editors). *Artificial Intelligence in Design '96.* Dordrecht: Kluwer Academic Publishers. 151-170.

Koza, John R., Bennett III, Forrest H, Andre, David, and Keane, Martin A. 1999. *Genetic Programming III: Darwinian Invention and Problem Solving.* San Francisco, CA: Morgan Kaufmann.

Koza, John R., Bennett III, Forrest H, Andre, David, Keane, Martin A., and Brave, Scott. 1999. *Genetic Programming III Videotape: Human-Competitive Machine Intelligence.* San Francisco, CA: Morgan Kaufmann.

Koza, John R., Keane, Martin A., Yu, Jessen, Bennett, Forrest H III, and Mydlowec, William. 2000. Automatic creation of human-competitive programs and controllers by means of genetic programming. *Genetic Programming and Evolvable Machines.* 1 (1 -2) 121 - 164.

Man, K. F., Tang, K. S., Kwong, S., and Halang, W. A. 1997. *Genetic Algorithms for Control and Signal Processing.* London: Springer-Verlag.

Man, K. F., Tang, K. S., Kwong, S., and Halang, W. A. 1999. *Genetic Algorithms: Concepts and Designs.* London: Springer-Verlag.

Marenbach, Peter, Bettenhausen, Kurt D., and Freyer, Stephan. 1996. Signal path oriented approach for generation of dynamic process models. In Koza, John R., Goldberg, David E., Fogel, David B., and Riolo, Rick L. (editors). *Genetic Programming 1996: Proceedings of the First Annual Conference, July 28-31, 1996, Stanford University.* Cambridge, MA: MIT Press. Pages 327 - 332.

Menon, P. K., Yousefpor, M., Lam, T., and Steinberg, M. L. 1995. Nonlinear flight control system synthesis using genetic programming. *Proceedings of 1995 Conference on Guidance, Navigation, and Control.* Reston, VA: American Institute of Aeronautics and Astronautics. Pages 461 - 470.

Quarles, Thomas, Newton, A. R., Pederson, D. O., and Sangiovanni-Vincentelli, A. 1994. *SPICE 3 Version 3F5 User's Manual.* Department of Electrical Engineering and Computer Science, University of California. Berkeley, CA. March 1994.

Sterling, Thomas L., Salmon, John, Becker, Donald J., and Savarese, Daniel F. 1999. *How to Build a Beowulf: A Guide to Implementation and Application of PC Clusters.* Cambridge, MA: MIT Press.

Sweriduk, G. D., Menon, P. K., and Steinberg, M. L. 1998. Robust command augmentation system design using genetic search methods. *Proceedings of 1998 Conference on Guidance, Navigation, and Control.* Reston, VA: American Institute of Aeronautics and Astronautics. Pages 286 - 294.

Sweriduk, G. D., Menon, P. K., and Steinberg, M. L. 1999. Design of a pilot-activated recovery system using genetic search methods. *Proceedings of 1998 Conference on Guidance, Navigation, and Control.* Reston, VA: American Institute of Aeronautics and Astronautics.

Teller, Astro. 1996a. *Evolving Programmers: SMART Mutation.* Technical Report CMU-CS-96. Computer Science Department, Carnegie Mellon University.

Teller, Astro. 1996b. Evolving programmers: The co-evolution of intelligent recombination operators. In Angeline, Peter J. and Kinnear, Kenneth E. Jr. (editors). *Advances in Genetic Programming 2.* Cambridge, MA: The MIT Press.

Teller, Astro, and Veloso, Manuela. 1995a. *Learning Tree Structured Algorithms for Orchestration into an Object Recognition System.* Technical Report CMU-CS-95-101. Computer Science Department, Carnegie Mellon University.

Teller, Astro, and Veloso, Manuela. 1995b. Program evolution for data mining. In Louis, Sushil (editor). Special Issue on Genetic Algorithms and Knowledge Bases. *The International Journal of Expert Systems.* JAI Press. (3) 216 - 236.

Teller, Astro, and Veloso, Manuela. 1995c. A controlled experiment: evolution for learning difficult problems. *Proceedings of Seventh Portuguese Conference on Artificial Intelligence.* Springer-Verlag. Pages 165-76.

Teller, Astro, and Veloso, Manuela. 1995d. Algorithm Evolution for Face Recognition: What Makes a Picture Difficult? *Proceedings of the IEEE International Conference on Evolutionary Computation.* IEEE Press.

Teller, Astro, and Veloso, Manuela. 1995e. Language Representation Progression in PADO. *Proceedings of AAAI Fall Symposium on Artificial Intelligence.* Menlo Park, CA: AAAI Press.

Teller, Astro and Veloso, Manuela. 1996. PADO: A new learning architecture for object recognition. In Ikeuchi, Katsushi and Veloso, Manuela (editors). *Symbolic Visual Learning.* Oxford University Press. Pages 81-116.

Whitley, Darrell, Gruau, Frederic, and Preatt, Larry. 1995. Cellular encoding applied to neurocontrol. In Eshelman, Larry J. (editor). *Proceedings of the Sixth International Conference on Genetic Algorithms.* San Francisco, CA: Morgan Kaufmann. Pages 460 -467.

Genetic Programming for Service Creation in Intelligent Networks

Peter Martin

Marconi Communications Limited, Poole, Dorset, UK
peter.martin@marconicomms.com

Abstract. Intelligent Networks are used by telephony systems to offer services to customers. The creation of these services has traditionally been performed by hand, and has required substantial effort, despite the advanced tools employed. An alternative to manual service creation using Genetic Programming is proposed that addresses some of the limitations of the manual process of service creation. The main benefit of using GP is that by focussing on what a service is required to do, as opposed to it's implementation, it is more likely that the generated programs will be available on time and to budget, when compared to traditional software engineering techniques. The problem of closure is tackled by presenting a new technique for ensuring correct program syntax that maintains genetic diversity.

1   Introduction to Intelligent Networks

Traditional telephony in the past 20 years has concentrated on delivering telephony services to customers by means of stored program switches. Customers have, until recently, been restricted to relatively crude terminal equipment that supports voice and Dual Tone Multi Frequency (DTMF) user controls. A simplified view of this traditional system is shown in Figure 1.

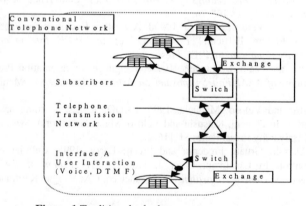

Figure 1 Traditional telephone system

The switches control the calls made by the subscribers. All the services such as ring back when free, are implemented in the switches. However, the time required to implement new services is in the order of 1 to 2 years (for example see [3]) because the services are tied closely to the switch development lifecycle.

As the number of services offered has grown and the sophistication of telephone equipment has risen, it has become clear that offering services via the traditional embedded switch technology does not scale well, and that other platforms for providing the services are required. The alternative is to move the services off the switches onto standard computing platforms. This is the Intelligent Network (IN) solution.

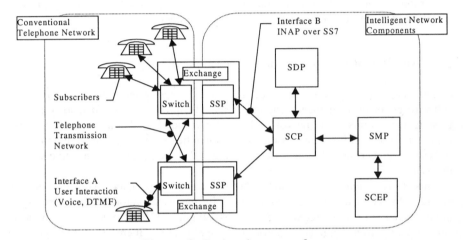

Figure 2 Basic elements of an Intelligent Network

The basic intelligent network is shown in Figure 2. Here it can be seen that the switches have been expanded to communicate with other network elements. The Service Switching Point (SSP) handles the interface between the telephone switch and the Intelligent Network. The interface uses a standardized Intelligent Network Application Part (INAP) which is carried over the standard Signaling System number 7 (SS7) network. In the Intelligent Network, the services are hosted on a Service Control Point (SCP). This is typically a number of high end UNIX servers. The SCP uses a Service Data Point (SDP) to store data, for example, numbers required to implement a FreePhone service. The system is managed by a Service Management Point (SMP). Finally, to enable the rapid creation and deployment of services, a Service Creation Environment (SCE) is used to build, test and deploy services to the network.

The primary objective of Intelligent Networks (IN) then is to move the service computation from the embedded switches to readily available computers to gain the benefits of mainstream IT techniques.

A secondary aim of introducing IN is to reduce the time required to develop and deploy new services. As already mentioned, traditional switch based solutions typically require 2 years from the initial requirements being specified until the service is in operation. In a highly competitive environment this is too long, and the market window will have disappeared by the time the services come into operation. IN aims to reduce this to around 6 months by exploiting mainstream IT techniques.

In order to achieve such a startling reduction in timescales, new methods of creating service applications were required. From this followed the introduction of the SCE, or Service Creation Environment Function.

Experience has shown that the time required to capture the requirements and create an outline of the service is relatively short, but the time required to implement and test the low level details of complex services can be several months. A typical non-trivial service can require several hundred icons, and results in dozens of valid traversals of the graph. A means of reducing the duration of the detailed engineering phase is therefore of benefit to the network and service operators.

1.1 An Alternative approach to Service Creation

The major problems encountered in the existing system are associated with software engineering management issues namely, productivity and quality control. Despite the promises of the early IN systems and the advanced tools available, complex services still take a considerable amount of time to develop using traditional software engineering techniques and there are still some defects found in the services themselves [2].

This work attempts to address the difficulties associated with the detailed engineering phase of service development, by means of automatically deriving an implementation from the requirements, or as Langdon [19] and others put it, by using Automatic Programming. This approach was hinted at by Boehm [5] Chap. 33 which mentions automatic programming. In 1981 the idea was considered interesting but 'somewhat beyond the current frontier of the state of the art'. This paper demonstrates that automatic programming by using Genetic Programming (GP) is now a viable alternative in the domain of IN.

To be able to judge whether an alternative approach to manual programming is worthwhile a number of questions need to be answered with regards to the alternative:

1. Has the alternative approach demonstrated that it can generate programs that perform as well as or better than a human?
2. In the domain being considered what are the observable and measurable attributes of the process of generating programs?
3. What are the observable and measurable attributes of the generated programs ?
4. Can it handle the range of program complexity that a human can; i.e.; is it scalable?

Firstly, GP has demonstrated that it can produce results that are at least as good as a human programmer and in some cases provide solutions to problems that a human has not been able to achieve as in the case of discovering an electronic circuit to yield a cube root function by Koza et al[15] and Sharman et al [25] has also shown that programs for Digital Signal Processors (DSPs) evolved using GP can outperform existing programs. Clearly then GP has the potential to generate programs that humans find hard.

Secondly, we can consider an existing service creation case study [2]. This study showed that for a complex service a team of engineers required 4.5 Man years of effort to analyse, design, code and test the service. A significant measurable attribute is therefore the elapsed time required to implement the service and this attribute will be quantified for GP by experimental data presented later. Other attributes are cost of equipment and the degree of human intervention but are not considered further in this work.

Thirdly, a key measurable attribute of the program is the level of defects. Broadly defects fall into one of two categories according to Sommerville [26]; errors due to incorrect requirements analysis and errors due to implementation deficiencies either by errors in programming or design. The first type is common to whatever method of programming is adopted. As summarised by Davis [8] the earlier that requirement related errors are found, the lower the cost to remedy the error. As will be seen later using GP forces the designer to consider requirements in more detail initially (for fitness evaluation) so the implication is that using GP will result in fewer errors introduced by faults in the requirements. Again the study by Boulton et al shows that even using advanced tools such as INventor, there were 15 failures associated with the service. Anecdotal evidence suggests that these were all implementation errors.

Lastly, the question of whether GP can scale can only be answered in full by analysing experimental data, but initial indications show that GP can create programs to solve complex problems in other domains.

2 Applying GP to Service Creation

2.1 Functions and Terminals

Classical tree based GP [16] requires a set of functions which form the non-leaf nodes in the tree and terminals which form the leaf nodes of the tree. The set of functions and terminals must satisfy the closure and sufficiency properties. Terminals may be side affecting or yield data. For this work, the functions were chosen to perform all external operations, while the terminals were chosen to yield data. In order to arrive at a sufficient set of data types, it is useful to consider what types of data are commonly encountered in telephony services. Table 1 summarizes these data types.

Table 1 Data types encountered in telephony services

Data Type	Comments
Telephone numbers	Strings of digits [0-9 # *] that can be dissected and concatenated. The string length may be up to 24 bytes.
Constant integral values	Used for counters and message parameter values
Boolean values	Flags and status values
Message types	An enumerated set used to distinguish messages

From this it is clear that restricting functions and terminals to use a single data type in order to satisfy the closure property is not feasible. In addition, since most IN services require some state information to be stored between messages, a mechanism for saving state information is required. The first approach to this requirement, Indexed Memory, was suggested by Teller [2] where he argues that in order for GP to be able to evolve any conceivable algorithm, GP needs to be Turing Complete and that addressable memory enables this. A useful side effect of this is that memory also allows state information to be explicitly saved and retrieved.

Of course other approaches to saving state information are possible as for example in the work by Angeline [1] that uses Multiple Interacting Programs (MIPS). However for the purposes of this work Indexed Memory was chosen since it was thought that it would be easier to analyse the operation of the evolving programs.

2.2 Achieving Closure

Several methods have been proposed to ensure that the closure property is maintained during initial creation and subsequent reproduction. This may be achieved in a number of ways. Firstly Koza [16] restricts the types of arguments and functions to compatible types. For instance, all floating point types as in the symbolic regression examples or logical in the Boolean examples. For simple problems with single data types this is sufficient.

Secondly, in strongly typed approaches such as those described by Montana [22] and Haynes et al [18] constraints are placed on the creation of individuals to satisfy the type rules. The advantage here is reducing the size of the search space by eliminating individuals that would fail due to syntax errors. Clack [6] extended this work to show that expression based parse trees can yield more correct programs, and introduced the idea of polymorphism into the data types.

An alternative to the strongly typed approach is proposed in this work, based on polymorphic data types with independent values for each type supported.

This approach was devised as an alternative to the strongly typed methods by making the observation that it is possible that the criteria used to decide what is a correct program has more to do with correctness as seen by a human programmer rather than any inherent property of GP. In other words, strong typing is a useful artifact of languages used by humans to help ease the burden on the programmer, by means of assisting machine interpretation. Perkis [2] has shown that an apparently haphazard mechanism in the form of a stack can yield useful results. Another objection to using a strongly based type system was that the potential number of solutions could be greatly diminished and biased , since the search space is constrained.

The work presented here uses a new data type termed Autonomous Polymorphic Addressable Memory (APAM). This consists of a set of memory locations $M=\{L_1,...L_n\}$ which can be addressed randomly or by name. Each location is a set of data items of different types $L=\{d_1,...d_n\}$. The values of $L_n.d_1$, $L_n.d_2$ etc are independent of each other. Selection of the correct type and therefore value is performed by any function that is passed a memory reference as an argument. An example of the use of this structure is when a program needs both integer and real values. Each location L would contain an integer data type and a floating point data type.

Memory M							
L_1			...		L_n		
d_1	d_2	d_3			d_1	d_2	d_3

Figure 3 Layout of Autonomous Polymorphic Addressable Memory

To support this memory architecture, the terminal set T consists of memory nodes $T=\{TVAR_1,...TVAR_n\}$. Each node returns a reference to memory location L_n. and can be passed as arguments to any function.

It should be noted that this is not the same as using a generic data type where a data item is coerced into the correct type at run time. A difficulty with coercion is that many automatic conversions are meaningless. For example, in the context of telephony it would be hard to imagine what the coercion of a Boolean value into a telephone number would mean.

2.3   Choosing a level of abstraction

Sufficiency is a problem specific attribute. In the domain of IN, there are three main levels of abstraction that can be considered. This list does not include low-level functions, for instance the UNIX API, or raw machine language, though the latter is clearly feasible as demonstrated by the use of Java byte code as the working set for C as described by Banzhaf et al [4]:

1. Icon level with attributes as terminals. This level is based on the set of functions offered to service creators using the GPT GAIN INventor ™ product [2]. Other service creation systems have similar or even higher level of abstraction. A subset of around twenty icons is sufficient to construct the majority of services encountered in existing networks.
2. Icon function level. This is the level used by the internal tools within GAIN INventor ™. Each icon typically makes use of between one and twenty functions. The total number of functions is around 200.

3. API level. This is the lowest practical level. This is the set of API functions offered by the target platform. In the case of the GPT GAIN INventor $^{(TM)}$ product, this is a set of over one hundred and fifty function calls designed to allow services and other applications to be constructed.

For these experiments, the level was initially pitched at the ICON level since this level allows humans to create production quality services. An attempt was made to see if this level of abstraction was optimal by carrying out additional experiments using a level closer to the API. Initial results indicate that using the ICON level may not be the most effective.

The functions chosen for the initial experiments are shown in Table 2

Table 2. Functions for high level abstraction

FSTART	Takes two arguments. It accepts an IDP message from the SSP and the calledDN value is stored at the location returned as a result of evaluating it's first parameter.
FDBREAD	This reads a data base, using the value of the first argument as a key and placing the result in the location returned as a result of evaluating it's second parameter
FROUTE	Evaluates the first argument which is used to furnish the new routed number for the connection message.
FEND	Sends a pre-arranged end to the SSP.
STRSUB	Performs a simple string shift operation to simulate real-life number manipulations.

For the lower level abstraction, the functions were:

Table 3. Functions for low level abstraction

ReadMSG:	Accepts a message from the SSP or SDP and places the parameters into variables.
SendMSG	Takes a number of parameters and builds a message which is then sent to the SSP or SDP.
STRSUB	Operates as already described

The system supports five message types analogous to the real world Intelligent Network Application Part (INAP) and SDF operations. These are InitialDP which is generated by the SSP as a result of a trigger detection point being activated by a call, DB_REQUEST which issues a database request, DB_RESPONSE which accepts a data base response, Connect which is an instruction from the SCF to the SSP to connect party A to party B, and END which terminates a service dialogue.

2.4 The Fitness Function for Service Creation

The decision was made to measure the fitness of the GP at the external INAP interface since this is a standardized external interface as described in Q.1211 [12] and would allow the specification of services to be performed at the network level.

This led to the use of Message Sequence Charts (MSCs) for deriving the fitness function. MSCs are commonly used in telecommunication system work for specifying system behaviour.

The Basic Call State Machine (BCSM) described in Q.1214 [13] is simplified, and called a Simple Call State Model (SCSM) in order to focus on the GP technique rather than being distracted by the complexities of the BCSM.

When running a fitness test, two related problem specific measures are used to determine how fit an individual is, as well as non-problem specific measures such as parsimony:

1. The number of correct state transitions made. Each correct transition is rewarded with a value of 100. Each incorrect transition is penalized with a value of -20. The reward and penalty values are summed. This value is called **s**.
2. The number of correct parameter values passed back to the SCSM. A correct parameter value is rewarded with a value of 100, and each incorrect value is penalized with a value of -20. The reward and penalty values are summed. This value is called **p**. These values are then used to compute a normalized fitness value as follows:

Raw fitness **r** is given by $\mathbf{r} = \mathbf{s} + \mathbf{p}$

Normalized fitness **n** is given by $\mathbf{n} = \mathbf{k} - \mathbf{r}$ where k is a constant that is dependent on the number of state transitions and message parameters in the problem being considered, such that for a 100% fit individual **n**=0.

A count is maintained of the number of correct and incorrect state transitions and correct and incorrect message parameter values to help with an analysis of the performance of GP.

2.5   Measuring performance and estimating effort

Additional information collected includes the total wall clock time taken for each run, the number of individuals processed, the number of unique individuals that were 100% fit, the number of 100% individuals at the final generation and details of the best individual of each run, including it's size.

3   Example of a Service - Complex Number Translation

We can now look at one of the experiments carried out as part of this research [21]. Number translation was chosen for this example since although it is one of the simpler services, it is also the most common service implemented using IN [9], and forms the basis of many more complex services such as Freephone, Premium Rate services and time based routing.

The experiment uses GP to evolve service logic for a number translation service where two data base lookups are required. This scenario occurs in the real world where a service requires two items of data in order to route a call. For example, a service may need to route to one number during working hours and another number during out of work hours.

For this experiment the population was set at 500, the number of generations was set at 200. These figures were arrived at after a number of trial runs using a wide range of values for the population size and number of generations. When this problem was run 50 times, GP created a 100% correct program 49 times. The probability of finding a correct solution at generation 200 was 72%.

Some interesting results were observed during this experiment. Firstly all 49 correct programs were different. The differences ranged from the selection of different variables to some counter intuitive program trees when compared to what a human programmer might have written. One of the less intuitive programs is shown in Figure 4. Note that the first function called is the route function, and the operation relies purely on the ordering of evaluation of function arguments, rather than a procedural sequencing more commonly found in human generated programs.

The number translation problem was performed twice, once using functions from the high level abstraction set as already described and once using functions from the lower level set. The use of the lower level abstraction system yielded solutions using fewer generations than the higher-level abstraction system. After 200 generations the probability of finding a 100% correct solution rose to 84%, compared to 74% for the high level functions. Weighed against this faster convergence is the fact for a given probability of finding a lower level abstraction, the low level functions required more time to evolve primarily due to the greater size of the program trees needed. An example of a 100% correct program is shown in Figure 5.

An interesting feature of this particular example is the regularity with which the pattern at nodes 3, 4, 5, 6 and 7 occur. This pattern is repeated at the subtrees rooted at nodes 10 and 18. It is likely that using Automatically Defined Functions (ADFs) [17] for this level of functions would be beneficial since there are repeating patterns emerging. A possibility that was not explored is that the common subtrees come from a common ancestor formed early in the run.

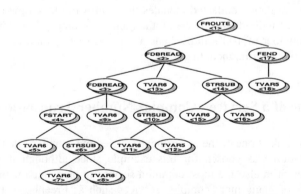

Figure 4 An example of a novel 100% correct program tree

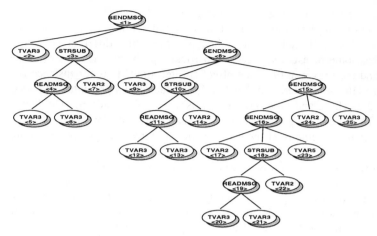

Figure 5 Example program tree using reduced complexity functions

3.1 Comparison of Performance Between High and Low Level Abstraction

Some additional measurements were taken to try to gauge the differences in the time required to find a solution and the resulting complexity of the 100% correct programs.

The average time for a run to complete is taken as the total wall clock time of the experiment divided by the number of runs, which was 50 in each case. The wall clock time is significant when trying to compare automatic programming as an alternative to human programming.

Table 4 Summary of experiments and results

Average time per run (secs)	Average Complexity of fittest	P(M,i) %	R(z)	ε
44	19	72	4	2,000
71	28	82	3	1,500

The average complexity is the sum of the complexity values of the fittest 100% correct individuals in each run, divided by the number of runs that produced a 100% correct individual. The complexity of an individual is simply the number of nodes in that individual.

4 Analysis and Discussion

During the early part of the work, considerable time was spent trying different combinations of the control parameters and the set arrived at for the experiments is probably not optimal.

Two questions arise from this:

1. Is there an envelope of operation that gives good results?
2. Is it possible to determine all environment control values by some method?

It should also be noted, that although studies into different control parameter values has some measurable effect on particular problems the scale of effect is often small, and the universality of the effect is often limited, as for instance reported by Goldberg [10] in his study on deme size, and the results presented as part of the GP kernel [28]. These and other questions raises the point made by Goldberg [11] that unlike GA there is no good theoretical basis for GP, and that until one is developed we are reliant on empirical methods for determining the operational parameters for GP.

The original choice of abstraction for the internal nodes gave satisfactory results, but as shown in the second experiment, a lower level of abstraction gives a better overall performance (higher probability of yielding a 100% correct program) using the same basic system architecture, but required approximately 40% more processing effort. Interestingly the average complexity of the reduced complexity experiment was also approximately 40% greater than the standard experiment. This suggests there may be a direct link between the two measures. Additionally, it is suggested that using ADFs could well be useful in this case. Clearly more work is required in order to arrive at an optimal level of abstraction.

The use of the Autonomous Polymorphic Addressable Memory System (APAMS) was very powerful. It meant that evolving programs were not constrained in the shape they took. The memory locations were used for several different purposes in the experiments – targets for storing message parameters, both string and integer, and a source for function arguments, and as a constant value as when used by some examples using the *FEQ* function. In the last experiment they also contained message types. Extension of APAMS would prove beneficial in future developments such as using it to hold partial or complete messages.

APAMS also contributed to the great variety seen in the 100% correct solutions by avoiding the need to restrict the semantic structure as in [22] and others. To examine this claim, a simple hypothetical case can be considered, such as the *FSTART* function. A strict typing of this by a human programmer during the early stages of building a GP system could define this function returning a status, or particular parameter to a calling function and having arguments of type DialledNumber for the first and some other type for the second. Immediately it can be seen that by adding these constraints, a human programmer imposes their own perceived structure on the function and therefore its place in any tree. Doing this would preclude the example solution illustrated in Figure 4 that started with the *FROUTE* function.

4.1 Performance of GP Compared to a Human Programmer

In terms of raw performance, GP service creation compares well to a human performing the same task, with GP taking minutes to find a 100% correct individual, and a human taking around one hour to hand code a similar program and test it. However a comparison made purely on time to complete a task does not tell the whole story. In the case of GP, one of the important tasks of the fitness function is to rank how well an individual is able to solve a problem. The fitness function could also be the test case for the solution so a correct program could well be classified as 100% tested, with the usual caveats concerning testing metrics. This is an appealing side effect from using GP.

4.2 Software Engineering Considerations

For many years automatic program generation has been a goal of software engineering. See for example [5]. GP at its current state of maturity goes some way to achieving this goal. This could be viewed as a threat to traditional software engineering, but such a narrow view misses the broader benefits of GP. GP offers an alternative to the traditional coding phase of software development, facilitating the creation of quality tested software. What remains of course are the essential activities of requirements capture and system specification.

A comment often raised when evolutionary techniques are discussed is how we can be sure that the evolved structures don't have any hidden surprises. For example, that a program may give erroneous results under a set of conditions that were not expected. Apart from the fact that programs written by humans are themselves often not free of hidden surprises, a strong argument against these objections is that every individual is tested against the fitness function. Since the fitness function in effect embodies the specification for the program, the fitness function can contain the test cases required to ensure that the evolved program performs as required.

The opaqueness of machine generated programs can of course be considered to be a positive attribute in that it forces the systems engineer to look more closely at the specification and the associated system testing. A consequence of this is that the systems engineer must specify *exactly* what the system should do, not as the introduction to Koza's third book [14] states '... a high level statement of the requirements ...'.

This question concerning the opaqueness of programs generated using GP or other EC techniques has inspired some work to try to address the perceived deficiency. For instance Pringle [2] suggests an approach that tries to create programs that look like those produced by a human programmer, while Langdon [19] has dedicated a whole book to automatic programming adopting techniques used by human programmers as building blocks. A potential flaw in this approach is that practices such as modularity, data hiding, object oriented disciplines, data structures and other 'good engineering practices' have been developed to aid human programmers in writing fault free and maintainable software. They are not of themselves required for a program to be correct and while the aforementioned work has delivered some useful techniques and insights it does not address any of the essential features of GP. A counter argument has been made by Blickle [3] pointing out that a clear structured program can give valuable insights into the problem being solved. For example when trying to find an analytical expression to difficult integral equations, a clear analytic expression would allow further investigation of the problem. However it is worth revisiting the original inspiration for this work and noting that Darwin observed 'nature cares nothing for appearances, except so far as they may be useful to any being' [7] (Chapter IV, 'Natural Selection').

Finally it is often remarked that understanding the evolved programs is often hard, since the programs do not always adhere to what a human programmer might consider good style. However, in a production environment it would be rare that an analytical understanding of the program is required. In any case, software engineering is currently happy to trust tools to generate code in other areas. A useful analogy here is to consider how CASE tools and high-level language optimizing compilers have replaced flow charts and hand crafted assembler code.

5 Conclusions

A new technique of achieving closure was developed that uses polymorphic data types. The principle advantage of using this approach is to facilitate a more complete search of the problem space by avoiding the selective search imposed by strongly typed techniques.

Choosing two different levels of abstraction for the function set indicated that selecting the optimum set of functions is not straightforward.

The level of defects in the generated application due to implementation errors is zero due to the fitness evaluation applied to the application. The level of defects due to errors in requirements should be reduced since more attention is needed at the specification stage.

GP still requires a significant amount of effort from the system designer in selecting a sufficient and appropriate set of functions and terminals, in selecting suitable run-time parameters and in fine-tuning the system during the development of useful programs.

This work has demonstrated that GP is a viable technique when applied to the problem of service creation. While there are some limitations with the present approach, ongoing research into GP is yielding better insights into the underlying theory of operation, and is delivering GP systems that can handle more complex tasks. Whether this application of GP is scalable to allow the creation of production quality services is still an open question.

6 Acknowledgements

I would like to thank Marconi Communications Limited for supporting this work and the encouragement and valuable comments from my colleagues Jeremy Bennett and Stuart Wray.

7 References

1   Angeline P. *An Alternative to Indexed Memory for Evolving Programs with Explicit State Representations.* pp 423-430 In Koza J, Deb K, Dorigo M, Fogel D.B, Garzon M, Iba H, and Riolo R (Eds.) Genetic Programming 1997. Proceedings of the Second Annual Conference July 13-16, 1997 Stanford University. Morgan Kaufmann Publishers, San Francisco, CA

2   Boulton C., Johnson W., and Prince M. 1997. *A Personal Number Service for British Telecom. (BT OneNumber)..* Unpublished paper by GPT Limited

3   Blickle T. *Evolving Compact Solutions in Genetic Programming: A Case Study.* In Voight H., Ebeling W., Recenberg I., Schwefel H., (Eds.): Parallel Problem Solving from Nature IV. Proceedings of the International Conference on Evolutionary, Berlin, September 1996. LNCS 1141, pp. 564-573, Heidelberg. Springer-Verlag.

4   Banzhaf W, Nordin P, and Olmer M. *Generating Adaptive Behaviour for a Real Robot using Function Regression within Genetic Programming.* pp 35-43 In Koza J, Deb K, Dorigo M, Fogel D.B, Garzon M, Iba H, and Riolo R (Eds.) Genetic Programming 1997. Proceedings of the Second Annual Conference July 13-16, 1997 Stanford University. Morgan Kaufmann Publishers, San Francisco, CA.

5   Boehm B. W. *Software Engineering Economics.* 1981 Prentice Hall

6   Clack T, and Yu T. *Performance Enhanced Genetic Programming.* In Angeline P., Reynolds R., McDonald J., and Eberhart R., Eds., Proceedings of the sixth conference on Evloutionary Programming, Volume 1213 of Lecture Notes in Computer Science, Indianapolis, Indiana, USA, 1997, Springer-Verlag.

7   Darwin, Charles. On the origin of Species by Means of Natural Selection, or the Preservation of Favoured Races in the Struggle for Life. 1st Edition. 1859.

8   Davis A M. *Software Requirements; Objects, Functions and States.* 1993. Prentice Hall

9   Eberhagen S. *Considerations for a successful introduction of Intelligent Networks from a marketing perspective.* In the Proceedings of the 5th International Conference on Intelligence in Networks, Bordeaux, France. 13/15 May 1998. Adera, France.

10   Goldberg, David E, Kargupta Hillol, Horn Jeffrey and Cantu-Paz Erik. *Critical Deme Size for Serial and Parallel Genetic Algorithms.* IlliGAL Report No. 95002. January 1995

11   Goldberg, David E., and O'Reilly, Una-May. *Where Does the Good Stuff Go, and Why? How Contextual semantics influences program structure in simple genetic programming,* in Banzhaf W., Poli R., Schoenauer M., and Fogarty T.C., (Eds.): First European Workshop, EuroGP'98, Paris, France, April 1998 Proceedings. LNCS 1391, Springer-Verlag.

12   ITU-T Q.1211. *Introduction to Intelligent Networks CS-1* 1994.6

13   ITU-T Q.1214. *Distributed Functional Plane for Intelligent Networks CS-1* 1994.

14   Koza John R. Andre David, Bennett Forret H and Keane, Martin. *Genetic Programming III* Unpublished draft version, available on the GP MAILING LIST

15   Koza J R, Bennett III, Forrest H, Andre D, Keane M. *Automated WYWIWYG design of both the topology and component values of analogue electrical circuits using genetic programming.* In Koza J R, Goldberg D E, Fogel D, and Riolo R. (Eds). Genetic Programming 1996: Proceedings of the First Annual Conference, July 28-31, 1996, Stanford University. Cambridge, MA. MIT Press.

16   Koza, John, R. *Genetic Programming, On the Programming of Computers by Means of Natural Selection.* 1st Ed. MIT Press 1992.

17   Koza John R. *Genetic Programming II. Automatic Discovery of Reusable Programs.* 1st Ed. MIT Press, 1994

18   Haynes T., Wainwright R., Sen S., and Schoenefeld D., *Strongly Typed Genetic Programming in Evolving Cooperation Strategies.* In Eshelman L., (Ed) Genetic Algorithms: Proceedings of the Sixth International Conference (ICGA95), pages 271-278, Pittsburgh, PA, USA, 15-19 July 1995. Morgan Kaufmann.

19   Langdon W B. *Genetic Programming and Data structures: Genetic Programming and Data structures = Automatic Programming!* 1st Edition. The Kluwer International Series in Engineering and Computer Science. Vol. 438. Kluwer Academic Publishers, Boston. 1998

20   Martin, Peter, N. *Service Creation for Intelligent Networks:* Delivering the Promise. Proceedings of the 4th International Conference on Intelligence in Networks, Bordeaux. 1996. ADERA.

21   Martin, Peter, N. *An investigation into the use of Genetic Programming Intelligent Network Service Creation.* MSc Dissertation, Bournemouth University, UK. http://www.martin21.freeserve.co.uk/gpproject.htm

22   Montana, David, J. *Strongly Typed Genetic Programming.* Evolutionary Computation, Volume 3, Issue 2, pp. 199-230. Summer 1995. MIT Press

23   Perkis, Timothy. *Stack Based Genetic Programming.* In the proceedings of the 1994 IEEE World Congress on Computational Intelligence 1994. Volume 1, pages 148-153, Orlando, Florida, USA, 27-29 June 1994. IEEE Press

24   Pringle W. ESP: *Evolutionary Structured Programming.* Technical Report, Penn State University, Greate Valley Campus, PA, USA, 1995.

25   Sharman K., Anna I. Esparcia A., and Yun Li.
Evolving signal processing algorithms by genetic programming. In A. M. S. Zalzala, editor, First International Conference on Genetic Algorithms in Engineering Systems: Innovations and Applications, (GALESIA), volume 414, pages 473--480, Sheffield, UK, 12-14 September 1995. IEE.

26   Sommerville I. *Software Engineering.* Fifth Ed. 1996. Addison Wesley Publishers Ltd.

27   Teller, Astro. *Turing Completeness in the Language of Genetic Programming with Indexed Memory.* Proceedings of the 1994 IEEE World Congress on Computational Intelligence., volume 1, Orlando, Florida, USA. June 1994. IEEE Press.

28   Weinbrenner, Thomas, *The genetic Programming Kernel,* Version 0.5.2: http://www..emk.e-tecknik.th-darmstadt.de/~thomasw/gp.html
Visited 12th Sept 1997

Cartesian Genetic Programming

Julian F. Miller[1], Peter Thomson[2]

[1] School of Computer Science, University of Birmingham, Birmingham, England, B15 2TT
j.miller@cs.bham.ac.uk

[2] School of Computing, Napier University, 219 Colinton Road, Edinburgh, Scotland, EH14
1DJ
p.thomson@dcs.napier.ac.uk

Abstract. This paper presents a new form of Genetic Programming called Cartesian Genetic Programming in which a program is represented as an indexed graph. The graph is encoded in the form of a linear string of integers. The inputs or terminal set and node outputs are numbered sequentially. The node functions are also separately numbered. The genotype is just a list of node connections and functions. The genotype is then mapped to an indexed graph that can be executed as a program. Evolutionary algorithms are used to evolve the genotype in a symbolic regression problem (sixth order polynomial) and the Santa Fe Ant Trail. The *computational effort* is calculated for both cases. It is suggested that *hit effort* is a more reliable measure of computational efficiency. A *neutral search* strategy that allows the fittest genotype to be replaced by another equally fit genotype (a neutral genotype) is examined and compared with *non-neutral* search for the Santa Fe ant problem. The neutral search proves to be much more effective.

1 Introduction

In its original form Genetic Programming (GP) [10][11] has evolved programs in the form of LISP parse trees. Usually large populations are used and crossover is used as the primary method of developing new candidate solutions from older programs. In contrast to this in Evolutionary Programming [4] [5] has tended to emphasize the importance of mutation operators.

Although GP and EP place different emphasis on the evolutionary operators they both tend to share the representation of programs as parse trees in which there is no distinction between genotype and phenotype. Trees are a special form of graphs in which two nodes must have at most one path between them (a *path* is a sequence of connected nodes). One of the original motivations for a tree-based approach was to allow a solution to the problem of applying crossover to variable length genotypes. In this work a new approach is proposed called *Cartesian Genetic Programming (CGP)* where the genotype is represented as a list of integers that are mapped to directed graphs rather than trees. One motivation for this is that it uses graphs which are more general than trees. However the original motivation came from the effectiveness of the approach in learning Boolean functions where it proved to be considerably more efficient than standard GP methods [14]. In CGP the genotypes are of fixed length but the phenotypes are of variable length according to the number of unexpressed genes. The importance and potential advantages of defining a genotype-phenotype mapping were discussed by Banzhaf [2] in Binary Genetic Programming (BGP) [1]. In BGP binary strings are translated and repaired to form valid programs. In CGP no repair is necessary (see section 2). Another potential advantage of employing a genotype-phenotype mapping is that it allows for the possibility of many genotypes mapping to the same phenotype and so explicitly allows *neutrality* to be present. Neutrality refers to the presence of genotypes with the same fitness. The importance of neutrality is widely recognized in modern theories of natural molecular evolution. Indeed, Kimura [8][9] and Ohta [15][16] maintains that evolution at the molecular level is mainly due to mutations that are nearly neutral with respect to natural selection. From the point of view of fitness landscapes in artificial evolution it has been suggested that neutrality may provide a route whereby adaptive evolution may cross regions with poor fitness [7]. This view is supported in a study of a genotype-phenotype mapping that was applied to a problem of constrained optimisation using genetic programming [2]. Harvey and Thompson, in an experiment that used artificial evolution to configure a programmable silicon chip, also found that the presence of neutrality in a genetically converged population could lead to further fitness improvement [6]. It is important to note that neutrality arising from genotype redundancy can only be useful when the neutral changes are likely to alter the potential effects of future genotypic change, merely adding unexpressed genes will not facilitate useful neutral evolution.

In CGP there are very large number of genotypes that map to identical genotypes due to the presence of a large amount of redundancy. Firstly there is *node redundancy* that is caused by genes associated with nodes that are not part of the connected graph representing the program. This redundancy is very large at the beginning of the evolutionary run as many nodes are not connected in the early populations. The node redundancy gradually reduces during the run to a level that is determined by average number of nodes required to implement a satisfactory program and the maximum allowed number of nodes. Another form of redundancy in CGP, also present in all other forms of GP is, *functional redundancy*. In this case, a number of nodes implement a sub-function that actually may be implemented with fewer nodes. This growth in the number of redundant nodes constitutes *bloat*. The third form of redundancy, called *input redundancy,* occurs when some node functions are not connected to some of the input nodes. For example, in section 3 it is seen that all the nodes used to evolve programs to solve the Santa Fe Ant Trail have three inputs and one output, despite the fact that only one function uses the three inputs. Node and input redundancy both have the potential of adding useful neutrality. An unconnected node may undergo neutral change and later become connected – this might be necessary to achieve a higher fitness. A node with a redundant input might, after a mutation that altered the arity of the node function, suddenly become useful. Functional redundancy probably positively contributes to the evolvability of the target function or program as it increases the number of ways that it might be built. The possibility of disconnecting nodes that CGP allows might have a useful role in keeping the attendant bloat in check.

The principal motivations behind the work reported in this paper were twofold, firstly, to introduce CGP as an alternative methodology to standard GP, and secondly, to extend CGP to non-Boolean problems and show that it is a useful method for evolving programs with other data types. It is important that researchers in Genetic Programming explores the advantages and disadvantages of many representations rather than confining themselves to one dominant form. Diversity is important in research just as it is in evolving populations.

The organization of the paper is as follows. In section 2 the method of CGP is explained. Section 3 describes the problems that CGP is tested on and the measure used to calculate computational effort. The test problems are: symbolic regression of a sixth order polynomial from input-output data, and the harder problem of evolving a program to control an artificial ant traversing the Santa Fe Ant Trail. In section 4 the results are presented and discussed and another useful measure of computational efficiency called the *hit effort* is defined. Finally in section 5 some conclusions are given.

2 Cartesian Genetic Programming

CGP is Cartesian in the sense that the method considers a grid of nodes that are addressed in a Cartesian coordinate system. CGP has a some of points of similarity with Parallel Distributed Genetic Programming (PDGP) [17][18] and the graph-based GP system PADO (Parallel Algorithm Discovery and Orchestration) [19]. In the former graphs were evolved *without* the use of a genotype-phenotype mapping and various sophisticated crossover operators were defined. In the latter, each program is represented as an arbitrary directed graph of N nodes, where each node may have up to N outputs. Moreover, in PADO each node possesses its own private stack based memory, and also access to a globally defined indexed memory.

A Cartesian program (CP) denoted P is defined as a set $\{G, n_i, n_o, n_n, F, n_r, n_c, n_o, l\}$ where G represents the genotype and is itself a set of integers representing the indexed n_i program inputs, the n_n node input connections and functions, and the n_o program output connections. The set F represents the n_f functions of the nodes. The number of nodes in a row and column are given by n_r, n_c respectively. Finally the program interconnectivity is defined by the levels back parameter l, which determines how many previous columns of cells may have their outputs connected to a node in the current column (the primary inputs are treated in the same way as node outputs). In this paper only feed-forward connectivity is considered. However a Cartesian program can be readily extended to include sequential processes (e.g. loops), by allowing the inputs to nodes to be connected to the program outputs through a clock. Note that the program inputs are allowed to connect to any node input. An example of the genotype and genotype-phenotype mapping is depicted in Fig. 1 for a program with six inputs and three outputs. The example shows a feed-forward program of three rows by four columns, with $l = 2$ and $n_n = 3$.

0 1 3 0   1 0 4 1   3 5 2 2   8 5 0 1   6 8 3 2   4 2 6 0
6 1 0 7 2   9 11 10 1   8 8 6 2   9 10 13 15   13 14 11 16   10 10 14 2   13 16 17

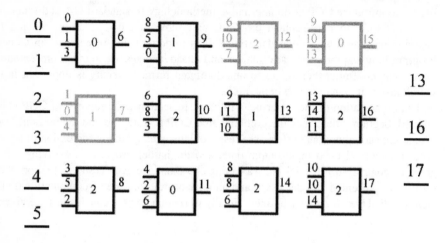

Fig. 1. Genotype-phenotype mapping a) Phenotype. b) Genotype. For a program with six inputs and 3 outputs, and three functions (0, 1, 2 inside square nodes, in italics in genotype). The grey squares indicate unconnected nodes.

Nodes in the same column are not allowed to be connected to each other, and any node may be either connected or disconnected. The CP with the most freely connected network is obtained when the $n_r=1$ and $l= n_c$, i.e. any node can be connected to any node on the right. The length of the genotype G that maps to a particular CP is fixed and is equal to $n_r n_c (n_n +1)+ n_o$ however this just means that the *maximum* size of the associated Cartesian program is fixed, the actual size may be anything from zero up to the maximum (in Fig. 1 only 9 nodes are connected).

When a population of genotypes is created or mutation is applied the genes must obey certain constraints in order for the genotype to represent a valid program. These are defined as follows:

Let c^n_{kj} represent the gene associated with the k-th input of a node in column j (the leftmost column is represented by 0) then it obeys the inequalities

$$c_{kj} < e_{max} , j < l$$
$$e_{min} \le c_{kj} < e_{max} , j \ge l$$
$$e_{min} = n_i+(j-l) n_r$$
$$e_{max} = n_i+j n_r .$$

(1)

Let c^o_k represent the gene associated with the k-th program output then it obeys the inequalities

$$h_{min} \le c^o_k < h_{max}$$
$$h_{min} = n_i+(n_c -l) n_r$$
$$h_{max} = n_i+(n_c -1) n_r .$$

(2)

Let c'_k represent the gene associated with the function of k-th node then it obeys the inequality

$$0 \le c'_k < n_f .$$

(3)

Provided the genotypes obey these constraints, crossover can be freely applied to produce a valid solution.

Modern forms of GP allow the use of Automatically Defined Functions (ADFs). In the form of CGP discussed here there are no ADFs, however as node outputs may be widely re-used one can consider it as employing Automatic Re-used Outputs (AROs). One of the attractive features of CGP is that no *explicit* encoding is required to facilitate this. A potential disadvantage is that AROs are not as general as ADFs as they can only re-use an output *with the same inputs*. One can control the amount of AROs by adjusting the levels-back parameter and the number of rows and columns. AROs are most strongly favoured in a configuration of nodes with one row, and in which $l = n_c$. At the other extreme AROs are forbidden when $n_c = 1$.

In this paper two evolutionary algorithms were applied. For the symbolic regression problem a generational Genetic Algorithm (GA) was used with uniform crossover, where each gene in the offspring is randomly picked up from the parents. Size two probabilistic tournament selection was used with the winner of the tournament being accepted with a probability of 0.7 (otherwise the loser was accepted). In the Santa Fe Ant Trail problem a simple form of $(1 + \lambda)$ Evolutionary Strategy (with $\lambda \approx 4$) was used. No crossover was applied. The algorithm was as follows:

Algorithm

1. Generate initial population at random (subject to constraints)
2. Evaluate fitness of genotypes in population
3. Promote fittest genotype to new population
4. Fill remaining places in the population with mutated versions of the fittest
5. Return to step 2 until stopping criterion reached

Note that at step 3. If no new genotype has greater fitness but other genotypes have the same fitness as the best then one of these would be randomly chosen and promoted to the new population. In this paper this is referred to as *neutral* search. If only better genotypes are chosen then the search is referred to as *non-neutral*. For the Santa Fe trail both methods were examined.

3   Test problems

Two problems were studied in this paper: Symbolic regression of a sixth order polynomial [11, pages 110-122]

$$x^6 - 2x^4 + x^2$$

(4)

The objective was to evolve a program that produces the given value of the sixth order polynomial above as its output, when the value of the one real-valued independent variable x is given as input. The program inputs were 1.0 and X. This departs from the definition given by Koza [11], who used *ephemeral constants* randomly defined over [-1.0, 1.0] with a given precision. Ephemeral constants are generated when the first program input is connected to a node in the initial population. From that point on in the evolutionary algorithm these constants are fixed and cannot be mutated. The function set used in this paper was {+, - *, div}, where div returns the numerator if the divisor is zero, otherwise it returns the normal result of division. The fitness cases were 50 random values of X from the interval [-1.0, +1.0]. These were fixed throughout the evolutionary run. The fitness of a program was defined as the sum over the 50 fitness cases of the sum of the absolute value of the error between the value returned by the program and that corresponding to the sixth order polynomial. The population size chosen was 10. The number of generations was equal to 8000. The crossover rate was 100% (entire population replaced by children). The mutation rate was 2%, i.e. on average 2% of the all the genes in the population were mutated. Other parameters were as follows: n_r =1, n_c =10, l =10. An evolved genotype with a high fitness is given below (with n_r =4, n_c =4, l =4):

$$
\begin{array}{llll}
1\ 0\ 3 & 1\ 1\ 3 & 1\ 1\ 2 & 0\ 0\ 2 \\
4\ 4\ 2 & 2\ 1\ 3 & 3\ 4\ 1 & 1\ 4\ 0 \\
6\ 3\ 2 & 1\ 8\ 2 & 0\ 8\ 2 & 8\ 6\ 2 \\
6\ 11\ 2 & \underline{11\ 11\ 2} & 5\ 3\ 3 & 13\ 13\ 3 \quad 15
\end{array}
$$

Fig. 2. The genotype for a program evolved to match a sixth order polynomial. The two inputs 0 and 1 refer to 1.0 and X respectively. Functions + , - , * , div are represented by integers 0, 1, 2, 3 respectively. The program output is taken from the node with output label 15 (underlined).

The other problem studied was the Santa Fe Ant Trail [10, pages 147-155]. An imaginary ant starts at position (0,0) (top-left corner) of a 32 x 32 toroidal grid of squares. The ant initially faces east. There is a trail of food covering 144 squares with may gaps and twists and turns containing 89 pieces of food. The ant can move one square in the direction it is facing or turn left or right on the square that is situated on. Each of these actions require one time step. The ant has a sensor that enables it to detect whether there is food on the square one position ahead in the direction it is facing. The ant eats any food on squares that it is occupying. The objective is to evolve a program that can successfully navigate the ant so that it consumes all 89 pieces of food (success predicate). Only 600 time steps are allocated to execute the ant control program[1]. If a program (as is very likely) finishes before 600 time steps have elapsed then it is repeated, starting at the current state of the ant and map, this process is repeated until the 600 time steps have elapsed. This is a much more testing problem for Genetic Pro-

---

[1] Although it appears that Koza [10] used 400 time steps. A personal communication from William B. Langdon assured me that the correct figure was 600 and that this figure is used by other investigators.

gramming than the sixth order polynomial because of the time dependent behaviour of the ant. Thus one is evolving a sequential program with states that depend on past history. The function set was as used by Koza { *if-food-ahead, prog2, prog3*} coded as functions 0, 1, 2 respectively. The program inputs were coded as 0, 1, 2 and represented *move, left, right* respectively.

The algorithm used to evolve the Cartesian programs was described in section 2. The specific parameters used were as follows: $\lambda = 4$. A range of mutation rates per individual genotype were investigated from 4%-40%. Population size was 10, 100 runs of 12,000 generations were carried out. Other parameters were as follows: $n_r = 1$, $n_c = 20$, $l = 20$. Thus with 81 genes and a mutation rate of 10%, 8 genes would be mutated in each genotype in the population.

4 Results

One method proposed for assessing the effectiveness of an algorithm, or a set of parameters was as follows [10, page 194]. It consists of calculating the number of individual chromosomes, which would have to be processed to give a certain probability of success. To calculate this figure one must first calculate the cumulative probability of success $P(M, i)$, where M represents the population size, and i the generation number. $R(z)$ represents the number of independent runs required for a probability of success (meeting success predicate), given by z, by generation i. $I(M, z, i)$ represents the minimum number of chromosomes which must be processed to give a probability of success z, by generation i. The formulae for these are given below, $N_s(i)$ represents the number of successful runs at generation i, and N_{total}, represents the total number of runs:

$$P(M,i) = \frac{N_s(i)}{N_{total}}, \quad R(z) = ceil\left\{ \frac{\log(1-z)}{\log(1-P(M,i))} \right\}, \quad I(M, i, z) = M\, R(z)(\, i+1) \tag{5}$$

Note that when $z = 1.0$ the formulae are invalid (all runs successful). In the expression for $I(M, i, z)$, $i+1$ is used to taken in account the initial population. In this paper z was chosen as 0.99 so that the computational effort represented the number of evaluations required to give a probability of success of 0.99.

4.1 Sixth order polynomial

One hundred runs were carried out with 61 runs successful. The minimum computational effort (see eqn. 5) was 90,060 that corresponded to 6 runs of 1,500 generations. It is not possible to compare this with the effort computed in [11] as the experimental conditions were not the same and employing the fixed constant 1.0 conferred a great advantage over randomly chosen ephemeral constants.

4. 2 Santa Fe Ant Trail

In Tables 1 and 2 are listed the results obtained after 100 runs for various mutation rates when the neutral and non-neutral strategies were employed. The number of solutions found with the maximum fitness is listed as #hits. The minimum computational effort I defined in equation 5 is listed in the third column (divided by 1000). The generation at which the minimum effort was observed is shown in the fourth column. Finally the number of hits that were observed at that generation is listed in the fifth column. The first thing to notice, is that some of the figures for the computational effort are calculated for a ridiculously small number of hits (less than 10, see column 5), and thus are instantly under suspicion, since in different batches of 100 runs these figures are likely to vary enormously (Actually most of figures in Table 2 are of this type). The computational effort is plotted against mutation rate in Fig. 3. It would be very easy to pick out a very good figure for I and compare it favourably with results found in the literature (see Table 3). It is quite difficult to define under what conditions the calculation of computational effort is reliable without undertaking many batches of 100 runs.

Table 1. Effort to solve the Santa Fe Ant Trail (neutral)

Mutation rate (%)	#hits	Min I /1000	Generation	#hits at generation
4	34	476	420	4
6	48	280	60	1
8	68	335	2580	30
10	83	252	2100	32
12	89	235	5880	70
14	93	173	1440	32
16	86	270	9000	79
18	85	193	2760	49
20	91	212	10,620	90
25	93	209	10,440	90
30	78	331	8280	69
40	63	511	1380	12

Table 2. Effort to solve the Santa Fe Ant Trail (non-neutral)

Mutation rate (%)	#hits	Min I /1000	Generation	#hits at generation
4	17	686	300	2
6	14	831	180	1
8	20	823	360	2
10	28	139	60	2
12	30	888	5220	24
14	39	458	300	3
16	42	276	120	2
18	50	204	180	4
20	48	441	900	9
25	52	324	1620	21
30	52	413	180	2
40	40	757	840	5

Table 3. Previously published effort to solve the Santa Fe Ant Trail [12]

Method	I/1000
Koza GP [10, page 202]	450
Size-limited EP [3]	136
Strict Hill Climbing [12]	186
PDGP [12]	336

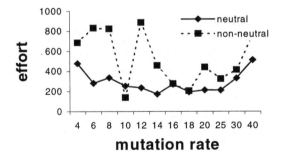

Fig. 3. The computational effort for neutral and non-neutral strategies for various mutation rates.

To assess the quality of an algorithm one could just look at the total number of hits (second column from left in Tables 1 and 2) rather than the computational effort. This is plotted against mutation rate in Fig. 4. Unfortunately this would give no information about the time taken by the algorithms to give these results.

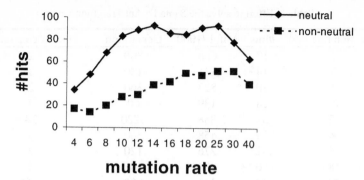

Fig. 4. The number of hits for neutral and non-neutral strategies for various mutation rates.

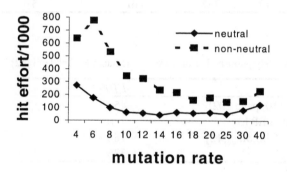

Fig. 4. The hit effort for neutral and non-neutral strategies for various mutation rates.

It is suggested that a more reliable measure of computational efficiency might be the *hit effort* which is defined here to be the total number of evaluations over the 100 runs divided by the number of hits. The is effectively the average number of evaluations required per hit. The hit effort is most reliable when a scenario is chosen that gives many successful runs as it is fairly sensitive to the number of hits. The advantage of using hit effort is that it is a figure that is measured over the *full set of runs* and is not an inference based on assumptions about likely outcome of another experiment. Qualitatively is satisfies the requirements of a good measure of computational efficiency and quality. In the experiments performed here it presented a much more regular behaviour as can be seen in Fig. 5 which shows the variation in hit effort with mutation rate.

Langdon and Poli [12] noted that genetic programming and other search techniques do not do much better than random search and that size of the allowed programs has a marked effect on performance, with larger programs fairing more badly (above some threshold). It can be seen from the results presented here that high mutation rates of about 14% are most effective with neutral search, and that a higher rate of 25% are better for non-neutral search. These results support the findings that the fitness space associated with the Santa Fe trail has a great deal of randomness associated with it [12]. with the search space. It should also be noted that the maximum length of CG programs used in this paper was 20, so the results may be slightly better because of this fact (see [12]).

5 Conclusions

This paper has presented a new form of genetic programming called Cartesian Genetic Programming. It has already been shown to be very effective for Boolean function learning [14] and here it is extended to problems with real-valued data and time dependent behaviour. It is certainly no less effective a method than other forms of GP when tested on the Santa Fe Ant problem.

The representation of genotypes in CGP has very large redundancy and it was shown that the neutrality in fitness that this allows can be used to improve the search. An important question that needs to be answered relates to which type of neutrality is most important. Other preliminary experiments for the problem of evolving binary arithmetic functions show that the search is impaired when either the node or functional redundancy is restricted when compared to a situation in which both are allowed. Other results indicate that provided a small but sufficient amount of node redundancy is present the search is most effective. Also experiments are underway that attempt to discover the relative importance of two types of neutrality- the first, genotypes with equal fitness that map to the same phenotype and second, genotypes with equal fitness that map to different phenotypes.

It was argued that the widely used measure of computational effort [10] is unreliable, and that a new measure called hit effort should be used instead. This gave much more stable results for experiments performed on the Santa Fe Ant Trail. Although the hit effort is still sensitive to the number of successful runs, at least it is a figure that is actually calculated for the entire set of runs, rather than being an inference.

There is much more work to be done and many questions remain. Crossover is easily accomplished with the Cartesian genotypes. But does it really confer a considerable advantage for other classes of problems? The mutation mechanism employed in this paper is extremely simple. In Evolutionary Programming this is generally more sophisticated. Could more involved methods of mutation improve the performance of algorithms based on Cartesian genotypes? Why is CGP effective is it due the presence and exploitation of neutrality? Would CGP be much more effective by allowing the possibility of true Automatically Defined Functions? How could this be accomplished?

References

1. Banzhaf, W.: Genetic Programming for Pedestrains, Technical Report No. 93-03, Mitsubishi Electric Research Labs, Cambridge, MA (1993)
2. Banzhaf, W.: Genotype-Phenotype-Mapping and Neutral Variation: A Case study in Genetic Programming. In: Davidor, Y., Schwefel, H.-P., R. Männer, R. (eds.): Proceedings of the conference on Parallel Problem Solving from Nature III. Springer-Verlag, Berlin Heidelberg New York(1994) 322-332.

3. Chellapilla K.: Evolutionary programming with tree mutations: Evolving computer programs without crossover. In: Koza, J. R., et al. (eds.): Genetic Programmimg 1997: Proceedings of the Third Annual Conference on Genetic Programming. Morgan Kaufmann, San Francisco, CA. (1998) 432-438.

4. Fogel, L. J., Owens, A. J., Walsh, M. J.: Artificial Intelligence through Simulated Evolution, Wiley (1996)

5. Fogel, D. B.: Evolutionary Computation: Towards a New Philosophy of Machine Intelligence, IEEE Press (1995)

6. Harvey, I., Thompson, A.: Through the labyrinth evolution finds a way: A silicon ridge. In: Higuchi, T., Iwata, M., Liu, W. (eds.): Proceedings of First International Conference on Evolvable Systems: From Biology to Hardware, Lecture Notes in Computer Science, Vol. 1259. Springer-Verlag, Berlin Heidelberg New York (1996) 406-422

7. Huynen, M. A., Stadler, P. F., Fontana, W.: Smoothness within ruggedness: The role of neutrality in adaptation. Proceedings of the National Academy of Science U.S.A. Vol. 93 (1996) 397-401

8. Kimura, M.: Nature, Vol. 217 (1968) 624-626.

9. Kimura, M.: The Neutral Theory of Molecular Evolution, Cambridge, Cambridge University Press (1983)

10. Koza, J. R.: Genetic Programming: On the programming of computers by means of natural selection. MIT Press (1992)

11. Koza, J. R.: Genetic Programming II: Automatic Discovery of Reusable Subprograms. MIT Press (1994)

12. Langdon, W. B., Poli, R.: Why Ants are Hard. In: Koza, J. R. et al. (eds.) Proceedings of the Third Annual Conference on Genetic Programming. Morgan Kaufmann, San Francisco, CA (1998) 193-201

13. Miller J. F., Thomson P.: Aspects of Digital Evolution: Geometry and Learning. In: Sipper, M. et al (eds.): Proceedings of the Second International Conference on Evolvable Systems: From Biology to Hardware, Lecture Notes in Computer Science, Vol. 1478. Springer-Verlag, Berlin Heidelberg New York (1998) 25-35

14. Miller, J. F.: An empirical study of the efficiency of learning Boolean functions using a Cartesian Genetic Programming Approach. In: Banzhaf, W. et al.: Proceedings of the Genetic and Evolutionary Computation Conference (GECCO'99), Morgan Kaufmann, San Francisco, CA (1999) 1135-1142

15. Ohta, T.: Slightly deleterious mutant substitutions in evolution. Nature Vol. 246 (1973) 96-97

16. Ohta, T.: The nearly neutral theory of molecular evolution. Annual Review of Ecology and Systematics Vol. 23 (1992) 263-286

17. Poli, T.: Parallel distributed genetic programming. Technical report CSRP-96-15, School of Computer Science, The University of Birmingham, UK (1996)

18. Poli, R.: Evolution of graph-like programs with parallel distributed genetic programming. In: Bäck, T. (ed): Genetic Algorithms: Proceedings of the Seventh International Conference, Morgan Kaufmann, San Francisco, CA (1997) 346-353

19. Teller, A., Veloso, M.: PADO: Learning tree structured algorithms for orchestration into an object recognition system. Technical Report CMU-CS-95-101, Dept. of Computer Science, Carnegie Mellon University, Pittsburg (1995)

Some Probabilistic Modelling Ideas for Boolean Classification in Genetic Programming

Jorge Muruzábal[1], Carlos Cotta-Porras[2] and Amelia Fernández[2]

[1] University Rey Juan Carlos, 28933 Móstoles, Spain
[2] University of Málaga, 29071 Málaga, Spain

Abstract. We discuss the problem of boolean classification via Genetic Programming. When predictors are numeric, the standard approach proceeds by classifying according to the *sign* of the value provided by the evaluated function. We consider an alternative approach whereby the *magnitude* of such a quantity also plays a role in prediction and evaluation. Specifically, the original, unconstrained value is transformed into a probability value which is then used to elicit the classification. This idea stems from the well-known logistic regression paradigm and can be seen as an attempt to squeeze all the information in each individual function. We investigate the empirical behaviour of these variants and discuss a third evaluation measure equally based on probabilistic ideas. To put these ideas in perspective, we present comparative results obtained by alternative methods, namely recursive splitting and logistic regression.

1 Introduction

Consider the problem of boolean classification via Genetic Programming (GP). The goal is to learn the relationship between a boolean response y and a set of predictors $x_1, x_2, ..., x_n$ on the basis of a body of known classifications. Let $\mathbf{x}$ and D denote the vector containing all predictors and the $M \times (n+1)$ training data matrix respectively. The GP approach proceeds by evolving a population of individuals or functions $\phi = \phi(\mathbf{x})$ according to some evaluation or fitness measure Θ and some set of basic functions Γ. All functions ϕ ever considered by the algorithm are combinations of members of Γ, inputs x_k and random constants; their fitness is given by $\Theta(\phi, D) \in \mathbb{R}$. The well-known procedures of selection and recombination are applied as usual for a fixed number of generations. Once this simulated evolution process is over, the best function found is ready for validation and, if appropriate, deployment on new cases $\mathbf{x}$.

The nature of predictors x_k is closely related to both the function set Γ and the evaluation method Θ. If all predictors are boolean, then one can naturally use logical operators in Γ [1]: all ϕ produce boolean values and no transformation is needed in Θ. In this paper, however, we consider the case of numeric predictors $x_k \in \mathbb{R}$. Now ϕ outputs real numbers and some kind of transformation is needed. The standard approach [2] proceeds by implementing a *wrapper* that classifies each $\mathbf{x}$ in D according to the *sign* of $\phi(\mathbf{x})$; the resulting predictions and evaluation measure are called *deterministic* for reasons that will become apparent below.

This idea returns to the previous boolean case but has a number of disadvantages. For example, it does not matter whether $\phi(\mathbf{x}) = +.1$ or $\phi(\mathbf{x}) = +1,000$, the prediction $\hat{y}$ will be 1 in either case. This is fine as long as $\hat{y} = 1$ is correct, but otherwise ϕ would seem much less valuable in the latter case. Another problem is that several unrelated functions may share the same signs over (the majority of) D. This may introduce a substantial amount of confusion in the evolutionary process. Finally, this approach yields no information about the *confidence* that we should place on our predictions; we can look at proximity to the border $\phi(\mathbf{x}) = 0$, but there is no absolute scale on which our assessment could be based. Overall, it seems a bit wasteful to carry out the search in the space of functions ϕ when all the system needs is a much simpler boolean decision criterion. It would be nice if we could profit from every bit of information in ϕ.

In this paper we study an alternative approach inspired by standard logistic regression modelling [3]. A distinctive feature of this approach is that the magnitude of $\phi(\mathbf{x})$ also plays a role. To define a new evaluation measure, the idea is to use first an appropriate function G to produce an estimate $\hat{\pi}(\mathbf{x}) = G \circ \phi(\mathbf{x})$ of the target conditional probability $\pi(\mathbf{x}) = P(Y = 1/X = \mathbf{x})$, then introduce a random mechanism to exploit the *probabilistic* information in $\hat{\pi}(\mathbf{x})$. Here X and Y denote the obvious random variables in the underlying statistical model. The random decision $\hat{y}$ is simply the result of a coin flip with probability of heads $\hat{\pi}(\mathbf{x})$. From this point on, the new evaluation measure works exactly as before. Note that constraints on G make only a few functions suitable for the present purpose. To illustrate their desired behaviour: if $\phi(\mathbf{x}) = +.1$, $\hat{\pi}(\mathbf{x})$ is slightly above .5 and will often yield $\hat{y} = 1$ *as well as* $\hat{y} = 0$; conversely, if $\phi(\mathbf{x}) = +1,000$, then $\hat{\pi}(\mathbf{x}) \approx 1$ and $\hat{y} = 1$ nearly always. This seems to be more in line with our intuition in this context.

We talk of *stochastic* predictions and evaluation in this case; let Θ_d and Θ_s respectively denote the deterministic and the stochastic evaluation measures. We provide some evidence below suggesting that Θ_s is better than Θ_d for some purposes and that the situation may be reversed for others. Specifically, Θ_s is better than Θ_d in the sense of providing more accurate results with "well-behaved" (or deterministic) training data; Θ_d seems more parsimonious when there is noise in the data.

As regards interpretation, $\hat{\pi}(\mathbf{x}) \in (0,1)$ is an extremely simple and informative summary for future $\mathbf{x}$. In particular, we shall explore below another way of exploiting the *whole set* $\{\hat{\pi}(\mathbf{x}), \mathbf{x} \in D\}$ for evaluation purposes. This will be seen as an attempt to quantify the overall *fit* of the selected ϕ in a well-defined sense.

The organization of the paper is as follows. Section 2 contains a brief description of the logistic regression paradigm. Section 3 presents our experimental setup and main empirical results. Section 4 links up with some related work. Section 5 closes with some discussion and an outline of future research.

2 A review of the logistic regression paradigm

In logistic regression (LR) modelling, the boolean random variable Y is assumed to depend on the available predictors x_1, x_2, ..., x_n in the following way. First, the contribution of these predictors is totally captured by a linear combination

$$f(\mathbf{x}, \boldsymbol{\omega}) = \omega_0 + \omega_1 x_1 + ... + \omega_m x_m.$$

Here we may have all or some of the original variables together with some other (simple) functions thereof (eg. squares). An important point is that the user needs to specify exactly what m variables come into play. Now, it is assumed that, for some *link* function G and some $\boldsymbol{\omega} \in \mathbb{R}^{m+1}$, Y follows, conditionally on $\mathbf{x}$, a Bernoulli distribution with parameter $G \circ f(\mathbf{x}, \boldsymbol{\omega})$. A link function is a one-to-one mapping from $\mathbb{R}$ onto the unit interval $(0, 1)$. While several choices for G are available, the most widely used is the inverse-logistic or sigmoid transformation

$$G(y) = 1/\left[1 + exp\left(-y\right)\right].$$

This particular choice will also be assumed throughout this paper. We note in passing that this model can be seen as a particular case (namely, the no-hidden-layer case) of the standard neural network model for classification tasks [5].

Jordan [6] has put together a number of useful results concerning this function. Most relevant for our purposes here is the fact that, if we model our classification problem in the usual way[3], then the posterior probability of either class can *always* be expressed as $G \circ \xi(x)$ for some function ξ. It follows that $G(y) = 1/\left[1 + exp\left(-y\right)\right]$ is the most natural choice for our transformation (which, while not bayesian, aims to provide estimated "posterior" probabilities).

Returning now to the LR paradigm, the key assumption is that the target function

$$\pi(\mathbf{x}) = P(Y = 1/X = \mathbf{x})$$

can be approximated by (some member of) the parametric form

$$h(\mathbf{x}, \boldsymbol{\omega}) = 1/\left[1 + exp\left(-f(\mathbf{x}, \boldsymbol{\omega})\right)\right].$$

It is then possible to formulate the likelihood function

$$L(\boldsymbol{\omega}/D) = \prod_{j=1}^{M} h(\mathbf{x}_j, \boldsymbol{\omega})^{y_j} \left[1 - h(\mathbf{x}_j, \boldsymbol{\omega})\right]^{1-y_j}$$

and estimate $\boldsymbol{\omega}$ by the method of maximum-likelihood. The LR algorithm implementing this method is well-studied and widely available [4]. While the resulting estimate $\hat{\boldsymbol{\omega}}$ is not guaranteed to maximize L, it typically provides sensible answers.

---

[3] That is, incorporating a priori probabilities for the two classes and conditional densities for the input vectors $\mathbf{x}$

The above problem may also be viewed alternatively by noting that $f(\mathbf{x}, \hat{\omega})$ aims to approximate the *logistic* transformation of $\pi(\mathbf{x})$, that is,

$$\lambda(\mathbf{x}) = logit(\pi) = log[\pi/(1 - \pi)] \in \mathbb{R}.$$

This *logit* function λ will also be the target for functions ϕ in our GP context. Indeed, in our simulations we specify λ and generate the response y associated to $\mathbf{x}$ on the basis of $G \circ \lambda(\mathbf{x})$.

Fitted probabilities for individual $\mathbf{x}_j \in D$ are given by $\hat{\pi}_j = h(\mathbf{x}_j, \hat{\omega})$. In ordinary LR, however, each prediction $\hat{y}_j \in \{0, 1\}$ simply reflects the sign of the corresponding $f(\mathbf{x}_j, \hat{\omega})$. Thus, following the terminology introduced earlier, the LR paradigm typically behaves deterministically (just like the standard GP approach).

As regards the practical aspects of the LR paradigm, it is clear that the analyst will rarely carry out a single fit (or estimation of ω). Upon examination of the first fit, the analyst may remove some variables and incorporate others. More often than not, however, we let the computing package carry out an *automatic* exploration of the space of all linear combinations of a specified set of variables along with their two-at-a-time products and squares [7]. For example, in S-PLUS this exploration is carried out via the `step` command. The algorithm then selects the "best-fitting" model and returns the associated vector of estimated coefficients $\hat{\omega}$ together with some indication of their significance given the remaining variables. This selection process is based on a scalar quantity that combines the quality of fit itself (or *deviance*, see [4]) with the number of terms (or ω's) in the model. However, this measure is less readily interpretable. Instead, the set of fitted probabilities $\hat{\pi}_j$ can be used to validate the model as shown next.

The Chapman data refer to past occurrence of cardio-vascular disease (CVD) amongst 200 males examined by certain hospital. There are six predictors including weight, height, age, blood pressure, etc. The percentage of positive response is only 13%; see [3] for further details. The preferred model is given in Table 1; thus, the key equation is given by

$$f(\mathbf{x}, \hat{\omega}) = -9.255 + 0.053 \ AGE + 0.018 \ WEIGHT + 0.007 \ CHOLESTEROL.$$

The interpretation is immediate: the older and heavier the patient, the higher the chance of having had CVD (and similarly for the cholesterol level, although the latter's coefficient is quite small). While the cholesterol variable is not overwhelmingly significant, it is seen to contribute to a sensible model as follows.

variable	coefficient	t-value
Intercept	−9.255	−4.49
Age	0.053	2.55
Weight	0.018	4.91
Cholesterol	0.007	0.79

Table 1. Selected model for the Chapman data, see [3].

Note first that the confusion matrix arising from this fit (given in Table 2) might lead us to think that the model is not good: $\hat{y}_j = 1$ in only two occasions, 24 positive cases go unpredicted! The following argument shows that this is not the case: the postulated model provides an excellent fit.

	zero	one
zero	174	0
one	24	2

Table 2. Confusion matrix for the training sample under the model shown in Table 1. Columns are predicted values $\hat{y}_j$; rows are observed values y_j.

Christensen [3] first groups individuals in categories according to their CVD probabilities $\hat{\pi}_j$, see Table 3. For example, the first category includes all individuals with $\hat{\pi}_j \in [0, 0.1)$; we find 99 cases here[4]. Now, if the midpoint .05 is selected as representative probability of this group, then we would naturally expect $99 \times 0.05 = 4.95$ positive cases. This figure is to be contrasted with the true number of CVD occurrences in the group: out of the 99 individuals with predicted probability below .1, we find 5 positive cases. Hence, the fit is very good in this first group. Indeed, the fit is rather good through all 6 groups.

Interval	[.0,.1)	[.1,.2)	[.2,.3)	[.3,.4)	[.4,.5)	[.5,.6)
E_i	4.95	9	5.5	3.5	3.15	1.1
O_i	5	10	2	5	2	2
Number of cases	99	60	22	10	7	2

Table 3. Groupwise expected (E_i) and true (O_i) number of CVD occurrences for the Chapman data based on the model shown in Table 1.

The explanation of this phenomenon is as follows. Since the LR method is based on a simple *linear* cut, the model can not tell whom exactly had CVD in each subinterval or risk group. However, the observed behaviour of each group as a whole is quite in line with the model's probabilistic assessment. This validation example is particularly interesting in that it refers to a case where the distribution of 0's and 1's is rather *skewed*. Skewed data sets (containing only a few 0's or 1's) are difficult for the conventional GP approach since any constant with the right (dominant) sign will have high fitness. Letting the evaluation function measure the overall fit in this way should be helpful to get rid of such free-riders.

We will return to Christensen's idea at the end of Section 3. A basic observation is that the previous assessment of fit depends in no way on the LR paradigm and can therefore be "exported" as soon as the analog of the $\hat{\pi}_j$'s are provided. It can also be used on test data as a validation tool.

---

[4] Of course, the fit can also be measured with a higher or lower number of subintervals.

3 Empirical work

In this Section we present the bulk of our empirical results. We first describe the basic experimental setting, then proceed to discuss some specific problems.

3.1 Experimental setup

We have extended the standard symbolic regression GP paradigm implemented in the `lil-gp` system [8]. Our modifications allow for more than one predictor at a time ($n > 1$) and incorporate all details required to test the ideas described above. The modified system currently runs on a LINUX-based PC; source code is available from the authors.

Due to the exploratory nature of this research, we choose to keep the situation as controlled as possible: all experiments reported below are based on simulated data. In our runs, we always use the same system parameters, see Table 4. Since the focus is on comparative results and not on optimal choices for these parameters, we have simply adopted `lil-gp`'s default options. In particular, we always use a single population and maintain the same reproductive plan throughout each run. We execute ten independent runs under each configuration; a summary of the runs is often given in the standard box-plot form. Either evaluation measure Θ_s or Θ_d is based on the number of correct predictions on the training sample as standardized fitness [2]:

$$\Theta(\phi, D) = M - \sum_{j=1}^{M} |y_j - \hat{y}_j|,$$

where $\hat{y}_j = \hat{y}_j(\phi, \mathbf{x}_j)$ may be computed either deterministically or stochastically as explained earlier.

System Parameter	Value
Maximum number of generations	200
Population size	500
Training sample size	500
Test sample size	500
Initial population generation	half-and-half
Maximum tree depth	5-10
Ephimeral random constant	Yes
Crossover/Reproduction	80/20
Mutation	None

Table 4. Execution parameters. See `lil-gp`'s documentation [8] for details.

Maximum tree depth is usually 10 but we sometimes switch to 5 in order to reduce the complexity of the output functions. We experiment with two function sets: Γ contains the arithmetic operators alone, $\Gamma = \{+, -, *, /\}$, where / stands for protected division as usual. Function set Γ^* contains also the protected logarithm $rlog(x)$, the exponential function $exp(x)$, and the trigonometric functions $sin(x)$ and $cos(x)$.

In our simulations, we work with $n = 5$ predictors; out of these, 2 or at most 3 are relevant, the rest contribute noise. To create our training sets D, we first draw, for each predictor, independent, uniformly distributed variates in $(-5, 5)$. Then we specify the "true" logit function $\lambda(\mathbf{x})$ and generate the associated responses accordingly. Just like in the case of predictions $\hat{y}$, these responses can be generated either deterministically (the "no noise" situation) or stochastically (the "noise" situation). The latter formulates a more realistic and typically harder problem. Results from both options are presented below.

As usual, for each problem we create a second sample to test generalization ability. This test sample itself can also be created either deterministically or stochastically; deterministic test samples are used except where otherwise noted. Lastly, test performance in turn can be determined in either of these two ways. Because the alternative techniques that we study make their predictions deterministically, so do we with our GP functions. This ensures the fairness of our comparisons throughout.

The outline of the rest of this Section is as follows. We first illustrate the "confusion of signs" problem mentioned in the introduction. Next, always under Θ_s and Θ_d, we analyze two problems well-suited for other techniques. We also approach these problems with their "natural" methods and discuss comparative performance. Specifically, we try the logistic regression and recursive splitting algorithms implemented in the S-PLUS system (V4.5), see [7]; all run parameters are fixed to their default values in either case. We only use deterministic training data so far. In Section 3.5 we introduce a third evaluation measure based on overall fit. Finally, we explore performance with noisy training data.

3.2   A basic confounding problem

Consider the problem of learning the logit function $\lambda_1(\mathbf{x}) = x_1 - 2x_2 + 5x_1x_2$. When both the generation of data D and the evaluation of functions ϕ are done deterministically, a basic confounding problem arises during learning[5]. As mentioned, this problem is due to the emergence of alternative functions which partially or totally match the signs of λ_1 over D. The result is that learning is impaired and we may end up with a totally mistaken function. For example, the system converged once to

```
(/ (rlog (exp x2))
   (* x1 0.86686)).
```

In another run, we found a function containing 10 instances of the "successful" string (/ x1 x2). Likewise, some variations of the sign-matching function x_1x_2 were discovered early and the system got stuck; this was the case of

---

[5] Actually, training data for this problem were generated stochastically (as shown in Fig. 1a). However, since the slopes near the axes are so sharp here, these data do not differ much from deterministically generated data.

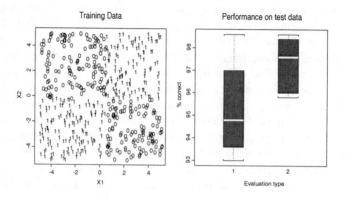

Fig. 1. Learning λ_1 with function set Γ^* and maximum depth set to 10. (a) Training data in the relevant $x_1 \times x_2$ space. (b) Boxplots (each based on ten runs) of performance results under the two evaluation measures Θ_d and Θ_s. In this and the following boxplots, standard GP (Θ_d) is identified as evaluation type 1. Recall that the box itself runs from the lower to the upper quartile, thus capturing central or typical behaviour. The white line represents the median. The "whiskers" stretch out to cover less typical values, whereas atypical values are shown individually (see eg. Fig. 2). Note also that the vertical scale changes from boxplot to boxplot.

```
(/ (* x2 x1)

    (exp x4)),
```

found at initialization and never improved upon! Naturally, this behaviour under Θ_d does not persist under Θ_s. In any case, performance on test data (see Fig. 1b) shows that Θ_s never declines below 95.8% and is thus more likely to prevent a poor fit.

3.3 The logistic regression problem

Consider now the logit function $\lambda_2(\mathbf{x}) = x_1 - 2x_2 + x_3 + 2x_1x_2 - 3x_2x_3$. This reflects exactly the type of structure that logistic regression (LR) searches for; it should therefore provide a nice basis for evaluation of the GP approach.

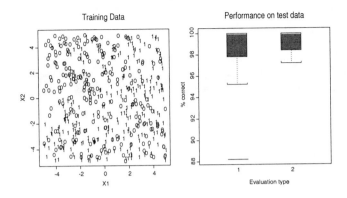

Fig. 2. Learning λ_2 with function set Γ and maximum depth set to to 5. (a) Partial view of the training data. (b) Boxplots of performance results (see Fig. 1). Recall that Θ_d and Θ_s are identified respectively with evaluation type 1 and 2.

variable	coefficient	t-value
Intercept	1.18	1.56
x_0	−0.44	−1.63
x_1	2.60	3.01
x_2	−4.36	−3.17
x_3	2.12	3.21
x_4	0.38	1.49
$x_2 \times x_3$	−7.32	−3.33
$x_1 \times x_2$	5.24	3.28
$x_2 \times x_4$	−0.36	−2.57
$x_0 \times x_2$	0.30	2.39
$x_0 \times x_1$	0.12	1.52

Table 5. Logistic regression fit for λ_2 (similar to Table 1).

Fig. 2b shows the results on the test sample. It appears that this problem is easy since the algorithm often finds the correct function (the median success rate is 100% in both cases). However, the lower tail of the distribution is again better for Θ_s.

As regards performance by the LR approach, the best fit (produced by the step command as discussed earlier) is summarized in Table 5. Note that we do not recover the desired function exactly. In particular, noise variables x_0 and x_4 show up several times. While the overall significance of such terms is not large in general, the −2.57 t-value of $x_2 \times x_4$ stands out. This is in contrast to the perfect match often achieved by the GP approach. On the other hand, all key terms are found significant, and indeed their coefficients are about 2.4 times their target values. Accordingly, performance on test data is 97.2%. Hence, as expected, the

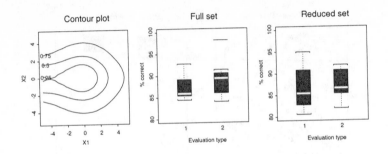

Fig. 3. Learning λ_3. (a) Contour plot of $G \circ \lambda_3$. (b) Performance results for the full function set Γ^* and maximum depth set to 10. (c) Performance results for Γ and maximum depth set to 5.

model fits the data well, yet the GP approach under Θ_s does better.

3.4 The recursive splitting problem

Let us now switch to the problem of learning

$$\lambda_3(\mathbf{x}) = 1.5 - \exp\{-\frac{x_1 + x_2^2}{5}\} + \log(\frac{1 + x_1^2 + x_2^2}{25}).$$

This function illustrates one of the worst-case scenarios for the LR method: while the basic structure is still a sum, summands comprise complicated functions of the input variables. However, the problem is relatively easy for the recursive splitting method. Indeed, the contour plot in Fig. 3a shows that parallel decision boundaries should not be too hard to find. But let us look first at performance by the GP algorithm.

To begin with, consider the case of Γ^* so that the exact target function can be found in principle. The results in Fig. 3b show that Θ_s yields again better results. Not only the best absolute result is achieved in this case, but the median percentage is about 4 points higher. Now we repeat the previous experiment with Γ instead of Γ^*; the system is thus somewhat handicapped but it is still interesting to compare performance. As shown in Fig. 3c, Θ_s leads once again to the best median, although the best overall individual was now evolved under Θ_d.

Consider now performance by the recursive splitting method. We find a relatively simple decision tree with 13 nodes achieving a 96.6% success rate on the test sample. Note that GP performs slightly worse, but it does provide a function! Obviously, things would be much harder for this method were the target region tilted towards, say, the main diagonal. We will examine below performance in a related problem of interest, namely, learning in the presence of noise.

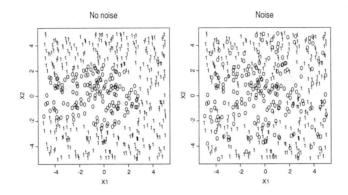

Fig. 4. Training data from λ_3.(a) Deterministic generation follows Fig. 3a closely. (b) Stochastic generation eliminates rigid borders (0's and 1's may occur anywhere).

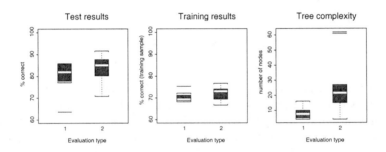

Fig. 5. Learning λ_3 with stochastically generated data (see Fig. 4b). (a) Performance on the test sample. (b) (Final) performance on the training sample. (c) Size of the best trees.

3.5 Measuring the overall fit

In the real world it is unlikely that we find perfectly separated classes as in Fig. 4a. The situation in Fig. 4b should be far more common. Recall that, under stochastic generation, we first compute the probabilities $\hat{\pi}_j$ and then obtain each y_j as a Bernoulli realization with probability $\hat{\pi}_j$. There are at least two issues of interest related to this type of training data. First, and most important, how well do the GP algorithm and its competitors perform here? Some results are reported in the next Section.

Second, we can naturally consider yet another evaluation measure, namely that hinted at in Table 3 above. Specifically, define the overall fit of a given

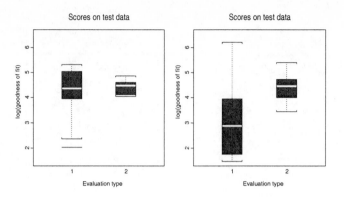

Fig. 6. Θ_f log-scores of individuals evolved earlier for logit functions λ_1 and λ_3 respectively. These numbers are based on a second stochastic test sample. In contrast to previous plots, the lower these values, the better.

individual ϕ as [3]

$$\Theta_f(\phi) = \sum_{i=1}^{10} \frac{[O_i(\phi) - E_i(\phi)]^2}{E_i(\phi)}.$$

Note that this measure is formally reminiscent of the standard χ^2 test for goodness of fit. Nevertheless, the idea is clearly different here since not only O_i but also E_i depend on ϕ. Note also that Θ_f does not make sense when predictions are deterministic. On the other hand, it is quite interesting for real data, particularly skewed data (see Section 2). However, unlike Θ_s, Θ_f does introduce an important computational overhead that needs careful assessment. Hence no GP runs based on it are reported here. To get some intuition for Θ_f, we have evaluated some functions evolved under Θ_s and Θ_d; these will be discussed in the next Section.

3.6 Learning in the presence of noise

We now return to $\lambda_3(\mathbf{x})$ and consider learning based on a *stochastic* sample, see Fig. 5. Θ_s provides once again slightly better results, although we now show that it requires twice as many nodes on average (note also that relatively complex functions emerge occasionally). To put these results in perspective, consider performance of the recursive splitting method on the same training set. We find a much bigger tree with 67 terminal nodes achieving a success rate of 86.4% on the training set but only 65.6% on the test sample. Thus, the GP algorithm does better in this case as well, particularly under Θ_s. Note also that, unlike the recursive splitting method, the GP algorithm generalizes correctly.

Finally, Fig. 6 shows (log) scores achieved under Θ_f by the 40 functions depicted in Figs. 1 and 5. Looking first at Fig. 6a, it appears that Θ_s and Θ_d are comparable as far as the median Θ_f is concerned. As usual, under Θ_s we get a tighter distribution avoiding very poor fits. On the other hand, the best fit is achieved by Θ_d and corresponds to the very simple $rlog(x_2) - .665$.

If we switch to Fig. 6b, the situation is markedly different: Θ_s can not compete now with Θ_d as far as the overall fit is concerned. Curiously enough, the absolute best fit is achieved by the nearly identical

```
(+ (rlog (cos (cos 0.59310)))
   (rlog x2)).
```

The fit, shown in Table 6, is indeed uniformly good. Overall, however, it would seem unlikely that the contribution of x_1 in either λ_1 or λ_3 could be found using Θ_f. On the other hand, Θ_f's apparent built-in bias for parsimony is worth investigating. We also note that the correlation between these Θ_f scores and the original performance scores (shown in Figs. 1 and 5) is low. Thus, the underlying search is quite different in each case.

Interval	[.0,.1)	[.1,.2)	[.2,.3)	[.3,.4)	[.4,.5)	[.5,.6)	[.6,.7)	[.7,.8)
E_i	0	2	7	16	18	39	84	121
O_i	1	3	9	14	18	30	90	127
Number of cases	9	16	29	46	39	71	129	161

Table 6. Groupwise expected (E_i) and true (O_i) number of 1's (analogous to Table 3) for the best fit under Θ_f in problem λ_3.

4   Some related research

The present approach concentrates on the quantitative nature of predictors and outputs real-valued functions. A completely different idea is to dismiss part of this detail and conduct the search over the space of boolean functions. The aim, of course, is to simplify interpretation. For example, Eggermont, Eiben and van Hemert [9] study a scheme based on "booleanization" of numeric predictors and inclusion of logical and comparison operators only. Booleanization is achieved by special functions $A_<(x_k, r)$ and $A_>(x_k, r)$, where r is some threshold, with the obvious interpretation $A_<(x_3, 32.5) = TRUE \iff x_3 < 32.5$. They use specialized mutation operators to modify the arguments of these functions. The overall results (based on several real data sets taken from [12]) are not quite satisfactory though: "Giving up the flexibility of numeric operators for the sake of transparency ... comes at costs of performance". A similar idea is advocated (although not tested) in [10].

Complexity in the resulting functions is indeed a recurrent area of research in GP. Cavaretta and Chellapilla [11] retain the set of mathematical functions but modify the usual fitness measure (Θ_d in our notation, *no-complexity-bias* in their terminology) by adding a term that penalizes functions with a large number of nodes. This is certainly not the first time that this idea is put to work in the GP context. Their results (based on a single test problem from the STATLOG suite [12]) indicate that "the models generated from the no-complexity-bias algorithm were very accurate, beating the best of the STATLOG algorithms, and were statistically significantly more accurate than the low-complexity-bias algorithm". These results are in line with the evidence provided above: under Θ_s, the returned functions are slightly more complex but perform more accurately on (deterministic) test samples.

Interestingly, these authors do not apply Θ_d over the *entire* training sample. As a basis for evaluation, they rather consider several subsamples from the training set, and they let these subsamples change with each generation. Although their main motivation is to reduce the chance of overfitting, this stochastic fitness measure can also be seen as an attempt to capture the overall fit of individual functions. Unfortunately, however, it is not entirely clear how this idea was implemented (see also below).

A different evaluation measure was tested out in [1]. If we label the entries of the confusion table as follows (see Table 2)

	zero	one
zero	tn	fp
one	fn	tp

Table 7. Generic confusion matrix. Columns are predicted values, rows are true values.

then their proposal is $\frac{tp}{tp+fn}\frac{tn}{tn+fp}$ (instead of the traditional $\frac{tn+tp}{tn+tp+fn+fp}$). This involves familiar quantities from the medical domain. The authors report that some comprehensible rules of interest were discovered by this measure.

Resampling underlies also the work of Iba [13]. This author believes that the new strategies called *boosting* and *bagging* may help to "control the bloating effect by means of the resampling techniques". In these variants, the population is divided into subpopulations and a different training sample is used in each subpopulation. Boosting places the emphasis on adaptive subsampling of *hard* cases, see also the cited [9]. This approach is closely related to various *co-evolutionary* schemes inspired by the seminal work of Hillis [14]. Bagging proceeds by simply counting votes from the independently evolved solutions. These strategies have proved useful in several machine learning tasks and show great potential in GP (at the expense of some added computational effort).

5 Summary and concluding remarks

We have presented a number of ideas and preliminary empirical results concerning the introduction of simple probabilistic machinery in the GP approach to boolean classification. The main focus has been placed on the computation of the predictive probabilities $\hat{\pi}$ and the associated stochastic evaluation of individuals Θ_s. The main ideas are central to the well-known method of logistic regression (LR). The added computational cost with respect to the more usual Θ_d is negligible and there is some indication that better results are obtained.

Endowed with this new evaluation method, the GP approach has been shown to provide competitive results with respect to standard methods. Further, the GP approach sometimes exhibits pleasant features that the alternative methods do not (eg., it finds the correct function in the LR problem), and is much better in the noisy recursive splitting problem.

We have also provided some insights into an alternative evaluation measure (Θ_f) based on the idea of overall fit. We believe that this is a promising alternative to tackle highly skewed data sets and we intend to investigate its performance in this direction. We have discussed evaluations of some of the functions found under Θ_d and Θ_s. As it turns out, the best-fitting functions under Θ_f are crude but useful versions of the target functions. This is positive in that a tendency towards parsimony may be in effect. On the other hand, it remains to be seen that fine detail is within the reach of the GP algorithm trained with Θ_f. In any case, further research is needed to ascertain its merit, to quantify the effect of the number of subintervals, etc.

Extensive experimentation with real data, particularly from the STATLOG project, should also be conducted to validate our findings. It would be equally interesting to compare the previous results to other established methods such as neural nets.

Finally, we have reviewed a number of recent developments in the GP literature. Our approach is based on maintaining quantitative detail throughout; exclusive consideration of boolean functions just precludes computation of any $\hat{\pi}$'s whatsoever. With this exception, we find it likely that portions of the present framework may cross-fertilize with some previously advocated ideas. Overall, the GP paradigm seems a worthwhile alternative for the problem of boolean classification. Extensions to deal with the general $(k-\text{leg})$ classification problem deserve attention as well.

Acknowledgement. We are grateful to Bill Punch for assistance with `lil-gp`. The first author is supported by grants HID98-0379-C02-01 and TIC98-0272-C02-01 from the spanish CICYT agency.

References

1. C. C. Bojarczuk, H. S. Lopes and A. A. Freitas (1999). Discovering Comprehensible Classification Rules Using Genetic Programming: A Case Study in a Medical Domain. Proceedings of the Genetic and Evolutionary Computation Conference (GECCO–99), Vol. 2.
2. J. R. Koza (1992). *Genetic Programming.* MIT Press.
3. R. Christensen (1997). *Log-Linear Models and Logistic Regression* (2nd Ed.). Springer.
4. McCullagh, P. and Nelder, J. A. (1983). *Generalized Linear Models.* Chapman & Hall.
5. C. M. Bishop (1995). *Neural Networks for Pattern Recognition.* Oxford University Press.

6. M. I. Jordan (1995). Why the Logistic Function? A Tutorial Discussion on Probabilities and Neural Networks. Computational Cognitive Science Technical Report 9503. M. I. T.

7. W. N. Venables and B. D. Ripley (1997). *Modern Applied Statistics with S-PLUS* (2nd Ed.). Springer.

8. See http://GARAGe.cps.msu.edu/software/lil-gp/lilgp-index.html.

9. J. Eggermont, A. E. Eiben and J. I. van Hemert (1999). A Comparison of Genetic Programming Variants for Data Classification. Proceedings of the Third Symposium on Intelligent Data Analysis (IDA–99).

10. A. A. Freitas (1997). A Genetic Programming Framework for Two Data Mining Tasks: Classification and Generalized Rule Induction. Proceedings of the Second Genetic Programming Conference (GP–97).

11. M. J. Cavaretta and K. Chellapilla (1999). Data Mining Using Genetic Programming: the Implications of Parsimony on Generalization Error. Proceedings of the 1999 Conference on Evolutionary Computation (CEC–99).

12. D. Michie, D. J. Spiegelhalter and C. C. Taylor (1994). *Machine Learning, Neural and Statistical Classification*. Ellis Horwood.

13. H. Iba (1999). Bagging, Boosting and Bloating in Genetic Programming. Proceedings of GECCO–99, Vol. 2.

14. W. D. Hillis (1991). Co-Evolving Parasites Improve Simulated Evolution as an Optimization Procedure. In *Artificial Life II, SFI Studies in the Science of Complexity*, C. G. Langton, C. Taylor, J. D. Farmer and S. Rasmussen, Eds., Addison-Wesley.

Crossover in Grammatical Evolution: A Smooth Operator?

Michael O'Neill & Conor Ryan

Dept. Of Computer Science And Information Systems
University of Limerick
Ireland
{Michael.ONeill|Conor.Ryan}@ul.ie

Abstract. Grammatical Evolution is an evolutionary algorithm which can produce code in any language, requiring as inputs a BNF grammar definition describing the output language, and the fitness function. The usefulness of crossover in GP systems has been hotly debated for some time, and this debate has also arisen with respect to Grammatical Evolution. This paper serves to analyse the crossover operator in our algorithm by comparing the performance of a variety of crossover operators. Results show that the standard one point crossover employed by Grammatical Evolution is not as destructive as it might originally appear, and is useful in performing a global search over the course of entire runs. This is attributed to the fact that prior to the crossover event the parent chromosomes undergo alignment which facilitates the swapping of blocks which are more likely to be in context.

1 Introduction

While crossover is generally accepted as an explorative operator in string based G.A.s [4] the benefit or otherwise of employing crossover in tree based Genetic Programming is hotly debated. Work such as [2] went as far as to dismiss GP as a biological search method due to its use of trees, while [1] presented results which suggested that crossover in GP provides little benefit over randomly generating subtrees. Langdon and Francone have also addressed this issue, the former on tree based GP and the latter on linear structures, and have both introduced new crossover operators in an attempt to improve exploration [3] [6].

Both of these works exploit the idea of a homologous crossover, which draws inspiration from the molecular biological crossover process. The principal exploited being the fact that in nature the entities swapping genetic material only swap fragments which belong to the same position and are of similar size, but which do not necessarily have the same functionality. This, it is proposed, will result in more productive crossover events. Indeed, results from both Langdon and Francone et. al. provide evidence to back up this claim. A consequence arising from these conservative crossover operators is the reduction of the bloat phenomenon, occurring due to the fact that these new operators are less destructive and therefore the production of increasingly longer genomes being deemed unnecessary. Suggestions have been made that destructive crossover events could be responsible for bloat, bloat arising as a mechanism to prevent destructive crossover events occurring by acting as buffering regions in which crossover can occur without harming functionality. As with many non-binary representations, it is often not clear how much useful genetic material is being exchanged during crossover, and thus not clear how much exploration is actually taking place.

Grammatical Evolution (GE) an evolutionary algorithm which can evolve code in any language, utilises linear genomes [12] [11] [9] [8] [7]. As with GP systems, GE has come under fire for its seemingly destructive crossover operator, a simple one-point crossover inspired from GA's. In this paper we address crossover in GE, and seek answers to the questions of how destructive our one-point crossover operator is, and could the system benefit from a homologous-like crossover operator as proposed for tree-based GP, and the linear AIM GP. Results suggest that GE's one-point crossover operator is in fact a form of an homologous crossover, which acts as a global search operator throughout entire runs. As regards the destructive nature of GE's crossover, at the beginning of runs results show that the number of crossover events which produce individuals which are better than those in the current population is very high. On average this ratio remains relatively consistent during the duration of runs, which tells us that crossover is in fact a useful operator in recombining individuals effectively, rather than causing mass destruction.

2 Grammatical Evolution

Grammatical Evolution (GE) is an evolutionary algorithm which can evolve computer programs in any language. Rather than representing the programs as parse trees, as in GP [5], we use a linear genome representation. Each individual, a variable length binary string, contains in its codons (groups of 8 bits) the information to select production rules from a Backus Naur Form (BNF) grammar. BNF is a notation which represents a language in the form of production rules. It is comprised of a set of non-terminals which can be mapped to elements of the set of terminals, according to the production rules. An example excerpt from a BNF grammar is given below. These productions state that S can be replaced with either one of the non-terminals $expr$, $if - stmt$, or $loop$.

```
S ::= expr       (0)
    | if-stmt    (1)
    | loop       (2)
```

In order to select a rule in GE, the next codon value on the genome is generated and placed in the following formula:

$Rule = Codon\ Integer\ Value\ MOD\ Number\ of\ Rules\ for\ this\ non-terminal$

If the next codon integer value was 4, given that we have 3 rules to select from as in the above example, we get 4 MOD 3 $=$ 1. S will therefore be replaced with the non-terminal $if - stmt$.

Beginning from the the left hand side of the genome codon integer values are generated and used to select rules from the BNF grammar, until one of the following situations arise: (a) A complete program is generated. This occurs when all the non-terminals in the expression being mapped are transformed into elements from the terminal set of the BNF grammar. (b) The end of the genome is reached, in which case the *wrapping* operator is invoked. This results in the return of the genome reading frame to the left hand side of the genome once again. The reading of codons will then continue unless an upper threshold representing the maximum number of wrapping events has occurred during this individuals mapping process. (c) In the event that a threshold on the number of wrapping events has occurred and the individual is still incompletely mapped, the mapping process is halted, and the individual assigned the lowest possible fitness value.

GE uses a steady state replacement mechanism [11], and the standard genetic operators of mutation (point), and crossover (one point). It also employs a duplication operator, which duplicates a random number of codons and inserts these into the penultimate codon position on the genome. A full description of GE can be found in [12] [9] [8] [7].

3 Crossover Operators in GE

By default, GE employs a standard one point crossover operator as follows: (i) Two crossover points are selected at random, one on each individual (ii) The segments on the right hand side of each individual are then swapped. As noted in Section 1, it has been suggested that the cost of exploration via crossover is the possible destruction of building blocks. Indeed, there has been work which shows that a reason for the increase in bloat in many GP runs is to protect individuals from destructive crossover. It is argued here that bloat occurs as a mechanism to prevent the disruption of functional parts of the individual arising from crossover events. To counteract the potentially destructive property of the crossover operator, and to indirectly reduce the occurrence of bloat, novel crossover operators have been developed [6][3]. Dubbed *homologous* crossover, these operators draw inspiration from the molecular biological process of crossover in which the chromosomes to crossover align and common crossover points are selected on each individual (i.e. at the same locus), and typically a two point crossover occurs.

This paper serves to investigate crossover in GE, and proposes a new form of operator inspired by the novel homologous crossover designed for GP. We then compare the standard GE one point crossover to two different versions of this homologous crossover, as well as two alternative forms of a two point crossover operator.

The standard GE homologous crossover proceeds as follows: (i) During the mapping process a history of the rules selected in the grammar is stored for each individual. (ii) The histories of the two individuals to crossover are aligned. (iii) Each history is read sequentially from the left while the rules selected are identical for both individuals, this region of similarity is noted. (iv) The first two crossover points are selected to be at the boundary of the region of similarity, these points are the same on both individuals. (v) The two second crossover points are then selected randomly from the regions of dissimilarity. (vi) A two point crossover is then performed. The reasoning behind this operator is to facilitate the recombination of blocks which are in context with respect to the current state of the mapping process. In this case these blocks are of differing lengths. The second form of the homologous crossover operator differs only in that is swaps blocks of identical size. In step (v) the two second crossover points are at the same locus on the individual. The two point crossover employed was the standard two point operator, and one in which the size of the fragments being swapped are the same.

Table 1. Grammatical Evolution Tableau for Symbolic Regression

Objective :	Find a function of one independent variable and one dependent variable, in symbolic form that fits a given sample of 20 (x_i, y_i) data points, where the target functions is the quartic polynomial $X^4 + X^3 + X^2 + X$
Terminal Operands:	X (the independent variable)
Terminal Operators	The binary operators $+, *, /$, and $-$ The unary operators Sin, Cos, Exp and Log
Fitness cases	The given sample of 20 data points in the interval $[-1, +1]$
Raw Fitness	The sum, taken over the 20 fitness cases, of the error
Standardised Fitness	Same as raw fitness
Hits	The number of fitness cases for which the error is less than 0.01
Wrapper	Standard productions to generate C functions
Parameters	$Population = 500$, $Generations = 20$ $pmut = 0.01$, pcross $= 0.9$

Table 2. Grammatical Evolution Tableau for the Santa Fe Trail

Objective :	Find a computer program to control an artificial ant so that it can find all 89 pieces of food located on the Santa Fe Trail.
Terminal Operators:	left(), right(), move(), food_ahead()
Fitness cases	One fitness case
Raw Fitness	Number of pieces of food before the ant times out with 615 operations.
Standardised Fitness	Total number of pieces of food less the raw fitness.
Hits	Same as raw fitness.
Wrapper	Standard productions to generate C functions
Parameters	$Population = 500$, $Generations = 20$ $pmut = 0.01$, pcross $= 0.9$

4 Experimental Approach

For each type of crossover 100 runs were carried out on the Santa Fe trail and a Symbolic regression problem. Performance of each operator was measured for the following (1) Cumulative frequency of success, (2) Average size of fragments being swapped at each generation, (3) Ratio of the average fragment size being swapped to the average genome length and, (4) Ratio of crossover events resulting in successful propagation of the individual to the next generation and the total number of crossover events. Measures (2) and (3) have been used previously in [10]. Tableau's describing the parameters and terminals are given in Tables 1 and 2.

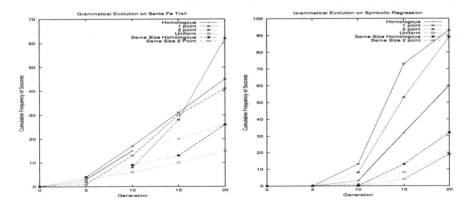

Fig. 1. Comparison of the cumulative frequencies of success for each crossover operator

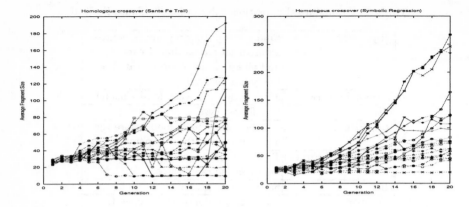

Fig. 2. Average fragment size being swapped each generation for Homologous crossover

5   Results

As can be seen in Figure 1, the cumulative frequencies of success clearly show that standard one and two point crossover are superior to the other operators on both problem domains. We will now describe the results for each operator under the other measures stated in Section 4.

Figure 2 shows the average fragment size being swapped at each generation with homologous crossover. Data can be seen for 20 separate runs, note that we are interested in general trends over the set of runs as opposed to the precise details for individual runs in each of these figures. Overall the fragment size increases as each generation passes, however the length of the chromosomes are also increasing, so a more useful measure of the ratio of average fragment size to the average chromosome length can be seen in Figure 3. The results for same size homologous, two point, same size two point, and one point can be seen in Figures 4, 5, 6, and 7 respectively. Figure 8 shows the ratio of individuals undergoing crossover which have been successfully propagated to the next generation and the total number of crossover events which have occurred. The results of this measure for all other operators can be seen in Figures 9,10,11,and 12.

6   Discussion

Examining figures 13 and 14, we can see the average over 20 runs of the ratio of individuals undergoing crossover which are propagated to the next generation to the total number of crossover events for each generation, and the ratio of the average crossover fragment size to the average chromosome length, respectively. In terms of the ratio of individuals being propagated to the next generation having undergone crossover, we can see for one point crossover in the case of the Symbolic Regression problem the rate of transfer remains relatively constant throughout the run around the value 0.35. A similar trend can be seen for one point crossover in the case of the Santa Fe trail problem although there is a slight

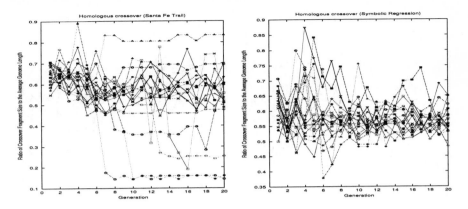

Fig. 3. Ratio of the average fragment size being swapped and the average chromosome length at each generation for Homologous crossover

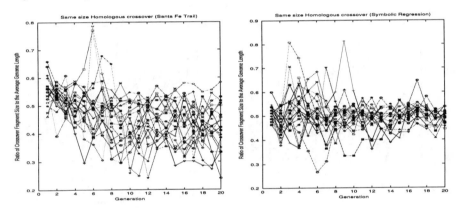

Fig. 4. Ratio of the average fragment size being swapped and the average chromosome length at each generation for Same size Homologous crossover

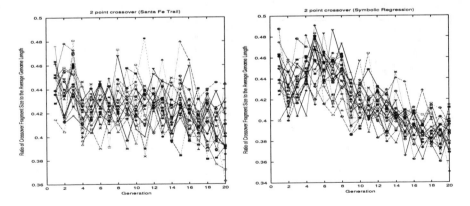

Fig. 5. Ratio of the average fragment size being swapped and the average chromosome length at each generation for two point crossover

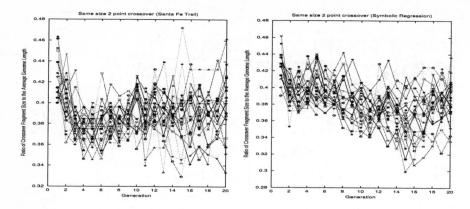

Fig. 6. Ratio of the average fragment size being swapped and the average chromosome length at each generation for same size two point crossover

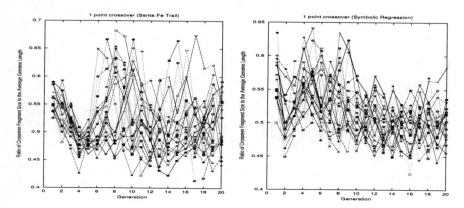

Fig. 7. Ratio of the average fragment size being swapped and the average chromosome length at each generation for one point crossover

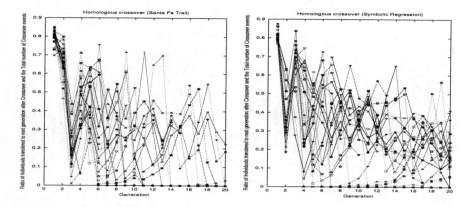

Fig. 8. Ratio of the number of individuals undergoing Homologous crossover which have been propagated to the next generation and the total number of crossover events occurring in that generation

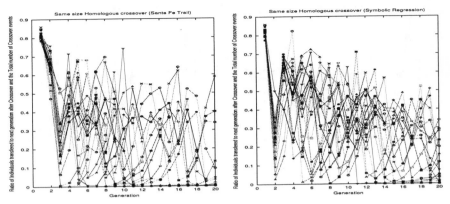

Fig. 9. Ratio of the number of individuals undergoing Same size Homologous crossover which have been propagated to the next generation and the total number of crossover events occurring in that generation

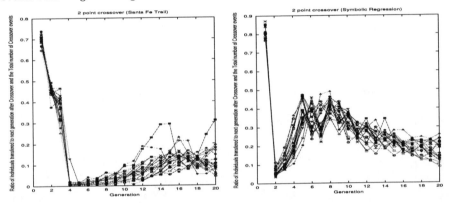

Fig. 10. Ratio of the number of individuals undergoing two point crossover which have been propagated to the next generation and the total number of crossover events occurring in that generation

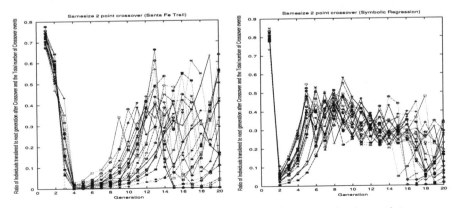

Fig. 11. Ratio of the number of individuals undergoing same size two point crossover which have been propagated to the next generation and the total number of crossover events occurring in that generation

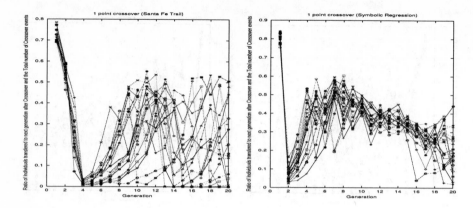

Fig. 12. Ratio of the number of individuals undergoing one point crossover which have been propagated to the next generation and the total number of crossover events occurring in that generation

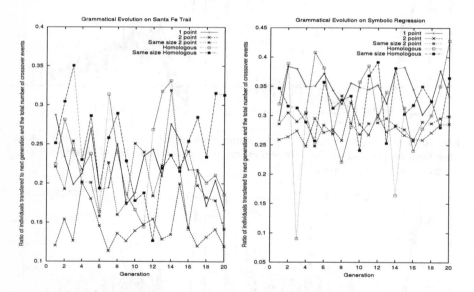

Fig. 13. Ratio of the number of individuals undergoing crossover which have been propagated to the next generation and the total number of crossover events occurring in that generation averaged over 20 runs

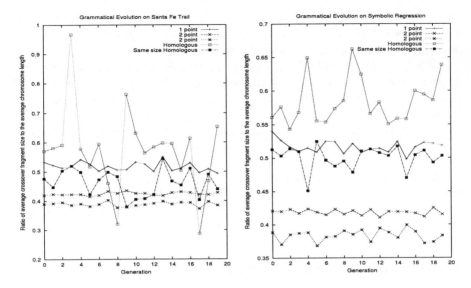

Fig. 14. Ratio of the average fragment size being swapped and the average chromosome length at each generation averaged over 20 runs

drop off as runs progress. Looking at the ratio of individuals transfered to the total number of crossover events for individual runs in figures 8, 9, 10, 11, and 12, we can see in all cases in the first few generations that this value is extremely high. Crossover then is acting in the initial stages of runs to aid in the removal of the poorer individuals randomly created in generation zero. These results for homologous crossover are not as consistent as those for one point, and directly compared appear erratic. The results show that the effort required to carry out our version of homologous crossover and the occasional peak it achieves in terms of individual transfer to each generation, is outweighed by the consistent results produced by the much simpler one point operator.

In general though, we can see the usefulness of this one point crossover operator by virtue of the fact it produces individuals which are capable of being propagated to successive generations given the steady state replacement strategy. This trend is observed over the course of entire runs.

Looking at the ratio of the average crossover fragment size to the average chromosome length, in the case of one point crossover we get a relatively smooth line around the 0.5 mark for both problems. Similar trends are observed for both types of two point crossover, however, these are localised to lower values. It can also be seen that the homologous operators are more erratic, although in the case of symbolic regression has a consistently higher value than all other operators. The experimental evidence shows us that crossover results in individuals which are being propagated to the next generation, and that the size ratio of these fragments to the actual genome length is consistent throughout the runs. This is in direct contrast to the results obtained in [10] which showed a drop off for all crossover operators except naturally for the uniform operator. These results showed how the operators examined in this case start off as global search operators but change rapidly to local search operators.

In light of the data obtained, namely the transfer of individuals having undergone crossover to the next generation, one could suspect that within GE individuals there exist useful building blocks which are being recombined to produce better performing individuals. The default one point crossover operator then is acting in the fashion of an homologous crossover operator by swapping blocks which are in fact in context due to the method in which the chromosomes are aligned prior to the crossover event.

With respect to the homologous operator described in [6] for tree based GP, a lot of effort is required to carry out this operation. Whereas in GE we can get a homologous type crossover when using standard one and two point operators, by virtue of the linear nature of our individuals.

7   Conclusions & Future Work

In conclusion, we have examined various crossover operators and their performance in the Grammatical Evolution evolutionary algorithm. It has been found that the standard one point crossover operator already employed in the algorithm is the most consistent of those examined, in general producing more successful runs. This one point crossover can result in a type of homologous crossover by virtue of the linear chromosome representation, while being computationally much simpler than other homologous operators for tree based GP. It also achieves a smoother and, therefore, more consistent graph for the ratio of individuals transferred to the next generation having undergone this operation to the total number of crossover events. Further investigation is required to establish if indeed building blocks do in fact exist in GE, to determine what form these blocks may take, and to see if GE's current one point crossover is exploiting these blocks.

Acknowledgement

This paper benefitted from conversations with Bill Langdon.

References

1. Angeline, P.J. 1997. Subtree Crossover: Building block engine or macromutation? In *Proceedings of GP'97*, pages 9-17.
2. Collins, R. 1992. Studies in Artificial Life. PhD thesis. University of California, Los Angeles.
3. Francone F. D., Banzhaf W., Conrads M, Nordin P. 1999. Homologous Crossover in Genetic Programming. In *Proceedings of the Genetic and Evolutionary Computation Conference, GECCO 99*, pages 1021-1038.

4. Goldberg, David E. 1989. Genetic Algorithms in Search, Optimization and Machine Learning. Addison Wesley.
5. Koza, J. 1992. *Genetic Programming*. MIT Press.
6. Langdon W.B. 1999. Size Fair and Homologous Tree Genetic Programming Crossovers. In *Proceedings of the Genetic and Evolutionary Computation Conference, GECCO 99*, pages 1092-1097.
7. O'Neill M., Ryan C. 1999. Genetic Code Degeneracy: Implications for Grammatical Evolution and Beyond. In *Proceedings of the Fifth European Conference on Artificial Life*.
8. O'Neill M., Ryan C. 1999. Under the Hood of Grammatical Evolution. In *Proceedings of the Genetic & Evolutionary Computation Conference 1999*.
9. O'Neill M., Ryan C. 1999. Evolving Multi-line Compilable C Programs. *Lecture Notes in Computer Science 1598, Proceedings of the Second European Workshop on Genetic Programming*, pages 83-92. Springer-Verlag.
10. Poli Riccardo, Langdon W.B. 1998. On the Search Properties of Different Crossover Operators in Genetic Programming. In *Proceedings of the Third annual Genetic Programming conference 1998*, pages 293-301.
11. Ryan C., O'Neill M. Grammatical Evolution: A Steady State Approach. In *Late Breaking Papers, Genetic Programming 1998*, pages 180-185.
12. Ryan C., Collins J.J., O'Neill M. 1998. Grammatical Evolution: Evolving Programs for an Arbitrary Language. *Lecture Notes in Computer Science 1391, Proceedings of the First European Workshop on Genetic Programming*, pages 83-95. Springer-Verlag.

Appendix

The grammars for the two problem domains addressed in this paper are as follows:

A   Symbolic Regression Grammar

$$N = \{expr, op, pre\_op\}$$

$$T = \{Sin, Cos, Exp, Log, +, -, /, *, X, ()\}$$

$$S =< expr >$$

And P can be represented as:

```
(1) <expr> ::= <expr> <op> <expr>      (A)
            | ( <expr> <op> <expr> )  (B)
            | <pre-op> ( <expr> )     (C)
            | <var>                   (D)
```

```
(2) <op> ::= + (A)
          | - (B)
          | / (C)
          | * (D)
```

```
(3) <pre-op> ::= Sin (A)
             | Cos (B)
             | Exp (C)
             | Log (D)
```

```
(4) <var> ::= X
```

B Santa Fe Trail Grammar

$$N = \{code, line, expr, if - statement, op, if - true, if - false\}$$

$$T = \{left(), right(), move(), food\_ahead(), else, if, \{, \}, (,)\}$$

$$S = < code >$$

And P can be represented as:

```
(1) <code> :: =  <line>              (A)
                |<code><line>        (B)
```

```
(2) <line> :: =  <expr>
```

```
(3) <expr> :: =  <if-statement>      (A)
                |<op>                (B)
```

```
(4) <if-statement> :: = if(food_ahead())<if-true><if-false>
```

```
(5) <op> :: =     left()             (A)
               | right()             (B)
               | move()              (C)
```

```
(6) <if-true>:={<expr>}
```

```
(7) <if-false>:=else{<expr>}
```

Hyperschema Theory for GP with One-Point Crossover, Building Blocks, and Some New Results in GA Theory

Riccardo Poli
(R.Poli@cs.bham.ac.uk)

School of Computer Science, University of Birmingham, Birmingham, B15 2TT, UK

Abstract. Two main weaknesses of GA and GP schema theorems are that they provide only information on the expected value of the number of instances of a given schema at the next generation $E[m(H, t + 1)]$, and they can only give a lower bound for such a quantity. This paper presents new theoretical results on GP and GA schemata which largely overcome these weaknesses. Firstly, unlike previous results which concentrated on schema survival and disruption, our results extend to GP recent work on GA theory by Stephens and Waelbroeck, and make the effects and the mechanisms of schema creation explicit. This allows us to give an exact formulation (rather than a lower bound) for the expected number of instances of a schema at the next generation. Thanks to this formulation we are then able to provide in improved version for an earlier GP schema theorem in which some schema creation events are accounted for, thus obtaining a tighter bound for $E[m(H, t+1)]$. This bound is a function of the selection probabilities of the schema itself and of a set of lower-order schemata which one-point crossover uses to build instances of the schema. This result supports the existence of building blocks in GP which, however, are not necessarily all short, low-order or highly fit. Building on earlier work, we show how Stephens and Waelbroeck's GA results and the new GP results described in the paper can be used to evaluate schema variance, signal-to-noise ratio and, in general, the probability distribution of $m(H, t + 1)$. In addition, we show how the expectation operator can be removed from the schema theorem so as to predict with a known probability whether $m(H, t + 1)$ (rather than $E[m(H, t + 1)]$) is going to be above a given threshold.

1 Introduction

Since John Holland's seminal work in the mid seventies and his well known schema theorem [1, 2], schemata are traditionally used to explain why GAs and more recently GP [3, 4, 5] work. Schemata are similarity templates representing entire groups of points in the search space. The schema theorem describes how schemata are expected to propagate generation after generation under the effects of selection, crossover and mutation.

The usefulness of the schema theorem has been widely criticised (see for example [6, 7, 8, 9]), and many people in the evolutionary computation field nowadays seem to believe that the theorem is nothing more than a trivial tautology of no use whatsoever. So, recently, the attention of GA theorists has moved away from schemata to land onto Markov chains [10, 11, 12, 13, 14, 15]. However, many of the problems attributed to the schema theorem are probably not due to the theorem itself, rather to its over-interpretations [16].

The main criticism for schema theorems is that they cannot be used easily for predicting the behaviour of a GA over multiple generations. One reason for this is that schema theorems give only a *lower bound* for the *expected value* of the number of instances of a schema H at the next generation $E[m(H, t + 1)]$ as a function of quantities characterising the schema at the previous generation. The presence of an expectation operator means that it is not possible to use schema theorems recursively to predict the behaviour of a genetic algorithm over multiple generations unless one assumes the population is infinite. In addition, since the schema theorem provides only a lower bound, some people argue that the predictions of the schema theorem are not very useful even for a single generation ahead.

Clearly there is some truth in these criticisms. However, this does not mean that schema theorems are useless. As shown by our and other researchers' recent work [17, 18, 19, 20, 21] schema theorems have not been fully exploited nor fully developed, and when this is done they become very useful.

This paper presents new theoretical results on GP and GA schemata which largely overcome some of the weaknesses of the schema theorem. Firstly, unlike previous results which concentrated on schema survival and disruption, our results extend to GP recent work on GA theory by Stephens and Waelbroeck, and make the effects and the mechanisms of schema creation explicit. This allows us to give an exact formulation (rather than a lower bound) for the expected number of instances of a schema at the next generation. Thanks to this formulation we are then able to provide in improved version for an earlier GP schema theorem in which some schema creation events are accounted for, thus obtaining a tighter bound for $E[m(H, t + 1)]$. This bound is a function of the selection probabilities of the schema itself and of a set of lower-order schemata which one-point crossover uses to build instances of the schema. This result supports the existence of building blocks in GP which, however, are not necessarily all short, low-order or highly fit. Building on earlier work, we show how Stephens and Waelbroeck's GA results and the new GP results described in the paper can be used to evaluate schema variance, signal-to-noise ratio and, in general, the probability distribution of $m(H, t + 1)$. In addition, we show how the expectation operator can be removed from the schema theorem so as to predict with a known probability whether $m(H, t + 1)$ (rather than $E[m(H, t + 1)])$ is going to be above a given threshold.

The structure of the paper is the following. Since the work presented in this paper extends our and other people's work, before we introduce it we will extensively review earlier relevant work on GP and GA schemata in Section 2. Then, in Section 3 we connect the recent GA results by Stephens and Waelbroeck to previous theoretical work on schemata, indicating how this leads to a series of new theoretical results for GAs. Section 4 presents perhaps the main contribution of this paper: the hyperschema theory for GP, its uses and interpretations. We draw some conclusions in Section 5.

2 Background

2.1 Holland's GA Schema Theory

In the context of GAs operating on binary strings, a schema (or similarity template) is a string of symbols taken from the alphabet $\{0,1,\#\}$. The character $\#$ is interpreted as a "don't care" symbol, so that a schema can represent several bit strings. For example the schema $\#10\#1$ represents four strings: 01001, 01011, 11001 and 11011. The number of non-$\#$ symbols is called the *order* $\mathcal{O}(H)$ of a schema H. The distance between the furthest two non-$\#$ symbols is called the *defining length* $\mathcal{L}(H)$ of the schema. Holland obtained a result (the schema theorem) which predicts how the number of strings in a population matching (or belonging to) a schema is expected to vary from one generation to the next [1]. The theorem can be reformulated as follows:

$$E[m(H, t+1)] \geq Mp(H, t) \cdot (1 - p_m)^{\mathcal{O}(H)} \cdot \left[1 - p_{xo} \frac{\mathcal{L}(H)}{N-1} (1 - p(H, t))\right] \quad (1)$$

where p_m is the probability of mutation per bit, p_{xo} is the probability of crossover, N is the number of bits in the strings, M is the number of strings in the population, $E[m(H, t+1)]$ is the expected number of strings matching the schema H at generation $t+1$, and $p(H, t)$ is the probability of selection of the schema H. In fitness proportionate selection this is given by $p(H, t) = \frac{m(H,t)f(H,t)}{M\bar{f}(t)}$ where $m(H, t)$ is the number of strings matching the schema H at generation t, $f(H, t)$ is the mean fitness of the strings matching H, and $\bar{f}(t)$ is the mean fitness of the strings in the population.[1]

2.2 GP Schema Theories

One of the difficulties in obtaining theoretical results on GP using the idea of schema is that the definition of schema is much less straightforward than for GAs and a few alternative definitions have been proposed in the literature. All of them define schemata as composed of one or multiple trees or fragments of trees. In some definitions [23, 24, 25, 26, 27] schema components are *non-rooted* and, therefore, a schema can be present multiple times within the same program. This, together with the variability of the size and shape of the programs matching the same schema, leads to considerable complications in the calculations necessary to formulate schema theorems for GP. In more recent definitions [3, 5] schemata are represented by *rooted* trees or tree fragments. These definitions make schema theorem calculations easier. We describe these schema definitions below.

---

[1] This is a slightly different version of Holland's original theorem which applies when crossover is performed taking both parents from the mating pool [2, 22].

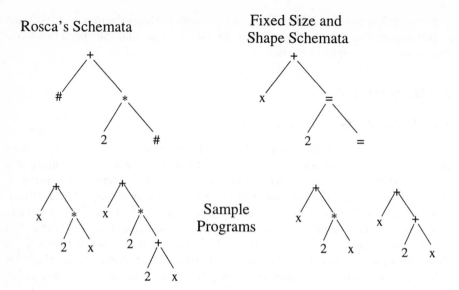

Fig. 1. Examples of rooted schemata (top) and some instances of programs sampling them (bottom).

Rosca's GP Schema Theory for Standard Crossover Rosca [5] has proposed a definition of schema, called *rooted tree-schema*, in which a schema is a rooted contiguous tree fragment. For example, the rooted tree-schema (+ # x) represents all the programs whose root node is a + the second argument of which is x. Another example of schema with some of its instances is shown in Figure 1 (left). With this definition, schemata divide the space of programs into subspaces containing programs of different sizes and shapes.

Rosca derived a schema theorem for GP with standard crossover which provided a lower bound for the expected number of instances of a schema at the next generation as a function of the schema order, and the fitness and size of its instances in the population.

GP Fixed-Size-and-Shape Schema Theory for One-point Crossover In [3] a simpler definition of schema for GP was proposed in which a *schema* is a tree composed of functions from the set $\mathcal{F} \cup \{=\}$ and terminals from the set $\mathcal{T} \cup \{= \}$, where $\mathcal{F}$ and $\mathcal{T}$ are the function set and the terminal set used in a GP run. The symbol = is a "don't care" symbol which stands for a *single* terminal or function. In line with the original definition of schema for GAs, a schema H represents programs having the same shape as H and the same labels for the non-= nodes. For example, if $\mathcal{F}=\{+, -\}$ and $\mathcal{T}=\{x, y\}$ the schema (+ (- = y) =) would represent the four programs (+ (- x y) x), (+ (- x y) y), (+ (- y y) x) and (+ (- y y) y). Another example of schema with some of its instances is shown in Figure 1 (right). This definition of schema partitions the program space into subspaces of programs of fixed size and shape. For this reason in the following we will refer to these schemata as *fixed-size-and-shape schemata*. The definition is lower-level than those described above, as a smaller number of trees can be represented by schemata with the same number of "don't care" symbols and it is possible to represent other types of schemata by using a collection of these. This is quite important because it makes it possible to export some results of the fixed-size-and-shape schema theory to other kinds of schemata.

The number of non-= symbols is called the *order* $\mathcal{O}(H)$ of a schema H, while the total number of nodes in the schema is called the *length* $N(H)$ of the schema. The number of links in the minimum tree fragment including all the non-= symbols within a schema H is called the *defining length* $\mathcal{L}(H)$ of the schema (see [3, 4] for more details). For example the schema (+ (- = =) x) has order 3 and defining length 2. These definitions are independent of the shape and size of the programs in the actual population.

In order to derive a GP schema theorem for these schemata non-standard forms of mutation and crossover, namely point mutation and one-point crossover, were used. *Point mutation* is the substitution of a node in the tree with another node with the same arity. *One-point crossover* works by selecting a common crossover point in the parent programs and then swapping the corresponding subtrees, like standard crossover, as illustrated in Figure 2(a). In order to account for the possible structural diversity of the two parents, one-point crossover analyses the two trees from the root nodes and considers for the selection of the crossover point only the parts of the two trees (common region) which have the same topology (i.e. the same arity in the nodes encountered traversing the trees from the root node) [3, 4] as illustrated in Figure 2(b).

The resulting schema theorem is the following:

$$E[m(H, t+1)] \geq Mp(H, t)(1 - p_m)^{\mathcal{O}(H)}. \tag{2}$$

$$\left\{ 1 - p_{xo} \left[p_{\text{diff}}(t) \left(1 - p(G(H), t) \right) + \frac{\mathcal{L}(H)}{(N(H) - 1)} \left(p(G(H), t) - p(H, t) \right) \right] \right\}$$

where p_m is the mutation probability (per node), $G(H)$ is the zero-th order schema with the same structure of H where all the defining nodes in H have been replaced with "don't care" symbols, M is the number of individuals in the population, $p_{\text{diff}}(t)$ is the conditional probability that H is disrupted by crossover when the second parent has a different shape (i.e. does not sample $G(H)$), and the other symbols have the same meaning as in Equation 1 (see [3, 4] for the proof). The zero-order schemata $G(H)$'s represent different groups of programs all with the same shape and size.

The probability $p_{\text{diff}}(t)$ is hard to model mathematically, but its expected variations during a run were analysed in detail in [3].

2.3 General Schema and Schema Variance Theorems in the Presence of Schema Creation

After the work on GP schemata in [3, 28, 4] we started asking ourselves whether the estimate (lower bound) for the expected value of the number of individuals

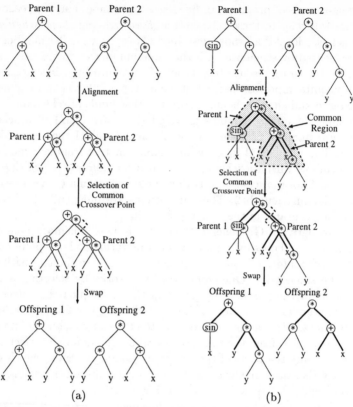

Fig. 2. (a) One-point crossover for GP when both parents have the same shape, and (b) general form of one-point crossover for GP. In (b) the links that can be selected as common crossover points are drawn with thick lines.

sampling a given schema at the next generation provided by the schema theorem is reliable. In order to investigate this issue in [29] we analysed the impact of variance on schema transmission and we obtained a schema-variance theorem which is general and applicable to GAs as well as GP. The analysis revealed the relative dependencies between schema transmission, population size, schema measured fitness, schema fragility and schema creation. In the rest of this subsection the main mathematical results of that work are summarised.

Crossover will cause some of the individuals matching H in the mating pool to produce offspring not sampling H and vice versa. Let us call $p_s(H,t)$ the probability that individuals in H will survive crossover (in the sense that their offspring will still sample H) and $p_c(H,t)$ the probability that offspring which sample H are *created* by parents not sampling H. Then the total schema transmission probability for the schema H is[2]

$$\alpha(H,t) = p_s(H,t)p(H,t) + p_c(H,t)(1 - p(H,t)).$$

Since we can see the selection/crossover process as a Bernoulli trial (each time an individual is produced, either the individual samples H or it does not), $m(H,t+1)$ can be seen as a binomial stochastic variable. Therefore,

$$\Pr\{m(H,t+1) = k\} = \binom{M}{k}\alpha(H,t)^k(1 - \alpha(H,t))^{M-k}, \tag{3}$$

whereby

$$E[m(H,t+1)] = M\alpha(H,t), \tag{4}$$

$$Var[m(H,t+1)] = M\alpha(H,t)(1 - \alpha(H,t)), \tag{5}$$

and

$$\left(\frac{S}{N}\right) \stackrel{def}{=} \frac{E[m(H,t+1)]}{\sqrt{Var[m(H,t+1)]}} = \sqrt{M}\sqrt{\frac{\alpha(H,t)}{1 - \alpha(H,t)}}, \tag{6}$$

which represents the signal-to-noise ratio of the schema. When the signal-to-noise ratio is large, the propagation of a schema will occur nearly exactly as predicted. When the signal-to-noise ratio is small, the actual number of instances of a schema in the next generation may be essentially random with respect to its current value.

From Equation 3 it is also possible to compute the probability of schema extinction in one generation:

$$\Pr\{m(H,t+1) = 0\} = (1 - \alpha(H,t))^M \tag{7}$$

It should be noted that although the results reported above include the probability of schema creation $p_c(H,t)$, such a quantity was not modelled mathematically in more detail. It is also worth noting that these results are valid independently of the representation adopted for the individuals in the population, the operators used, and the definition of schema. Therefore, they are valid for GAs as well as for GP.

2.4 General Schema Theorems without Expected Values

Building on the work described in the previous section in [20] two new schema theorems were introduced in which expectations are not present. The theorems were obtained using Chebychev inequality in conjunction with Equations 4 and 5. These theorems are a first step towards removing one of the most criticised components of the schema theorem: the expectation operator. One of the theorems states that for any given constant $k > 0$

$$\Pr\{m(H,t+1) > M\alpha(H,t) - k\sqrt{M\alpha(H,t)(1 - \alpha(H,t))}\} \geq 1 - \frac{1}{k^2}. \tag{8}$$

[2] The total schema transmission probability α at time t corresponds to the expected proportion of population sampling H at generation $t+1$. Thus, Equation 4 can be seen as a reformulation of the "gain and losses" equation in [30].

This result is necessarily weaker than the obvious consequence of Equation 3

$$\Pr\{m(H, t+1) \geq h\} = \sum_{k=h}^{M} \binom{M}{k} \alpha(H, t)^k (1 - \alpha(H, t))^{M-k},\qquad(9)$$

but it is easier to handle mathematically when $\alpha(H, t)$ is unknown [19].

Interestingly, Equation 8 can be used to provide an upper bound for the probability of schema extinction in one generation as a simple function of the signal-to-noise ratio. This is achieved by solving for k the equation $M\alpha(H, t) - k\sqrt{M\alpha(H, t)(1 - \alpha(H, t))} = 0$ obtaining $k = \sqrt{M}\sqrt{\frac{\alpha(H,t)}{1-\alpha(H,t)}} = \left(\frac{S}{N}\right)$. This, substituted into Equation 8, after simple calculations leads to

$$\Pr\{m(H, t+1) = 0\} \leq \frac{1}{\left(\frac{S}{N}\right)^2}.\qquad(10)$$

(This is a small new result, but we report it here to make it easier to see where this comes from.)

Again, these results are valid for GAs as well as for GP.

2.5 Stephens and Waelbroeck's GA Schema Theory

The results in the previous two sections require the knowledge of the total transmission probability α, i.e. they require that not only schema survival and schema disruption events be modelled mathematically but also that schema creation events are. This is not be an easy task if one wants to do that using only the properties of the schema H (such as the number of instances of H and the fitness of H) and those of the population when expressing the quantity α. Indeed, to the best of our knowledge, none of the schema theorems presented to date in the literature have succeeded in doing this. This is the reason why all of schema theorems provide upper bounds. However, thanks to the recent work of Stephens and Waelbroeck [17, 18] it is now possible to express exactly $\alpha(H, t)$ for GAs operating on fixed-length bit strings by using properties of lower-order schemata which are supersets of the schema under consideration.

For a GA with one point crossover applied with a probability p_{xo}, $\alpha(H, t)$ is given by the following equation:

$$\alpha(H, t) = (1 - p_{xo})p(H, t) + \frac{p_{xo}}{N-1} \sum_{i=1}^{N-1} p(L(H, i), t)p(R(H, i), t)\qquad(11)$$

where $L(H, i)$ is the schema obtained by replacing with "don't care" symbols all the elements of H from position $i + 1$ to position N, $R(H, i)$ is the schema obtained by replacing with "don't care" symbols all the elements of H from position 1 to position i, and i varies over the valid crossover points. The symbol L stands for "left part of", while R stands for "right part of". For example, if H =1*111, $L(H, 1)$ =1****, $R(H, 1)$ =**111, $L(H, 3)$ =1*1**, $R(H, 3)$ =***11.

It should be noted that Equation 11 is in a considerably different form with respect to the equivalent results in [17, 18]. This is because we developed it using our own notation following a simpler approach with respect to that used by Stephens and Waelbroeck. In our approach we assume that while producing each individual for a new generation one flips a biased coin to decide whether to apply selection only (probability $1 - p_{xo}$) or selection followed by crossover (probability p_{xo}). If selection only is applied, then there is a probability $p(H, t)$ that the new individual created sample H (hence the first term in Equation 11). If instead selection followed by crossover is selected, we use the unusual idea of first choosing the crossover point and then the parents. When selecting the crossover point, one has to choose randomly one of the $N - 1$ crossover points each of which has a probability $1/(N - 1)$ of being selected. Once this decision has been made, one has to select two parents. Then crossover is executed. This will result in an individual that samples H only if the first parent has the correct left-hand side (with respect to the crossover point) *and* the second parent has the correct right-hand side. These two events are independent because each parent is selected with an independent Bernoulli trial. So, the probability of the joint event is the product of the probabilities of the two events. Assuming that crossover point i has been selected, the first parent has the correct left-hand side if it belongs to $L(H, i)$ while the second parent has the correct right-hand side if it belongs to $R(H, i)$. The probabilities of these events are $p(L(H, i), t)$ and $p(R(H, i), t)$, respectively (whereby the terms in the summation in Equation 11, the summation being there because there are $N - 1$ possible crossover points). Combining the probabilities all these events one obtains Equation 11.

By substituting Equation 11 into Equation 4 and performing some minor additional calculations[3] the GA schema theorem described in [17, 18] is obtained. Stephens and Waelbroeck reported a number of other important ideas (including a refinement of the concept of effective fitness firstly defined and applied to GP in [31, 32, 33]) and results on the behaviour of a GA over multiple generations in the assumption of infinite populations.

2.6 Building Block Hypothesis in GAs and GP

The *building block hypothesis* [2] is typically stated informally by saying that a GA works by combining short, low-order schemata of above average fitness (building blocks) to form higher-order ones over and over again until it converges to an optimum or near-optimum solution.

The building block hypothesis and the related concept of deception have been strongly criticised in [34], in the context of binary GAs, where it was suggested that these ideas can only be used to produce first-order approximations of what really happens in a GA. In [25] the criticisms to the building block hypothesis have been extended to the case of GP with standard crossover, arguing that the hypothesis is even less applicable to GP because of the highly destructive effects

---

[3] In [17, 18] the summation in Equation 11 is performed only over the crossover points between the extreme defining bits in a schema.

of crossover. The analysis of Equation 2 in [3, 4] and the experimental results in [28] suggest that the latter criticism does not apply to GP with one-point crossover. Therefore, we cannot exclude that the building block hypothesis might be a good first approximation of the behaviour of GP with one-point crossover.

Stephens and Waelbroeck analysed their schema theorem (equivalent to Equation 11) and stated that the presence of the terms (equivalent to) $p(L(H,i),t)p(R(H,i),t)$ is a clear indication of the fact that, indeed, GAs build higher-order schemata (H) by juxtaposing lower-order ones (the $L(H,i)$'s and $R(H,i)$'s) [17, 18]. However, these building blocks are not necessarily all fitter than average, short or even low-order.

3 New Results on GA Schemata

The availability of Equation 11 makes it possible and easy to produce a series of new results which go beyond the schema theorem obtained by Stephens and Waelbroeck and provide additional statistical information on schema behaviour for finite populations. This can be done by simply substituting the expression of $\alpha(H,t)$ given in Equation 11 into Equations 3, 5, 6, 7, 8, 9 and 10.

We believe that the equations resulting from Equations 8 and 9 are particularly important, at least in principle, because they show that it is possible to remove the expectation operator from exact schema theories without having to assume infinite populations. We have started studying the consequences of this fact in [19].

4 GP Hyperschema Theory

A question that immediately comes to mind is whether it would not be possible to extend Rosca's or the fixed-size-and-shape GP schema theories to emulate what Stephens and Waelbroeck have done for GAs. This is important because it would make it possible to exploit the recent and new GA results summarised above in GP, too. As described below we were able to extend the fixed-size-and-shape GP schema theory for one-point crossover, but that required to generalise the definition of GP schema.

4.1 Hyperschemata

In order to do obtain for GP with one-point crossover results similar to those obtained Stephens and Waelbroeck for binary GAs we need to introduce a new, more general definition of schema. The definition is as follows:

Definition 1 (GP hyperschema). *A GP hyperschema is a rooted tree composed of nodes from the set $\mathcal{F} \cup \mathcal{T} \cup \{=, \#\}$, where $\mathcal{F}$ and $\mathcal{T}$ are the function set and the terminal set used in a GP run, and the operators = and # are polymorphic functions with as many arities as the number of different arities of the elements of $\mathcal{F} \cup \mathcal{T}$, the terminals in $\mathcal{T}$ being 0-arity functions. The operator = is*

*a "don't care" symbols which stands for exactly one node, while the operator #
stands for any valid subtree. The internal nodes of a hyperschema can be chosen
from $\mathcal{F} \cup \{=\}$, while the leaves of a hyperschema are elements of $\mathcal{T} \cup \{=, \#\}$.*

This definition presents some of the features of the schema definition we used
for our GP schema theory [4]: hyperschemata are rooted trees, and they include
"don't care" symbols which stand for one node only. However, Definition 1 also
includes one of the features of the schema definitions proposed by other authors [25, 5]: hyperschemata also include "don't care" symbols which stand for
entire subtrees. Indeed, the notion of hyperschema is a generalisation of both
Rosca's schemata (which are hyperschemata without = symbols) and fixed-size-
and-shape schemata (which are hyperschemata without # symbols). So, a hyperschema can represent a group of schemata in essentially the same way as such
schemata represent groups of program trees (hence the name "hyperschema").
An example of hyperschema is (* # (+ x =)). This hyperschema represents all
the programs with the following characteristics: a) the root node is a product, b)
the first argument of the root node is any valid subtree, c) the second argument
of the root node is +, d) the first argument of the + is the variable x, e) the
second argument of the + is any valid node in the terminal set.

4.2 Theory for Programs of Fixed Size and Shape

If one had a population of programs all having exactly the same size and shape,
thanks to the definition of hyperschema, it would be possible to express the total
transmission probability of a fixed-size-and-shape schema using the properties
of lower-order hyperschemata, in the presence of one-point crossover, in exactly
the same way as in Equation 11, i.e.

$$\alpha(H,t) = (1 - p_{xo})p(H,t) + \frac{p_{xo}}{N(H) - 1} \sum_{i=1}^{N(H)-1} p(L(H,i),t)p(U(H,i),t) \quad (12)$$

where: $N(H)$ is the number nodes in the schema H (which is assumed to have
exactly the same size and shape of the programs in the population); $L(H,i)$ is
the hyperschema obtained by replacing with = nodes all the nodes on the path
between crossover point i and the root node, and with # nodes all the subtrees
connected to the nodes replaced with =; $U(H,i)$ is the hyperschema obtained
by replacing with a # node the subtree below crossover point i; i varies over
the valid $N(H) - 1$ crossover points. The symbol L now stands for "lower part
of", while U stands for "upper part of". For example, if $H = (* = (+ x =))$
and the crossover points are numbered as in Figure 3 (top left) then $L(H,1)$
is obtained by first replacing the root node with a = symbol and then replacing the subtree connected to the right of root node with a # symbol obtaining (= = #), as indicated in the second column of Figure 3. The schema
$U(H,1)$ is instead obtained by replacing the subtree below the crossover point
with a # symbol obtaining (* # (+ x =)), as illustrated in the third column
of Figure 3. The fourth and fifth columns of Figure 3 show how $L(H,3) = (=$

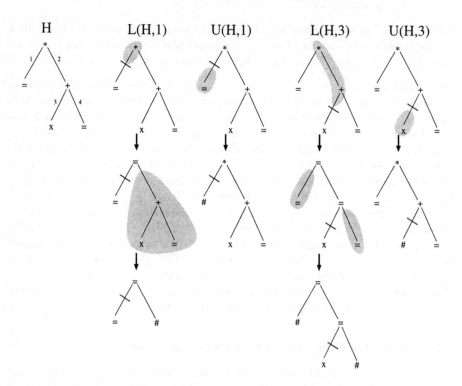

Fig. 3. Example of hyperschema and some of its potential building blocks.

(= x #)) and $U(H,3) =(*$ = (+ # =)) are obtained. The remaining build-
ing blocks for the schema H are: $L(H,2) =(=$ # (+ x =)), $U(H,2) =(*$ = #),
$L(H,4) =(=$ # (= # =)), and $U(H,4) =(*$ = (+ x #)).[4]

Formally this result can be obtained by proceeding like in Section 2.5. Again
we assume that while producing each individual for a new generation one first
decides whether to apply selection only (probability $1-p_{xo}$) or selection followed
by crossover (probability p_{xo}). If selection only is applied, the new individual created samples H with a probability $p(H,t)$ (hence the first term in Equation 12).
If selection followed by crossover is selected, we first choose the crossover point
(to be interpreted as a link between nodes) randomly out of one of the $N(H)-1$
crossover points available. Then we select two parents and perform crossover.

---

[4] One might think that the definitions of $L(H,i)$ and $U(H,i)$ given above are overly
complicated with respect to their GA counterparts. Indeed, if all the programs in the
population have the same size and shape, one could equivalently define $L(H,i)$ as
the schema obtained by replacing with = nodes all the nodes above crossover point
i and $U(H,i)$ as the schema obtained by replacing with a = nodes all the nodes
below crossover point i. So, it would not be necessary to introduce the notion of
hyperschema altogether. However, as clarified below, the notion of hyperschema can
be used to produce more general results which apply when the population contains
programs of different size and shape.

This will result in an individual that samples H only if the first parent has the correct lower part (with respect to the crossover point) *and* the second parent has the correct upper part. Assuming that crossover point i has been selected,[5] the parents recreate H if they belong to $L(H,i)$ and $U(H,i)$, respectively (whereby the terms in the summation in Equation 12). Combining the probabilities all these events one obtains Equation 12.[6]

4.3 General Case

Alternatively, Equation 12 can be obtained by specialising the following general result which is valid for populations of programs of any size and shape:

Theorem 1 (Exact GP Schema Theorem). *The total transmission probability for a fixed-size-and-shape GP schema H under one-point crossover and no mutation is*

$$\alpha(H,t) = (1 - p_{xo})p(H,t) \tag{13}$$
$$+p_{xo}\sum_{h_1}\sum_{h_2}\frac{p(h_1,t)p(h_2,t)}{NC(h_1,h_2)-1}\sum_{i\in C(h_1,h_2)}\delta(h_1 \in L(H,i))\delta(h_2 \in U(H,i))$$

where the first two summations are over all the individuals in the population, $NC(h_1,h_2)$ is the number of nodes in the tree fragment representing the common region between program h_1 and program h_2, $C(h_1,h_2)$ is the set of indices of the crossover points in such a common region, and $\delta(x)$ is a function which returns 1 if x is true, 0 otherwise.

The theorem can easily be proven by: a) considering all the possible ways in which parents can be selected for crossover and in which the crossover points can be selected in such parents, b) computing the probabilities that each of those events happens *and* the first parent has the correct lower part (w.r.t. the crossover point) to create H while the second parent has the correct upper part, and c) adding up such probabilities (whereby the three summations in Equation 14). Obviously, in order for this equation to be valid it is necessary to assume that if a crossover point i is in the common region between two programs but outside the schema H under consideration, then $L(H,i)$ and $U(H,i)$ are empty sets (i.e. they cannot be matched by any individual).

Equation 14 allows one to compute the exact total transmission probability of a GP schema. Therefore, by proceeding as indicated in Section 3 a series of new results for GP schemata can be obtained which provide important new statistical information on schema behaviour (e.g. schema probability distribution, schema variance, schema signal-to-noise ratio, extinction probability, etc.). In this paper

---

[5] Any numbering scheme for the crossover points produces the same result.

[6] A simpler version of this equation, applicable when crossover is applied with 100% probability, was presented in [21] where we failed to state the fact that, in this form, the equation is only applicable to populations of programs of fixed size and shape.

we will not study these results. Instead we will concentrate on understanding Equation 14 in greater depth.

If one restricts the first two summations in Equation 14 to include only the individuals having the same shape and size of the schema H, i.e. which belong to $G(H)$, one obtains:

$$\alpha(H,t) \geq (1 - p_{xo})p(H,t)+$$

$$\frac{p_{xo}}{N(H)-1} \sum_{i=1}^{N(H)-1} \sum_{h_1 \in G(H)} p(h_1,t)\delta(h_1 \in L(H,i)) \sum_{h_2 \in G(H)} p(h_2,t)\delta(h_2 \in U(H,i))$$

where we used the following facts: $NC(h_1, h_2) = N(H)$ since $h_1, h_2 \in G(H)$, all common regions have the same shape and size as H, and so $C(h_1, h_2) = \{1, 2, \ldots, N(H) - 1\}$ (assuming that the crossover points in H are numbered 1 to $N(H) - 1$). This equation can easily be transformed into the following

Theorem 2 (GP Schema Theorem with Schema Creation Correction).
A lower bound for the total transmission probability for a fixed-size-and-shape GP schema H under one-point crossover and no mutation is

$$\alpha(H,t) \geq (1-p_{xo})p(H,t)+\frac{p_{xo}}{N(H)-1} \sum_{i=1}^{N(H)-1} p(L(H,i) \cap G(H),t)p(U(H,i) \cap G(H),t),$$

(14)

the equality applying when all the programs in the population sample $G(H)$.

Let us compare this result to the one described in Section 2.2.

It is easy to see that by dividing the right-hand side of Equation 2 by M one obtains a lower bound for $\alpha(H,t)$. If we consider the case in which $p_m = 0$ and we assume the worst case scenario for $p_{\text{diff}}(t)$, i.e. $p_{\text{diff}}(t) = 1$, we obtain[7]

$$\alpha(H,t) \geq p(H,t)\left\{1 - p_{xo}\left[1 - p(G(H),t) + \frac{\mathcal{L}(H)}{(N(H)-1)}(p(G(H),t) - p(H,t))\right]\right\}$$

By subtracting the right-hand side of this equation from the lower bound provided by Equation 14 one can show that the difference is:

$$\Delta\alpha(H,t) = \frac{p_{xo}}{N(H)-1}\left(\sum_{i \in B(H)} p(L(H,i),t)p(U(H,i),t) - \mathcal{L}(H)p(H,t)^2\right),$$

where $B(H)$ is the set of crossover points in the tree fragment used in the definition of $\mathcal{L}(H)$. Since such tree fragment contains exactly $\mathcal{L}(H)$ crossover points, we can rewrite this equation as

$$\Delta\alpha(H,t) = \frac{p_{xo}}{N(H)-1} \sum_{i \in B(H)} [p(L(H,i),t)p(U(H,i),t) - p(H,t)^2].$$

---

[7] We are forced to assume $p_{\text{diff}}(t) = 1$ to maintain the generality of our results.

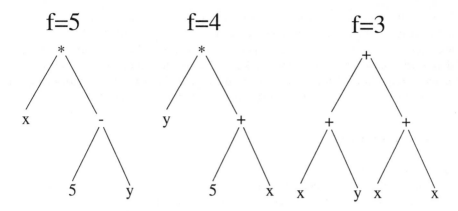

Fig. 4. Example population of programs with their fitnesses.

Since, for $i \in B(H)$, $L(H,i)$ and $U(H,i)$ are supersets of H, it follows that $p(L(H,i),t) \geq p(H,t)$ and $p(U(H,i),t) \geq p(H,t)$, whereby we see that $\Delta\alpha(H,t) \geq 0$, i.e. Theorem 2 provides a better estimate of the true transmission probability of a schema in the assumption that $p_{\text{diff}}(t) = 1$. This is why Theorem 2 is called "GP Schema Theorem with Schema Creation Correction".

4.4 Example

Let us consider an example. Let us imagine that the population contains 10 copies of each of the 3 programs in Figure 4, and let us consider the propagation of the schema $H' = (= = (= = =))$. Since $\mathcal{L}(H') = 0$ and $B(H') = \emptyset$ then $\Delta\alpha(H',t) = 0$. So, the new GP schema theorem cannot provide a better bound than the previous one.

However, if one considers the schema $H'' = (* = (- 5 =))$, with simple calculations (assuming fitness proportionate selection) one obtains $\Delta\alpha(H',t) \approx 0.07 \times p_{xo}$. This might look like a small difference, but the new schema theorem can provide a lower bound for $E[m(H,t+1)]$ of up to about 2.1 higher than the old theorem, the maximum value being reached when $p_{xo} = 1$. This is a big difference considering that the correct value for $E[m(H,t+1)]$ obtained from Equation 14 is 10.5 (for reference: the value obtained if selection only was acting is 12.5), and the lower bound provided by the old schema theorem is 7.3. So, in this example the "schema creation correction" improves the estimate by nearly 30% providing a very tight lower bound of 9.4 for $E[m(H,t+1)]$.

4.5 Hyper Building Blocks?

The presence of the terms $p(L(H,i),t)p(U(H,i),t)$ in the results presented in this section seems to suggest that some hyperschemata are a form of building blocks for fixed-size-and-shape schemata and that, indeed, also GP with one-point crossover builds higher-order schemata by juxtaposing lower-order ones.

The results also suggest that for this to happen the building blocks need not necessarily be all fitter than average, short or even low-order. The question is whether this is true for general hyperschemata (remember that fixed-size-and-shape schemata are only a special kind of hyperschemata, since they do not include # symbols). More work is needed to see if Equation 14 provides support for the existence of natural building blocks for general hyperschemata.

5 Conclusions

In this paper, for the first time, we provide an exact schema theorem for genetic programming which models the effects of schema creation due to crossover, in addition to schema survival and disruption. This theorem is the result of extending recent GA theory to GP through the definition of a new, more general notion of GP schema: the hyperschema.

In the paper we have also derived a simplified version of the exact GP schema theorem: the GP schema theorem with schema creation correction. This provides a simple way of calculating a lower bound for the total transmission probability of a schema which is significantly more accurate then the one provided by earlier schema theorems.

In addition, the paper indicates how recent general results reported in the GP literature can be applied to extend the most advanced results reported to date on GA schemata as well as the exact GP schema theorem presented in this paper. This extension provides much more information on the probabilistic behaviour of the number of instances of a schema at the next generation $m(H, t + 1)$.

This paper shows how the theory of genetic algorithms and theory of genetic programming are converging more and more and how they can mutually support the development of each other.

Acknowledgements

The author would like to thank the members of the EEBIC (Evolutionary and Emergent Behaviour Intelligence and Computation) group at Birmingham for useful discussions and comments.

References

[1] J. Holland, *Adaptation in Natural and Artificial Systems.* Cambridge, Massachusetts: MIT Press, second ed., 1992.

[2] D. E. Goldberg, *Genetic Algorithms in Search, Optimization, and Machine Learning.* Reading, Massachusetts: Addison-Wesley, 1989.

[3] R. Poli and W. B. Langdon, "A new schema theory for genetic programming with one-point crossover and point mutation," in *Genetic Programming 1997: Proceedings of the Second Annual Conference* (J. R. Koza, K. Deb, M. Dorigo, D. B. Fogel, M. Garzon, H. Iba, and R. L. Riolo, eds.), (Stanford University, CA, USA), pp. 278–285, Morgan Kaufmann, 13-16 July 1997.

[4] R. Poli and W. B. Langdon, "Schema theory for genetic programming with one-point crossover and point mutation," *Evolutionary Computation*, vol. 6, no. 3, pp. 231–252, 1998.

[5] J. P. Rosca, "Analysis of complexity drift in genetic programming," in *Genetic Programming 1997: Proceedings of the Second Annual Conference* (J. R. Koza, K. Deb, M. Dorigo, D. B. Fogel, M. Garzon, H. Iba, and R. L. Riolo, eds.), (Stanford University, CA, USA), pp. 286–294, Morgan Kaufmann, 13-16 July 1997.

[6] L. Altenberg, "The Schema Theorem and Price's Theorem," in *Foundations of Genetic Algorithms 3* (L. D. Whitley and M. D. Vose, eds.), (Estes Park, Colorado, USA), pp. 23–49, Morgan Kaufmann, 31 July–2 Aug. 1994 1995.

[7] W. G. Macready and D. H. Wolpert, "On 2-armed gaussian bandits and optimization." Sante Fe Institute Working Paper 96-05-009, March 1996.

[8] D. B. Fogel and A. Ghozeil, "Schema processing under proportional selection in the presence of random effects," *IEEE Transactions on Evolutionary Computation*, vol. 1, no. 4, pp. 290–293, 1997.

[9] D. B. Fogel and A. Ghozeil, "The schema theorem and the misallocation of trials in the presence of stochastic effects," in *Evolutionary Programming VII: Proc. of the 7th Ann. Conf. on Evolutionary Programming* (D. W. V.W. Porto, N. Saravanan and A. Eiben, eds.), pp. 313–321, 1998.

[10] A. E. Nix and M. D. Vose, "Modeling genetic algorithms with markov chains," *Annals of Mathematics and Artificial Intelligence*, vol. 5, pp. 79–88, 1992.

[11] T. E. Davis and J. C. Principe, "A markov chain framework for the simple genetic algorithm," *Evolutionary Computation*, vol. 1, no. 3, pp. 269–288, 1993.

[12] G. Rudolph, "Stochastic processes," in *Handbook of Evolutionary Computation* (T. Baeck, D. B. Fogel, and Z. Michalewicz, eds.), pp. B2.2–1–8, Oxford University Press, 1997.

[13] G. Rudolph, "Genetic algorithms," in *Handbook of Evolutionary Computation* (T. Baeck, D. B. Fogel, and Z. Michalewicz, eds.), pp. B2.4–20–27, Oxford University Press, 1997.

[14] G. Rudolph, "Convergence analysis of canonical genetic algorithm," *IEEE Transactions on Neural Networks*, vol. 5, no. 1, pp. 96–101, 1994.

[15] G. Rudolph, "Modees of stochastic convergence," in *Handbook of Evolutionary Computation* (T. Baeck, D. B. Fogel, and Z. Michalewicz, eds.), pp. B2.3–1–3, Oxford University Press, 1997.

[16] N. J. Radcliffe, "Schema processing," in *Handbook of Evolutionary Computation* (T. Baeck, D. B. Fogel, and Z. Michalewicz, eds.), pp. B2.5–1–10, Oxford University Press, 1997.

[17] C. R. Stephens and H. Waelbroeck, "Effective degrees of freedom in genetic algorithms and the block hypothesis," in *Proceedings of the Seventh International Conference on Genetic Algorithms (ICGA97)* (T. Bäck, ed.), (East Lansing), Morgan Kaufmann, 1997.

[18] C. R. Stephens and H. Waelbroeck, "Schemata evolution and building blocks," *Evolutionary Computation*, vol. 7, no. 2, pp. 109–124, 1999.

[19] R. Poli, "Probabilistic schema theorems without expectation, left-to-right convergence and population sizing in genetic algorithms," Tech. Rep. CSRP-99-3, University of Birmingham, School of Computer Science, Jan. 1999.

[20] R. Poli, "Schema theorems without expectations," in *GECCO-99: Proceedings of the Genetic and Evolutionary Computation Conference* (W. Banzhaf, J. Daida, A. E. Eiben, M. H. Garzon, V. Honavar, M. Jakiela, and R. E. Smith, eds.), (Orlando, Florida, USA), Morgan Kaufmann, 13-17 July 1999. Forthcoming.

[21] R. Poli, "Schema theory without expectations for GP and GAs with one-point crossover in the presence of schema creation," in *Foundations of Genetic Programming* (T. Haynes, W. B. Langdon, U.-M. O'Reilly, R. Poli, and J. Rosca, eds.), (Orlando, Florida, USA), 13 July 1999.

[22] D. Whitley, "A genetic algorithm tutorial," Tech. Rep. CS-93-103, Department of Computer Science, Colorado State University, August 1993.

[23] J. R. Koza, *Genetic Programming: On the Programming of Computers by Means of Natural Selection.* MIT Press, 1992.

[24] U.-M. O'Reilly, *An Analysis of Genetic Programming.* PhD thesis, Carleton University, Ottawa-Carleton Institute for Computer Science, Ottawa, Ontario, Canada, 22 Sept. 1995.

[25] U.-M. O'Reilly and F. Oppacher, "The troubling aspects of a building block hypothesis for genetic programming," in *Foundations of Genetic Algorithms 3* (L. D. Whitley and M. D. Vose, eds.), (Estes Park, Colorado, USA), pp. 73–88, Morgan Kaufmann, 31 July–2 Aug. 1994 1995.

[26] P. A. Whigham, "A schema theorem for context-free grammars," in *1995 IEEE Conference on Evolutionary Computation*, vol. 1, (Perth, Australia), pp. 178–181, IEEE Press, 29 Nov. - 1 Dec. 1995.

[27] P. A. Whigham, *Grammatical Bias for Evolutionary Learning.* PhD thesis, School of Computer Science, University College, University of New South Wales, Australian Defence Force Academy, 14 Oct. 1996.

[28] R. Poli and W. B. Langdon, "An experimental analysis of schema creation, propagation and disruption in genetic programming," in *Genetic Algorithms: Proceedings of the Seventh International Conference* (T. Back, ed.), (Michigan State University, East Lansing, MI, USA), pp. 18–25, Morgan Kaufmann, 19-23 July 1997.

[29] R. Poli, W. B. Langdon, and U.-M. O'Reilly, "Analysis of schema variance and short term extinction likelihoods," in *Genetic Programming 1998: Proceedings of the Third Annual Conference* (J. R. Koza, W. Banzhaf, K. Chellapilla, K. Deb, M. Dorigo, D. B. Fogel, M. H. Garzon, D. E. Goldberg, H. Iba, and R. Riolo, eds.), (University of Wisconsin, Madison, Wisconsin, USA), pp. 284–292, Morgan Kaufmann, 22-25 July 1998.

[30] D. Whitley, "An executable model of a simple genetic algorithm," in *Foundations of Genetic Algorithms Workshop (FOGA-92)* (D. Whitley, ed.), (Vail, Colorado), July 1992.

[31] P. Nordin and W. Banzhaf, "Complexity compression and evolution," in *Genetic Algorithms: Proceedings of the Sixth International Conference (ICGA95)* (L. Eshelman, ed.), (Pittsburgh, PA, USA), pp. 310–317, Morgan Kaufmann, 15-19 July 1995.

[32] P. Nordin, F. Francone, and W. Banzhaf, "Explicitly defined introns and destructive crossover in genetic programming," in *Proceedings of the Workshop on Genetic Programming: From Theory to Real-World Applications* (J. P. Rosca, ed.), (Tahoe City, California, USA), pp. 6–22, 9 July 1995.

[33] P. Nordin, F. Francone, and W. Banzhaf, "Explicitly defined introns and destructive crossover in genetic programming," in *Advances in Genetic Programming 2* (P. J. Angeline and K. E. Kinnear, Jr., eds.), ch. 6, pp. 111–134, Cambridge, MA, USA: MIT Press, 1996.

[34] J. J. Grefenstette, "Deception considered harmful," in *FOGA-92, Foundations of Genetic Algorithms*, (Vail, Colorado), 24–29 July 1992. Email: gref@aic.nrl.navy.mil.

Use of Genetic Programming in the Identification of Rational Model Structures

Katya Rodríguez-Vázquez[1] and Peter J. Fleming[2]

[1]DISCA-IIMAS-UNAM, P.O. Box 20-726, Alvaro Obregón, 01000 México, D.F.,
MEXICO
katya@uxdea4.iimas.unam.mx
[2]Department of Automatic Control and Systems Engineering, University of Sheffield,
Mappin Street S1 3JD, UNITED KINGDOM
P.Fleming@sheffield.ac.uk

Abstract. This paper demonstrates how genetic programming can be used for solving problems in the field of non-linear system identification of rational models. By using a two-tree structure rather than introducing the division operator in the function set, this genetic programming approach is able to determine the "true" model structure of the system under investigation. However, unlike use of the polynomial, which is linear in the parameters, use of rational model is non-linear in the parameters and thus noise terms cannot be estimated properly. By means of a second optimisation process (real-coded GA) which has the aim of tunning the coefficients to the "true" values, these parameters are then correctly computed. This approach is based upon the well-known NARMAX model representation, widely used in non-linear system identification.

1. Introduction

In Koza (1992), the identification process was presented as a symbolic regression problem where GP was used to find a mathematical expression able to reproduce a functional relationship based on given training data.

Iba *et al.* (1993) developed a software system called "STROGANOFF" for system identification based upon the Group Method of Data Handling (GMDH). In later work, they defined a fitness function based upon a Minimum Description Length (MDL) principle that had the aim of controlling the tree growth (Iba *et al.*, 1994; 1996). However, when a priori knowledge about the system is available, it is not easy to include this information when using GMDH.

McKay *et al.* (1995) and Schoenauer *et al.* (1996) also proposed the application of similar approaches for the modelling of chemical process and mechanical materials, respectively, represented via symbolic formulations.

Marenbach *et al.* (1997) presented an identification approach where the function set, in addition to the basic arithmetic functions, considers transfer blocks such as dynamics elements (time delays), non-linear elements and domain specific elements (from the industrial biotechnological fed-batch fermentation process). Models obtained using this method were expressed as block diagrams rather than mathematical functions and provided the basis for long-term prediction of the process behaviour.

Previous work by the authors formulated a non-linear system identification tool based upon the NARMAX model that used GP for determining the appropriate non-linear model structure and has been successfully applied to simulated and real identification problems (Rodríguez-Vázquez, et al., 1997a; Rodríguez-Vázquez and Fleming, 1998b).

Although, the polynomial model was successfully applied in this earlier work, the rational model has the advantage of modelling certain types of discontinuous functions and even severe non-linearities using only a very few parameters. The disadvantage of this rational representation is that it is non-linear in the parameters, thus making the identification a more complex process. Hence, this work attempts to extend the work on the identification of polynomial models to the use of the rational model. The rest of the paper is structured as follows. Section 2 introduces the rational model representation proposed for this work and Section 3 describes the GP encoding of the rational model. Section 4 then presents the GP operators used in this approach, while Section 5 describes the algorithm used for the estimation of parameters associated with the rational model. Section 6 presents results of the rational model system identification and, finally, section 7 draws conclusions and future developments.

Special features of the research are i) the use of a two-tree structure in the GP approach, ii) the use of a real-coded GA to address the "non-linear in the parameters problem" and iii) the use of multiple objectives in the GP context.

2. The Rational NARMAX Model

The model representation of this approach arises out of the NARMAX (Non-linear AutoRegressive Moving Average with eXtra inputs) model (Leontaritis and Billings, 1985). The general NARMAX model is an unknown non-linear function of degree ℓ defined as

$$y(k) = F^{\ell}(y(k-1),..., y(k-n_y), u(k-1),..., u(k-n_u), e(k-1),..., e(k-n_e)) + e(k) \qquad (1)$$

where n_y, n_u and n_e are the maximum lags considered for the output, input and noise terms, respectively. Moreover, $\{y(k)\}$ and $\{u(k)\}$ denotes a measured output and input of the system, and $\{e(k)\}$ is an unobservable zero mean independent sequence. As has been stated in Chen and Billings (1989), the most typical choice for $F^{\ell}\{\bullet\}$ in equation (1) is a polynomial expansion. The polynomial NARMAX model can be expressed as a linear regression model as

$$y(k) = P_i(k)\theta_i + \varepsilon(k) \quad k = 1, \cdots, N \qquad (2)$$

where N is the number of input-output data, $P_i(k)$ are possible regressors up to degree ℓ, θ_i are the associated coefficients, and M is the number of differents regressors. $P_i(k) \equiv l$ corresponds to the constant term. Thus, the polynomial model becomes linear in the parameters.

However, different representations for the NARMAX model exist. The rational representation is an alternative that contains as a special case, the polynomial representation when the polynomial denominator is equal to the constant term.

Classical approximation theory (Sontag, 1979) shows that rational functions are very efficient descriptors of the non-linear effect. This suggests that an extension of these concepts to the non-linear dynamic case may provide useful alternative models for non-linear systems compared with polynomial system expansions.

The rational model can then be derived and expressed by

$$y(k) = \frac{A(\bullet)}{B(\bullet)} + e(k) \tag{3}$$

where $A(\bullet)$ and $B(\bullet)$ are polynomial models of the form of equation (2), and $e(k)$ is an unobservable independent noise with zero mean and finite variance. Equation (3) is then rewritten as

$$y(k) = \frac{A(k)}{B(k)} = \frac{\sum\limits_{i=1}^{num} P_{n_i}(k)\theta_{n_i}}{\sum\limits_{j=1}^{den} P_{d_j}(k)\theta_{d_j}} + e(k) \tag{4}$$

where the numerator $\sum\limits_{i=1}^{num} P_{n_i}(k)\theta_{n_i}$ and the denominator $\sum\limits_{j=1}^{den} P_{d_j}(k)\theta_{d_j}$ are non-linear polynomial expressions.

3. Hierarchical Representation of the NARMAX Model

In previous work (Rodríguez-Vázquez, *et al.*, 1997a; Rodríguez-Vázquez and Fleming, 1998b), a population of non-linear polynomial structures, expressed as hierarchical trees, has been evolved in order to determine the correct model of the system under investigation, or at least, a model which is able to describe the dynamics of the non-linear system. The rational model can also be expressed in a similar way as described below.

3.1. Two-Tree Structures

This section introduces the concept of two-tree and how it can be used for encoding rational models.

A way of representing a rational model as a hierarchical tree is illustrated in Figure 1a. The tree representation of the rational structure may be implemented in a straightforward manner just by including the division "DIV" function. However, an alternative is a two-tree encoding. That means, a hierarchical tree arranged in a pair for each potential individual of the genetic population (Figure 1b). Tree expressions from Figure 1 represent the rational model given as

$$y(k) = \frac{\theta_{n_1} y(k-1) + \theta_{n_2} y(k-2) + \theta_{n_3} y(k-1)u(k-1) + \theta_{n_4} y(k-2)u(k-1)}{\theta_{d_1} + \theta_{d_2} y(k-1)} \quad (5)$$

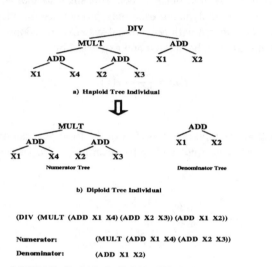

a) Haploid Tree Individual

b) Diploid Tree Individual

(DIV (MULT (ADD X1 X4) (ADD X2 X3)) (ADD X1 X2))

Numerator: (MULT (ADD X1 X4) (ADD X2 X3))

Denominator: (ADD X1 X2)

c) Reverse Polish Notation (Posfix) for the Haploid and Diploid Examples, Respectively.

Fig. 1. Hierarchical representation of a rational NARMAX model.

4. Two-Tree Genetic Programming Operators

In the case of n-tree encoding, Langdon (1995) has proposed a polyploid genetic programming population for evolving data structures (stack). In his approach, the crossover is performed in one tree selected randomly from the polyploid individual, and the other trees are copied from the parents to the offsprings. For the purpose of evolving rational models, the same criterion has been considered. That means, the crossover is restricted to take place between the numerators, or alternatively, the denominators of parent hierarchical trees.

This criterion for crossing over two-tree structures also applies to mutation. The mutation process is therefore performed over only one chromosome (tree) of the two-tree individual.

For the purpose of this type of structures (GP) proposed in this work, the switch_mutation operator is also introduced.

Angeline (1996) describes the switch_mutation operator as the selection of two subtrees of the same parent (haploid representation) which switches their positions creating a new individual. Because the rational model is the ratio of two polynomials, the switch operator here works by switching a selected subtree from each of the trees of the individual to be mutated; in others words, it exchanges a randomly selected subtree from the numerator (tree_1) by a random selected subtree from the denominator (tree_2). This switch_mutation is exemplified in Figure 2. Summarising, the genetic programming operators used in the identification of rational models are crossover, mutation and switch_mutation operators.

As mentioned earlier, rational NARMAX models are non-linear models in the input, output and also in the parameters. The next section gives a general description of the extended Least-Squares algorithm used for computing the set of parameters associated with the candidate rational models.

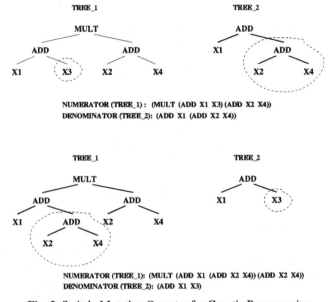

Fig. 2. Switch_Mutation Operator for Genetic Programming

5. Non-Linear Parameters Estimation

The main disadvantage of the rational model is the fact that it is non-linear in the parameters and the estimation becomes a complex process. Nevertheless, there have been developments on how to estimate this set of parameters. Billings and Zhu (1991) have suggested an extended Least Squares algorithm which is used in this work. Briefly, this algorithm is described as follow.

The mathematical expression of the rational NARMAX model given by equation (4) can be translated into a linear regression model by multiplying this expression by $B(k)$ (the polynomial denominator) on both sides and then, all the terms, except $y(k)P_{d_1}(k)\theta_{d_1}$, are moved to the right-hand side of the equation. This gives rise to the following expression

$$y(k)B(k) = A(k) + e(k)B(k) \tag{6}$$

$$Y(k) = A(k) - y(k)B(k) + y(k)P_{d_1}(k)\theta_{d_1} + \varepsilon(k) \tag{7}$$

$$= \sum_{i=1}^{num} P_{n_i}(k)\theta_{n_i} - \sum_{j=2}^{den} y(k)P_{d_j}(k)\theta_{d_j} + \varepsilon(k)$$

where

$$Y(k) = y(k)P_{d_1}(k)\theta_{d_1}\Big|_{\theta_{d_1}=1} = y(k)P_{d_1}(k) \tag{8}$$

$$= P_{d_1}(k)\frac{A(k)}{B(k)} + P_{d_1}(k)e(k)$$

Notice that

$$\varepsilon(k) = B(k)e(k) = \left(\sum_{j=1}^{den} P_{d_j}(k)\theta_{d_j}\right)e(k) \tag{9}$$

$$= P_{d_j}e(k) + \left(\sum_{j=2}^{den} P_{d_j}(k)\theta_{d_j}\right)e(k)$$

where $E[\varepsilon(k)] = E[B(k)]E[e(k)] = 0$.

Hence, $e(k)$ is reduced to an uncorrelated sequence as defined in equation (4). The disadvantage of this approach is now clearly seen. All the denominator terms $y(k)Pd_j(k)$ implicitly include a current noise term $e(k)$ which is highly correlated with $\varepsilon(k)$. This will introduce bias in the parameter estimation even when $e(k)$ is a zero-mean white-noise sequence.

The linear-in-the-parameters expression of the rational model (7) can be rewritten in a vector form as follows

$$Y(k) = P(k)\theta + \xi(k) = \hat{P}(k)\theta + P_{d_1}(k)e(k) \tag{10}$$

where

$$P(k) = \left[P_{n_1}(k),...,P_{n_{num}}(k),-P_{d_2}(k)(y(k)),...,P_{d_{den}}(k)y(k)\right] \tag{11}$$

$$\theta^T = [\theta_n \quad \theta_d] = \left[\theta_{n_1},...,\theta_{n_{num}},\theta_{d_2},...,\theta_{d_{den}}\right] \tag{12}$$

5.1. Parameter Estimation

Expanding the rational model to be linear in the parameters allows the application of the extended least square algorithm to estimate the coefficients of the rational structure. This algorithm has extensively been studied by several authors (Ljung, 1987). The main characteristic of the algorithm is the fact that the correlated noise is reduced to a white noise sequence by incorporating a noise model. Then, the algorithm proceeds in an iterative manner by which the process model and noise model parameters are estimated until the bias is reduced to zero.

This Extended Least Squares comprises the following step.

Step 1. An ordinary Least Squares algorithm is used to compute the initial parameters. In this step, *process model terms* are only considered. Thus,

$$\theta = \left[P^T P \right]^{-1} P^T Y \tag{13}$$

Step 2. The one-step ahead prediction error and the noise variance are estimated, where

$$\varepsilon(k) = y(k) - \hat{y}(k) \tag{14}$$

where $y(k)$ is the measured output and $\hat{y}(k)$ is the estimated output of the system.

Step 3. The *noise process terms* are now incorporated into the rational model structure by considering the noise sequence computed in the previous step.

Step 4. The *process and noise model terms* parameter are then estimated. The new noise sequence $\varepsilon(k)$ and noise variance are also computed.

Step 5. Go back to step 2 and repeat until the parameters estimates and the noise variance converge to constant values.

Based upon this Extended Least Squares algorithm, an example is presented in the next section to illustrate the use of a two-tree genetic programming approach to deal with the problem of identification of rational model structures.

6. Simulation Results

In order to test the use of genetic programming for rational model identification, a simulated example is given. An input-output sequence of 1000 points was generated using the model (Billings and Chen, 1989)

$$y(k) = \frac{y(k-1) + u(k-1) + y(k-1)u(k-1) + y(k-1)e(k-1)}{1 + y^2(k-1) + u(k-1)e(k-1)} + e(k) \tag{15}$$

The input was an independent sequence of uniform distribution with mean zero and variance 1.0. The noise was a Gaussian white noise sequence with zero mean and variance 0.01.

In this rational NARMAX identification example, linear output, input and noise terms with $n_y = n_u = n_e = 3$ maximum lag were considered. The GP population was set up to consist of 200 two-tree individuals (potential rational models) and the crossover, mutation and switch-crossover probabilities were 0.8, 0.1 and 0.1, respectively. The fitness was defined in terms of a multiobjective function where attributes of the identification process such as model complexity, performance and validation, were simultaneously evaluated. A multi-objective fitness ranking assignment was used as described in previous work (Rodríguez-Vázquez, et al., 1997b; Rodríguez-Vázquez and Fleming, 1998a). These three system identification attributes were measured based upon different objectives which are defined and classified as shown in Table 1. For all the correlation validity tests mentioned in Table 1, confidence intervals indicate whether the correlation between variables is significant or not. If N (the number of data points) is large, the standard deviation of the correlation estimate is $1/N$, where the 95% confidence limits are therefore approximately $\pm 1.96/\sqrt{N}$, assuming a Gaussian distribution.

Table 1. Description of objectives considered in the MOGP identification of rational model procedure

Attribute	Objective	Description
Model complexity	Model size	Number of process and noise terms
	Model degree	Maximum order term
	Model Lag	Maximum lagged input, output and noise terms
Model Performance	Residual variance	Variance of the predictive error between the OSAPE
	Long-term prediction error	Variance of the LTPE
Model validation	$\Phi_{ee}(\tau)$ (ACF) $\Phi_{ue}(\tau)$ (CCF) $\begin{bmatrix} \Phi_{\varepsilon(u\varepsilon)}(\tau) \\ \Phi_{u^2\varepsilon^2}(\tau) \\ \Phi_{u^2\varepsilon}(\tau) \end{bmatrix}$	ACF, CCF and higher-order correlation functions for the statistical validation of non-linear system models[1]

Models arising from several runs of the algorithm were rational representation which kept, in common, the process terms (see Table 2). Nevertheless these rational structures differed in the fact that noise terms were different in each case, the performance exhibited by them was, otherwise, similar.

From these results, it is clear that GP was able to identify the "true" structure of the system. Parameter estimation of the process terms using the linear-in-the-parameters mapping of the rational model and applying the extended LS algorithm did not correctly estimate the noise term coefficients. Nevertheless, the validity correlation

---

[1] The higher-order correlation functions between the input and residuals were evaluated based on the fact that the linear correlations (the autocorrelation of the residuals ACF and the crosscorrelation between the input and residuals ccf) are necessary but not sufficient conditions in order to state that a non-linear model is statistically valid.

functions were all satisfied. The "true" model structure and associated coefficients (model 4) obtained by using GP is shown in Table 3.

Table 2. Structure and performance of rational NARMAX model identified by using MOGP.

	Model 1	Model 2	Model 3	Model 4	Model 5
Numerator					
y(k-1)	x	x	x	x	x
u(k-1)	x	x	x	x	x
y(k-1)u(k-1)	x	x	x	x	x
y(k-1)e(k-1)				x	
u(k-1)e(k-1)			x		
Denominator					
C	x	x	x	x	x
y(k-1)2	x	x	x	x	x
u(k-1)e(k-1)		x	x	x	x
y(k-1)e(k-1)					x
Performance					
OSAPEx10^{-4}	4.7107	4.3116	4.3176	3.9781	4.2947
LTPEx10^{-4}	5.3000	4.4086	4.1928	4.2835	4.2700

6.1. A Second Search Procedure

From the results presented in the previous section, it is observed that using evolutionary techniques, such as genetic programming, optimal rational model structures can be generated. It is also noted that the correct parameter values of the process terms can be obtained by means of the extended LS algorithm. However, it is not the case for the estimation of noise term coefficients.

An alternative to overcome this drawback in the parameter estimation could be the use of genetic algorithm. Given the model structure, a GA would be used as a parameter optimiser. Hence, a population of a possible set of parameters would evolve and the single objective fitness function would be expressed as the minimisation of either the one-step ahead mean square error or the long-term mean square error.

Therefore, a GA optimiser using real-valued encoding was proposed. The chromosome was then defined as a $(P_d + P_n)$ elements vector of real values. That is

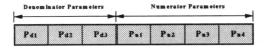

Fig. 3. Fixed-length real-valued chromosome enconding for the GA parameters optimiser.

where P_d and P_n are the number of terms in the denominator and numerator polynomials, respectively. Given a real-valued chromosome encoding, Mühlenbein and Schlierkamp-Voosen (1993) proposed a set of genetic operators for working on real encoding.

Having defined a real-valued chromosome representation, the second optimisation was performed where the objective function was defined as the minimisation of the residual variance estimated based upon the one-step ahead prediction error[2].

A population of 50 individuals was run for 50 generations. The crossover and mutation probabilities were 0.8 and $1/(P_d + P_n)$, respectively. The results show an improvement in the coefficients of the noise terms which are closer to the system parameters and a slight improvement in the one-step ahead and long-term predictive response as seen in Table 3.

Table 3. Comparison of extended LS and second parameter optimisation method.

Terms	Extended LS	GA Optimiser
Numerator		
y(k-1)	0.9987	0.9902
u(k-1)	0.9991	0.9915
y(k-1)u(k-1)	1.0000	0.9906
y(k-1)e(k-1)	-0.3705	1.0000
Denominator		
c	1.0000	0.9910
y(k-1)2	1.0001	0.9899
u(k-1)e(k-1)	-0.3295	0.9581
Performance		
OSAPEx10^{-4}	3.9781	3.7544
LTPEx10^{-4}	4.2835	4.1486

7. Conclusions

In system identification, GP has also been demonstrated to be useful for the determination of these rational structures. Although the main results demonstrate the use of GP in this rational identification case, it was tested on simulated data. However, these results are promising and future studies could contemplate the study of the identification of rational models based upon real data.

It is also relevant that a second optimisation procedure that was based upon a simple genetic algorithm was introduced. This second optimiser was necessary in order to tune the parameter of the rational structure by simply minimising a single objective function. Although the parameter values obtained after running the GA optimiser were closer to the "true" model parameters, the performance criteria did not show a

---

[2] The routines for the implementation of the GA parameters optimiser were all used from the GA Toolbox developed by Chipperfield *et al.* (1994).

great variation from the performance given by the model with the parameters previous to tune.

In view of the results, the identification of rational and polynomial models by means of evolutionary techniques is, therefore, a promising research area to be explored further.

Acknowledgments

Katya Rodríguez-Vázquez gratefully acknowledges the financial support of Consejo Nacional de Ciencia y Tecnología (CONACyT), México, and the tecnhical support made by Ing. Eliseo Díaz.

References

ANGELINE, P.J. (1996) An Investigation into the Sensitivity of Genetic Programming to the Frequency of Leaf Selection During Subtree Crossover. In: *Annual GP Conference* (Koza et al., 1996), pp. 21-29.

BILLINGS, S.A. AND S. CHEN (1989) Identification of Non-Linear Rational Systems Using a Prediction Error Estimation Algorithm. *Int. J. Systems Sci.,* 20(3), pp. 467-494.

BILLINGS, S.A. AND Q.M. ZHU (1991) Rational Model Identification Using an Extended Least-Squares Algorithm. *Int. J. Control,* **54**(3), pp. 529-546.

CHEN, S. AND S.A. BILLINGS (1989) Representation of Non-Linear Systems: the NARMAX Model. *Int. J. Control,* 49(3), pp. 1013-1032.

IBA, H. T. KURITA, H. DE GARIS AND T. SATO (1993) System Identification Using Structured Genetic Algorithms. In *Proc. of the 5th ICGA*, pp. 467-491.

IBA, H., H. DE GARIS AND T. SATO. (1994) Genetic Programming Using a Minimum Description Length. In: *Advance in Genetic Programming* (Kinnear, Ed.). pp. 265-284.

KOZA, J.R. (1992) *Genetic Programming: On the Programming of Computers by Means of Natural Selection.* MIT Press.

LANGDON, W.B. (1995) Evolving Data Structures with Genetic Programming. In: *6th ICGA* (Eshelman, Ed.), pp. 295-302.

LEONTARITIS, I.J. AND S.A. BILLINGS (1985) Input-Output Parametric Models for Non-Linear Systems. Part I and Part II. *Int. J. Control,* 41(2), pp. 304-344.

LJUNG, L.S. (1987) *System Identification:* Theory for the User. Prentice Hall.

MARENBACH, P., K.D. BETTENHAUSEN AND S. FREYER. (1996) Signal Path Oriented Approach for Generation of Dynamic Process Models. In (Koza et al., 1996), pp. 327-332.

MCKAY, B., M.J. WILLIS AND G.W. BARTON. (1995) Using a Tree Structures Genetic Algorithm to Perform Symbolic Regression. In *1st. Int. Conf. on Genetic Algorithms in Engineering Systems: Innovations and Applications.* IEE, pp. 487-492.

MÜHLENBEIN, H. AND D. SCHLIERKAMP-VOOSEN (1993) Predictive Models for the Breeder Genetic Algorithm. *Evolutionary Computation,* 1(1), pp. 25-49.

RODRÍGUEZ-VÁZQUEZ, K., C.M. FONSECA AND P.J. FLEMING. (1997a) An Evolutionary Approach to Non-Linear Polynomial System Identification. In *11th IFAC Symposium on System Identification,* Vol. 3, pp. 1395-1400.

RODRÍGUEZ-VÁZQUEZ, K., C.M. FONSECA AND P.J. FLEMING (1997b) Multiobjective Genetic Programming: A Non-Lineat System Identification Application. *Late Breaking Paper at the GP'97 Conference*. pg. 207-212.

RODRÍGUEZ-VÁZQUEZ, K. AND P.J. FLEMING (1998a) Multiobjective Genetic Programming for Non-Linear System Identification. *Ellectronics Letters*, **34**(9), p. 930-931.

RODRÍGUEZ-VÁZQUEZ, K. AND P.J. FLEMING (1998b) Multiobjective Genetic Programming for a Gas Turbine Engine Model Identification. *International Conference on Control'98*, pg. 1385-1390.

SCHOENAUER, M., M. SEBAG, T. JOUVE, B. LAMY AND H. MAITOURAM. (1996) Evolutionary Identification of Macro-Mechanical Models. In: *Adavances in Genetic Programming 2* (Angeline and Kinnear, Eds.). pp. 467-488.

SONTAG, E.D. (1979) *Polynomial Response Maps. Lectures Notes in Control and Information Sciences*, Vol. **13**. Springer-Verlag.

Grammatical Retina Description with Enhanced Methods

Róbert Ványi[1], Gabriella Kókai[2], Zoltán Tóth[1] and Tünde Pető[3]

[1] Institute of Informatics, József Attila University
Árpád tér 2, H-6720 Szeged, Hungary
e-mail: h531774|h531714@stud.u-szeged.hu

[2] Department of Computer Science, Programming Languages
Friedrich-Alexander University of Erlangen-Nürnberg
Martensstr. 3. D-91058 Erlangen, Germany
e-mail: kokai@informatik.uni-erlangen.de

[3] Department of Opthalmology
Albert Szent-Györgyi Medical University
Korányi f. 12, H-6720 Szeged, Hungary
e-mail: peto@opth.szote.u-szeged.hu

Abstract. In this paper the enhanced version of the *GREDEA* system is presented. The main idea behind the system is that with the help of evolutionary algorithms a grammatical description of the blood circulation of the human retina can be inferred. The system uses parametric *Lindenmayer systems* as description language. It can be applied on patients with diabetes who need to be monitored over long periods of time. Since the first version some improvements were made, e.g. new fitness function and new genetic operators. In this paper these changes are described.[1]

1 Introduction

In this paper the enhanced version of the *GREDEA* (*Grammatical Retina Description with Evolutionary Algorithms* [5]) system is presented. This system is being developed to describe retina images. The main idea behind this system is to work out patient-specific monitoring programs for examining the blood circulation of the human retina. The system can be used on patients with diabetes who need to be monitored over long periods. It is essential to check the eyesight of patients with this disease, because the deterioration of the vascular tree caused by diabetes has a direct effect on the vision quality.

The first version *GREDEA* system had some restrictions: Only simple images were allowed and the convergence speed was low. The solutions of these problems are discussed in the following sections.

We proceed as follows. In Section 2 the description of the GREDEA system is given. Remember that the system is still under development. Here only some

[1] This work was supported by the grants of Bayerischer-Habilitationsförderpreis 1999

approvements are described. First in Section 3 the new image processing algorithms are summarized. In Section 4 the modified genetic operators, in Section 5 the new fitness functions are presented. In Section 6 the test results are discussed. Finally in Section 7 the conclusions and some future plans are mentioned.

2   The GREDEA system

The *GREDEA* system (see in Figure 1) is designed to describe the blood vessels of the human retina using some kind of grammars. The first version of the system is described in this section. In *GREDEA* an individual description of the blood circulation of the human retina is created for each patient. To obtain the description, the process starts from preprocessed fundus images taken with a fundus camera. It is assumed that the background is black (0 values) and the vascular tree is white (1 values). Later these images are referred to as *destination images*.

Then a parametric *Lindenmayer system* (*L-system*) – creating the pattern closest to the vascular tree of the patient – is developed. This *L-system* can be stored (this storage method needs less storage than bitmap images) and used later to make comparisons. If the patient returns for a check-up, new pictures are recognised and preprocessed, but the previously generated L-system is used for starting the evolution process.

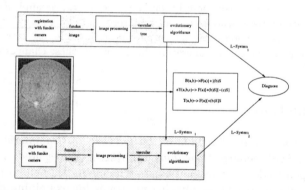

Fig. 1. Structure of the GREDEA system

L-systems are parallel grammatical rewriting systems. With the help of them complex objects can be described by starting from the axiom and using the productions [13]. When choosing the representation form, the fact that *L-systems* can be evolved using evolutionary algorithms [6] was taken into consideration.

Since parametric L-systems are evolved, two kinds of evolutionary algorithms are applied: *genetic programming* [3] is used on the rewriting rules of the evolved L-systems, and *evolution strategies* [12] are applied on the parameters of these L-systems.

The searching for string-representation of an L-system is called the inverse problem of L-systems, which is a problem that has already been described in several papers [4], [8] and [9]. But the primary task was to find an evolutionary algorithm system to describe and classify retina images. It is an image processing problem, where branching structures and curves in different length have to be described. For this reason the task is to construct appropriate opreators and a fitness function.

The L-systems and the evolution process are described in the following.

2.1 Lindenmayer systems

Lindenmayer introduced special kinds of descriptive grammars in order to describe the natural development processes of plants [7]. These are called Lindenmayer grammars or *Lindenmayer systems* (*L-systems*). The most important characteristic of *L-systems* compared with Chomsky grammars is *parallel rewriting*. In Lindenmayer systems, rewriting rules are applied in parallel and all occurances of the left side of the chosen rewriting rule are replaced simultaneously.

Different *L-system* classes are introduced, such as deterministic or nondeterministic *L-systems*. Distinction can be made between context free and context sensitive *L-systems*. In this paper only *deterministic context free L-systems* (*D0L-systems*) are used. In order to enlarge the number of described objects, *L-systems* can also be associated with parameters.

2.1.1 Parametric $D0L$-system

With the idea of parametric *L-systems*, the concept of parallel rewriting to parametric words instead of strings of symbols can be applied. Following the notation of Prusinkiewicz [10] the word "module" as a synonym for "letter with associated parameters" is used.

The real-valued *actual* parameters appearing in words have a counterpart in the *formal* parameters which may occur in the specification of *L-system* productions. Let Π be the set of formal parameters. The combination of formal parameters and numeric constants using arithmetic $(+, -, *, /)$, relational $(<, <=, >, >=, ==)$, logical operators $(!, \&\&, ||)$ and parentheses $(())$ generates the set of all correctly constructed logical and arithmetic expressions with parameters from Π. These are denoted by $\mathcal{C}(\Pi)$ and $\mathcal{E}(\Pi)$, respectively. A *parametric D0L-system* is an ordered quadruple $G = (\Sigma, \Pi, \alpha, P)$, where

- $\Sigma = s_1, s_2, ..., s_n$ is the alphabet of the system,
- Π is the set of formal parameters,
- $\alpha \in (\Sigma \times \mathbb{R}^*)^+$ is a nonempty parametric word called the axiom,
- $P \subset (\Sigma \times \Pi^*) \times \mathcal{C}(\Pi) \times (\Sigma \times \mathcal{E}(\Pi)^*)^*$ is a finite set of productions.

There is no essential difference between *D0L-systems* that operate on strings with or without brackets. In our implementation brackets are used. However, in this case productions involving brackets have to the following forms:

- *pred* | *cond* → *succ*, where *pred* and *cond* are as before, but $succ \in ((\Sigma \times \mathcal{E}(\Pi)^* \cup \{[,]\})^*$, and *succ* is well nested;
- [→ [, or] →].

where $pred \in \Sigma \times \Pi^*$ denotes the *predecessor*, $succ \in ((\Sigma \times \mathcal{E}(\Pi)^*)^*$ is the *successor* and $cond \in \mathcal{C}(\Pi)$ describes the *conditions*.

A string S is called well-nested if and only if for all S' prefix of S, the number of '[' symbols is not less than the number of ']' symbols, and the number of these symbols is equal in S. More details and examples of parametric *D0L-systems* can be found in [11].

When evolving L-systems, one mustn't forget to mention the stability (or instability) of these structures and the evolution process. The GA part of the GREDEA system (which evolves rewriting rules) is not stable; for example, putting a pair of brackets into a rewriting rule can generate a substantially different image: consider $F + FF$ and $F + [F]F$. Fortunately, this part is the less important one, it is used for founding an L-system which broadly covers the target image; and it is used only at the first examination of the patient, in the case of check-ups only ES is applied. The ES part (which evolves the parameters of the L-systems) is stable, because a small distortion in a pitch generates large distortion in the resulting L-system only in larger iteration steps, and the GREDEA system uses only small iteration numbers. This is an advantage of the L-systems over the iterated function systems (IFSs), because a small change in a parameter of an IFS can cause a completely different image.

To display an image defined by an *L-system*, the well-known turtle graphics is used [1]. The turtle is an object in space such that its position is defined by a vector and its orientation is defined by a coordinate-system. The coordinate-system determines the heading (h), the left (l) and the up (u) directions. Using these vectors, the turtle can be moved in space. There are several turtle operations, also known as turtle commands. The most important ones are drawing ('F') and turning left and right ('$+$' and '$-$'). There are special turtle operations, called stack operations. These are denoted by '[' and ']', and used to save and restore turtle position and orientation. Using stack operations, branching structures can be produced.

To use turtle graphics with *L-systems*, turtle commands as the elements of the alphabet are applied.

2.2 Evolution process

During the evolution process the individuals are represented by *parametric L-systems*. According to the definition of the D0L-systems, parameters and constant values may appear in the rewriting rules. The *constant values* are considered as the *parameters* of the whole L-system. For example, if an $X \rightarrow Y(10)$ is present, it is assumed that an $X \rightarrow Y(A)$ rule is used and the L-system has a parameter called 'A' with a current value of 10. It is also assumed that the axiom is S.

Because the shape of the generated image depends both on the rewriting rules and on the parameters, the representation can be divided into two main parts. The right sides of the rules are represented as strings, while the parameters are described as vectors of real values. Usually it is more important to find the appropriate right sides. Changing the parameters is used for the fine-tuning of the image.

L-systems with the same rules are considered as an *L-system group*. In an *L-system group* the rules are constant, but the parameters can be varied. To find the best parameters *evolutionary strategies (ES)* are used inside the *L-system group*. Therefore an *L-system* group works as a *subpopulation*. A subpopulation can be represented by rewriting rules. Because the number of the rules, and also the left sides are constant, an *L-system group* can be represented by an array of strings. To find the rewriting rules, a *genetic algorithm (GA)* is applied on the string arrays representing an *L-system group*. So the *L-system groups* make up the population. The structure of the population can be seen in Figure 2.

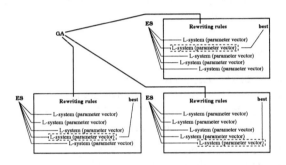

Fig. 2. Structure of the population

2.3.1 The structure of the population

To calculate the fitness value, the images described by the L-systems are created, then the distance between these generated images and the destination image is calculated with a suitable function. The original fitness function will be discussed in section 5. Also the disadvantages and possible solutions are described there. Using this fitness value, *evolutionary strategies* are applied on the parameter vectors several times. The number of the evolution steps (denoted by S_{ES}) is defined at the very beginning of the process. When the ES steps are finished, the best individuals in each subpopulation are selected. The parameters and the fitness values of these *L-systems* will act as the parameters and the fitness value for the whole subpopulation. Using this fitness value, several subpopulations can be selected and then they are used to create new subpopulations with *genetic algorithms (GA)*. Because ES is usually applied on real valued vectors and GA is used with words over a finite alphabet, in this process both of them are used. Because of the data structure, the GA has been changed to GP.

2.3.2 The form of an individual

Using that the retinal blood vessel can only contain bending (B), branching (Y) and forking (T), the form of an individual is the following *parametric L-system*:

$$\Sigma = \{S, F(a), +(a), -(a), [,], B(a,b), Y(a,b,c), T(a,b)\}$$
$$\alpha = S$$

rewriting rules:

- $S \to v$, where $v \in \{S, B(a,b), Y(a,b,c), T(a,b), [,]\}^*$ is a well-nested string, with constant parameters.
- $B(a,b) \to S+(a)F(b)$
- $Y(a,b,c) \to F(a)[+(b)S][-(c)S]$
- $T(a,b) \to F(a)[+(b)S]S$

The form of the first rule can be similar to the following:

$$S \to B(A,B)Y(C,D,E)Y(F,G,H)T(I,J)B(K,L).$$

In this case the L-system has twelve parameters: from 'A' to 'L'.

Now the description of the original system is finished. In the following sections the solutions of the problems and the enhancements are discussed.

3   Preprocessing images

The first part of the GREDEA system preprocesses fundus images. It was aided to the system previously, but for example noise cleaning had to be done by hand. Now the process takes a fundus image (Figure 3), and its output is the vascular tree. The algorithm has the following parts:

Median filtering for noise correction.
Gradient computing to find the blood vessels. (Figure 4)
Thresholding to create a binary image.

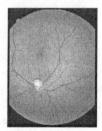

Fig. 3. Original fundus image   **Fig. 4.** Result of the edge detection

The most important improvement is that instead of using a grayscaled image, the green component of the color fundus image is used. This results in better image quality, because the blood vessels are usually red, thus they don't contain green component, while the background is orange, which does contain green component. The preprocessed grayscaled image and green component can be seen in Figures 5 and 6, respectively.

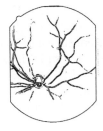

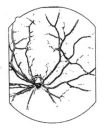

Fig. 5. Preprocessed grayscaled image   **Fig. 6.** Preprocessed green component

4   Genetic operators

The main algorithm of the evolution process, as described previously in Section 2.2, was not changed. The evolution strategies to find the best possible L-system parameters remained unchanged, but but the genetic algorithms to find the best L-system were modified. During the evolution, one must operate with well-nested strings and their substrings. It is much more easy to handle them as trees instead of strings. But the structure of the L-systems, as defined in Section 2.1, implies the use of strings. That's why in the new version the operators are defined on trees, but in fact applied on strings. Finally, the L-systems are drawn, so the operators should work correcly on this level as well. In short, there are 3 levels of the operations:

- theoretical level: the operations are defined on this level
- physical level: the operations are carried out on this level
- effective level: the result appears on this level

Previously mutation changed a single character and crossover replaced substrings. Now they are slightly redefined to work on trees and subtrees. The definitions of these operators can be found in the following subsections.

Originally, the operators were applied on well-nested strings. Now they work with trees, as a tree can be assigned to all well-nested strings. An example tree for the string $a[b[c]d]ef[g][hi]jk$ can be seen in Figure 7. Another example is in Figure 8.

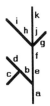

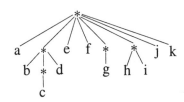

Fig. 7. Tree for a well-nested sting   **Fig. 8.** Syntactic tree for a well-nested string

The first example represents the string drawn by a turtle. In this case, the edges are labelled with the characters, and well-nested substrings are the subtrees. The main problem is that this representation is not bijective. For example, we get the same tree if we replace any x character with $[x]$. The latter one is more like syntactical, so it is called *syntactical tree*. Here the characters are the leafs, and there is exactly one internode for each well-nested substring. This representation is bijective. In GREDEA, the second method is used. But because the first one – which represents the structure – will appear in the generated image, this type also has to be taken into consideration. We can conclude the followings:

- the theoretical level is the level of the syntactical trees
- the physical level is the level of well-nested strings
- the effective level is the level of the structural (or drawn) trees

The formal definition of this syntactical tree is the following:

Definition 1 Let us take a well-nested string. Index each character with its position in order to make difference between them. We say that a *well-nested string* is *unseparable* iff it begins with a '[' and its length is minimal. The syntactic tree for a well-nested string is the following $G = G(V, E)$ graph:

- $V = V_c \cup V_s$, where V_c is the set of the non-bracket characters and V_s is the set of all unseparable well-nested substrings plus the whole string.
- $xy \in E$ iff $x \in V, y \in V_s, x \subset y$ and $(\forall y')(x \subset y' \Rightarrow y \subseteq y')$

For both the mutation and the crossover, several tree operations are needed, such as cutting a subtree. These are described in the following subsection.

4.1 Tree operations

The used tree operations are the following: random tree generation, finding a subtree, deleting a subtree and inserting a subtree. When creating a random tree or finding a subtree, it is usually important to know the depth of the (sub)tree. In the case of L-systems, the tree is stored as a string, so the corresponding string operations have to be found. The depth of a subtree can also be calculated within the string. The leafs are represented with characters in the string, their depth is 0. The internodes are represented with the right bracket of the corresponding substring. Their depth can be calculated in the following way:

- Set variable D and D_{max} to 0
- Go through the string from left to right
- For each character x,
  - if x is a left bracket, increase D and D_{max}
  - if x is a right bracket, decrease D and label x with $D_{max} - D$. If $D = 0$ after decreasing, then set D_{max} to 0.
  - otherwise label x with 0 and do not change D and D_{max}

It is easy to see that this algorithm labels the brackets with the appropriate values.

4.2 GP Mutation

The mutation, using the previously defined tree operations, is the following:

- Find a subtree with depth less than or equal to R_{mut}. Its root will be the chosen node.
- Delete the subtree at the chosen node.
- Create a random subtree the depth of which is less than or equal to R_{mut}.
- Insert this subtree at the chosen node.

When using strings, the algorithm is the following:

- Find a character which is not a left bracket and its label is less than or equal to R_{mut}. If the chosen character is not a right bracket, then it will be the chosen substring. Otherwise find the corresponding left bracket, and the string between the brackets will be the chosen substring, including the brackets.
- Cut the chosen substring from the string.
- Randomly create a well-nested substring the "depth" of which is less than or equal to R_{mut}. The "depth" of a string is the depth of the represented tree. If the generated string is not a single non-bracket character, then enclose it into brackets.
- Insert this string at the place of the chosen substring.

In Figures 9 through 11, the result of this operation on a syntactical tree, a structural tree and also on a well-nested string can be seen.

Fig. 9. Mutation of a syntactical tree   **Fig. 10.** Mutation of a structural tree

a[b[c]d]e[f]   a[b[c]d]e[g[hi]j]

Fig. 11. Mutation of a string

4.3 GP Crossover

The crossover operation uses similar tree and string functions as mutation:

- Find a subtree in both trees such that their depths are less than or equal to R_{cross}. Their root will be the chosen nodes.

- Delete the subtrees at the chosen nodes.
- Insert these subtrees at the chosen nodes in the other trees.

When using strings, the algorithm is the following:

- Find a character which is not a left bracket and its label is less than or equal to R_{cross} in both strings. If the chosen character is not a right bracket, then it will be the chosen substring. Otherwise find the corresponding left bracket, and the string between the brackets will be the chosen substring including the brackets.
- Cut the chosen substrings from the strings.
- Insert these strings at the place of the chosen substrings in the other strings.

The effect of this operator on well-nested strings and structural trees is similar to the effect of mutation shown previously.

5   Fitness functions

There are many possibilities to define the distance between two images. A very simple method to calculate the distance d_q between two images I_1 and I_2 is the quadratic error of the images. During the definitions, it is assumed that the sizes of the pictures are the same: $N \times M$. Originally, the quadratic error was used in GREDEA.

5.1   Quadratic error

The distance is the sum of quadratic errors of two corresponding pixels:

$$d_q(I_1, I_2) = \sum_{x=0}^{N-1} \sum_{y=0}^{M-1} (I_1(x, y) - I_2(x, y))^2$$

For binary images, where $I_1^B(x, y)$ and $I_2^B(x, y)$ are two pixels, the quadratic error is:

$$d_p(I_1^B(x, y), I_2^B(x, y)) = \begin{cases} 1 & \text{if } I_1^B(x, y) \neq I_2^B(x, y) \\ 0 & \text{otherwise} \end{cases}$$

So let the distance between two binary pictures be defined like this:

$$d_q(I_1, I_2) = \sum_{x=0}^{N-1} \sum_{y=0}^{M-1} d_p(x, y)$$

In GREDEA an extended version of this function was used: Let us denote the pixels with $p_d = I_d^B(x, y)$ and $p_g = I_g^B(x, y)$, where p_d means that the pixel value on the destination image I_d and p_g means the value of the pixel on the generated image I_g. In this case the weighted distance of two pixels can be defined:

$$d_p'(x, y) = \begin{cases} W_0 & \text{, if } p_d = p_g = 0 \\ W_1 & \text{, if } p_d = p_g = 1 \\ -W_2 & \text{, if } p_d < p_g \\ -W_3 & \text{, if } p_d > p_g \end{cases}$$

So the fitness function φ is the weighted distance between two images:

$$\varphi_q = d'(I_1, I_2) = \sum_{x=0}^{N-1} \sum_{y=0}^{M-1} d'_p(x, y)$$

Using this definition, a missing point will cost more than an unnecessary one if W_3 is set to greater than W_2. These parameters can also be used to fine-tune the fitness function. Note that W_0 is usually 0.

5.2 Multi-level fitness

The quadratic error is easy to calculate, but has several disadvantages. It says nothing about the image structure, thus a matching point anywhere will increase fitness with the same value, and similarly any unnecessary point will cost the same, regardless to its distance. So a new version of the fitness function was developed, which is related to the so called distance images descriebed for example in [2]. In the new version, the destination image is preprocessed in the following way:

- Parts of the vascular tree get higher value
- Parts of the background gets lower values depending on their distance from the vascular tree

The fitness is simply the sum of these values for the points of the generated image. As in the quadratic error, the missing pixels can also be added to this fitness with a negative value. Note that the values of the background pixels must be negative, otherwise the L-system covering the whole image would get the highest fitness. The most simple multi-level fitness is the following:

$$d_p^m(x, y) = \begin{cases} W_f & \text{, if } p_g = p_d = 1 \\ -D_{x,y} & \text{, if } p_d = 0 \text{ and } p_g = 1 \\ -W_m & \text{, if } p_d = 1 \text{ and } p_g = 0 \\ 0 & \text{otherwise} \end{cases}$$

where the distance of (x, y) and the vascular tree is $D_{x,y}$. The distance of pixels (x_1, y_1) and (x_2, y_2) is usually the euclidian norm, that is $\sqrt{(x_2 - x_1)^2 + (y_2 - y_1)^2}$, but it can be simply $|x_2 - x_1| + |y_2 - y_1|$ here. The distance of a pixel and a multi-point object is the smallest distance between the pixel and any point of the object. The multi level fitness function is the following:

$$\varphi_m = \sum_{x=0}^{N-1} \sum_{y=0}^{M-1} d_p^m(x, y)$$

5.3 General fitness

Examining the previous fitness function, one can easily find a more general fitness for binary images. Let the distance of two pixels be modified as follows:

$$d_p^g(x, y) = \begin{cases} W(D_{x,y}) & \text{, if } p_g = 1 \\ -W_m & \text{, if } p_d = 1 \text{ and } p_g = 0 \\ 0 & \text{otherwise} \end{cases}$$

Thus we get a fitness function for each $W(d)$ function. It is easy to see that the previously definied fitness functions are the special cases of this general fitness function:

Quadratic error: $W(0) = W_1, W(d) = -W_3 \ (d > 0), W_m = W_2$
Multi-level fitness: $W(0) = W_f, W(d) = -d \ (d > 0)$

5.4 Calculating the fitness

Comparing two images pixel by pixel is very slow, even with the fast computers used nowadays. Well, it is not important if this has to be done only a few times. But during the evolution process, fitness has to be calculated many times. So it is important to have a fast fitness calculating procedure. Notice that black pixels (0 values) usually do not count in the fitness value. So leaving the black pixels can increase calculation speed. The best way to do this is by combining the turtle with the fitness calculation, resulting in an object called *fitness-calculating turtle* (*FCT*).

5.4.1 FCT for quadratic error
When the FCT draws an object, it also examines the corresponding pixels of the destination image. According to the result, the pixel is classified as *fpixel* if the pixel is present in the destination image or *upixel* if not. The number of these pixels is denoted by N_f and N_u. Before the evolution process, the number of black pixels (N_0) and white pixels (N_1) in the destination image can be calculated. Using these values, the number of pixels which appear in the destination image but not in the generated image can be determined. This kind of pixels is called *mpixels*. The number of these pixels is denoted by N_m and can be calculated in the following way:

$$N_m = N_1 - N_f$$

The not examined pixels are called *npixels*. The number of these pixels is:

$$N_n = N_0 - N_u$$

Assuming that $W_0 = 0$ and denoting the constant values with c_i, the result is an easy-to-use formula:

$$\varphi_q = c_{13} N_f - c_{02} N_u - c, \ c_{13} = W_1 - W_3, \ c_{02} = W_0 + W_2 = W_2, \ c = W_3 N_1$$

This turtle was already used in GREDEA but it had to be extended to use the new fitness functions.

5.4.2 FCT for general fitness
For the general fitness function, it is not enough to classify pixels depending on whether they are part of the vascular tree or not. In this case the classification

has to be made by considering the distance from the vascular tree.

$$\varphi_g = \sum_{x=0}^{n-1}\sum_{y=0}^{n-1} d_p^g(x,y) =$$

$$\sum_{fpixel} d_p^g(x,y) + \sum_{upixel} d_p^g(x,y) + \sum_{mpixel} d_p^g(x,y) + \sum_{npixel} d_p^g(x,y) =$$

$$\sum_{fpixel \cup upixel} d_p^g(x,y) + (N_1 - N_f)W_m = \sum_{fpixel \cup upixel} d_p^g(x,y) + N_1 W_m - N_f W_m$$

The $fpixels$ and $upixels$ are those which has $p_g = 1$, so $d_p^g(x,y) = W(D_{x,y})$. Using this, the sum in the previous formula can be modified as follows:

$$\sum_{f \cup u} d_p^g(x,y) = \sum_{d=0}^{\infty}\sum_{D_{x,y}=d} W(D_{x,y}) = \sum_{d=0}^{\infty}\sum_{D_{x,y}=d} W(d) = \sum_{d=0}^{\infty} N(d)W(d),$$

where $N(d)$ is the number of the points in d distance from the vascular tree. Note that $N_f = N(0)$. In real applications, there is an upper bound for d, namely $\sqrt{N^2 + M^2}$ in the case of the euclidian norm, or $N + M$ when using the more simple distance measurement. Remember that the size of the image is $N \times M$. Pixels outside the image can be farther than $N + M$ from the vascular tree, but in this case their distance can be considered $N + M$. So the general fitness is:

$$\varphi_g = \sum_{d=0}^{N+M} N(d)W(d) - N(0)W_m - c,$$

where $c = N_1 W_m$. So the FCT for general fitness simply calculates N by classifying the drawn pixels.

6   Experimental results

In order to visualize the fitness landscape, a simple test image and L-system was used. The image was a 200×200 image with a $(100,0) - (100,99)$ line. The L-system had only one $S \to +(a)F(b)$ rule.

For **quadratic error** or binary fitness, the parameters were $W_0 = W_2 = W_3 = 0$ and $W_1 = 100$, which means $c_{13} = 100$, $c_{02} = 0$, and $c = 0$. The fitness was calculated for $a = -180\ldots180$ and $b = 0\ldots100$. The result can be seen in Figure 12. The first part is the slice for $b = 100$.

Note that by giving c_{02} a non-zero value, the landscape can change: Assume that x is the number of the found pixels (so $fitness/100$). If y is the length, then

$$\varphi = xc_{13} - (y - x)c_{02} = xc_{13} + xc_{02} - yc_{02} = xc' - yc_{02}$$

This means that the fitness landscape will be multiplied by a scalar (c'/c_{13}) and it will be slanted around the *angle* axis, but the high peak remains.

For **multi-level fitness**, the parameters were $W_f = 100$ and $W_m = 0$. The fitness was calculated with the same parameters. The result can be seen in Figure 13.

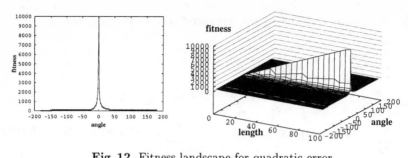

Fig. 12. Fitness landscape for quadratic error

Just like for the binary fitness, the W_m value can also be set to a non-zero value. Doing so, the fitness landscape can be slanted, thus fitness values for longer lines can be higher. The breakpoint at $a = -90$ and $a = 90$ is caused by the edge of the picture. For $a < -90$ and $a > 90$, the line is out of the image, so its distance is calculated by a formula. By changing this calculation, the local peak at $a = 180, b = 100$ can be eliminated.

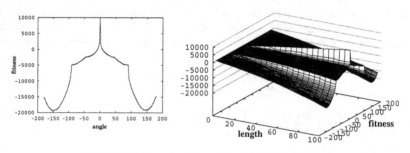

Fig. 13. Fitness landscape for Multi-Level fitness

To estimate the behaviour of the evoltion process with multi-level fitness is difficult from these results, because L-systems use much more parameters and with brackets they describe more complex objects. The experiments carried out so far shows that the evolution process finds better individuals using the multi-level fitness, though some experiments still needed to find the correct evoluiton parameters. In Figure 14 the comparison of the two fitness functions can be seen. The binary fitness shows the fitness values of the best individual using binary fitness function. The multi-level fitness shows the fitness values of the best individual using multi-level fitness. Note, that these values are divided by 10. These two values cannot be compared, because the binary and multi-level fitness values of the same individual can be completely different. To make the comparison the binary fitness values of the best individuals from the second run are also calculated. This is displayed as multi-binary fitness.

A test was done to examine the fitness calculation times for different fitness functions as well. The results can be seen in Figure 15. According to the results, there is no significant difference between running times. That's because the number of comparisons is the same, only some calculations changed. Thus

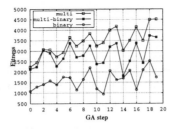

Fig. 14. Fitness values with different fitness functions

without additional computational cost many kind of distance functions can be used, for examle quadratic or logarithmic distances.

For the multi-level and the general fitness, preprocessing has to be done before the evolution process. The required time can be seen in Figure 16. The worst case is when the image is black, the best is when it is white. The general case was tested using a random image with about 10% coverage. Note that this process has to be done only once.

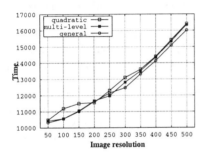

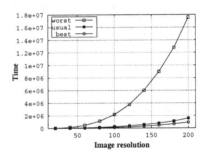

Fig. 15. Fitness calculation times

Fig. 16. Preprocessing times

7 Conclusion, future plans

In this paper the improvements of the GREDEA system is presented. The idea of the system that it can be applied in diabetic retinopathy for diabetic retinal disease. It solves the problem of describing the blood vessels of the eye using parametric *Lindenmayer systems* (*L-systems*). To find the most suitable description, evolutionary algorithms are applied. The development of the whole system is a long time period project. In this paper we restricted only on several enhancements and new solution:

- Image preprocessing is fully automatized
- Genetic operators are redefined:
  1. Being tree-operations, they are much easier to control,
  2. By constructing their string counterparts, it is still possible to use them on L-systems

3. When drawing the object defined by the L-system, the operations still carry out the desired operation
- New fitness function is used to increase convergence speed
- A general fitness function was specified:
  1. Other fitness functions are easy to add by defining a cost-matrix
  2. Fitness Calculating Turtle (FCT) is still usable, thus changing fitness do not cause significant speed change

It is among our future plans to develop a method to compare the two different L-systems generated by *GREDEA*. Our idea is that from both L-systems graphs can be generated and the transformation between these graphs can be found. After that, with the help of a classification process, a decision can be made whether the patient needs a therapy or not. The detection of other illnesses using the structure of the L-systems is also planned.

References

1. H. Abelson and A. A. diSessa. *Turtle Geometry*. MIT Press, Cambridge, 1982.
2. G. Borgefors. Distance transformations in arbitrary dimensions. *Computer Vision, Graphics, and Image Processing*, 27(3):321–345, Sept. 1984.
3. D. Goldberg. *Genetic Algorithms in Search, Optimization, and Machine Learning*. Addison-Wesley Reading, MA, 1989.
4. C. Jacob. Mathevolvica simulierte evolution von entwicklungsprogrammen der natur. Ph.d thesis, Arbeitsberichte des Instituts für Mathematische Maschinen und Datenverarbeitung (Dissertation), Vol. 28, No. 10, Universität Erlangen-Nürnberg,, 1995.
5. G. Kókai, Z. Tóth, and R. Ványi. Modelling blood vessel of the eye with parametric L-systems using evolutionary algorithms. In *Artificial Intelligence in Medicine, Proceedings of the Joint European Conference on Artificial Intelligence in Medicine and Medical Decision Making, AIMDM'99*, volume 1620 of *Lecture Notes of Computer Science*, pages 433–443. Springer-Verlag, 1999.
6. J. R. Koza. *Genetic Programming: On the Programming of Computers by Means of Natural Selection*. MIT Press, Cambridge, Massachusetts, 1992.
7. A. Lindenmayer. Mathematical models for cellular interaction in development. *Journal of Theoretical Biology*, 18:280–315, 1968.
8. K. J. Mock. Wildwood: The evolution of l-system plants for virtual environments. In *Proceedings ICEC 98*, pages 476–480. IEEE-Press, 1998.
9. G. Ochoa. On genetic algorithms and lindenmayer systems. In *Proceedings of the Parallel Problem Solving from Nature, International Conference on Evolutionary Computation, PPSN V*, volume 1498 of *Lecture Notes of Computer Science*, pages 335–344. Springer-Verlag, Berlin, 1998.
10. P. Prusinkiewicz and M. Hammel. Language-restricted iterated function systems, Koch constructions and L-systems. In *New Directions for Fractal Modeling in Computer Graphics*. ACM Press, 1994. SIGGRAPH'94 Course Notes.
11. P. Prusinkiewicz and A. Lindenmayer. *The Algorithmic Beauty of Plants*. Springer-Verlag, New York, 1990.
12. I. Rechenberg. *Evolutionsstrategien: Optimierung Technischer Systeme nach Prinzipen der Biologischen Evolution*. Fromman-Holzboog, Stuttgart, 1973.
13. G. Rozenberg and A. Salomaa. *The book of L*. Springer-Verlag, 1985.

Intraspecific Evolution of Learning by Genetic Programming

Yoshida Akira(akira-yo@is.aist-nara.ac.jp)

Graduate School of Information Science, Nara Institute of Science and Technology
8916-5 Takayama, Ikoma, Nara, 630-0101 Japan

Abstract. Spatial dynamic pattern formations or trails can be observed in a simple square world where individuals move to look for scattered foods. They seem to show the emergence of co-operation, job separation, or division of territories when genetic programming controls the reproduction, mutation, crossing over of the organisms. We try to explain the co-operative behaviors among multiple organisms by means of density of organisms and their environment. Next, we add some interactions between organisms, and between organism and their environment to see that the more interaction make the convergence of intraspecific learning faster. At last, we study that MDL-based fitness evaluation is effective for improvement of generalization of genetic programming.

1 Introduction

Genetic Programming [1,2] is an interesting extension of Genetic Algorithms. In GP, a population of programs, that are represented by trees in a task specific language, evolve to accomplish the desired task. One of the advantages of genetic programming over genetic algorithms is noted as its dynamic tree representation. The interpretation of a particular expression tree is performed without consideration to the position of the allele in genetic programming. On the contrary, in genetic algorithm, the typical interpretation function uses a positional encoding where the semantics of the interpretation are tied to the position of the bit in the genotype.

Koza's artificial ant problems were (1) to find a computer program to control an artificial ant so that it can find all foods located on the Santa Fe trail or San Mateo Trail, and (2) multiple ants collect scattered or concentrated piles of foods and put them in their nest site or some other place. His conclusion was that we can see the emergence of beneficial and interesting higher-level collective behavior similar to that observed in nature, although he did not decide complex overall rules in his programs.

But our objective is to see how multiple-ants can generate the emergence of co-operation, job separation, or division of territories by means of genetic programming, and also referencing results by Biophysics and Population Ecology[3]. The reason why the behavior of multiple organisms should attract our attention is that a special feature of all social insects is their living in a group or forming colonies. Spatial dynamic pattern formations or trails by organisms attract us, which remind us chaos and fractal.

Co-operation, job separation, and territory are the results of a particular observations for organisms by human beings. We would like to distinguish co-operation from competition if a computing or evaluating our fitness function could give a clue, since we can not understand the motive of organisms for their behaviors. The selection of fitness function depends on the particular way of a person. Another person may select another fitness function and have different interpretation even when he looks at the same behavior of organisms. Darwinian fitness explains the fundamental theorem of natural selection but, in Japan, Imanishi K had proposed "Habitat segregation", and Kimura M argued "The neutral theory of molecular evolution"[5].

We had been studying adaptive organisms in evolving populations, and evolution of learning using fully connected neural network and genetic algorithms. We introduced the concept of "Mutual Associative Memory" to use as a memory for virtual organism[8]. Next, we started a new step to use genetic programming instead of genetic algorithms to study interactions, learning and evolution of organisms[9].

Here, first, we explain basic methods of simulation using genetic programming to observe evolution and learning of collective behavior of organisms, and the process that much interaction makes intraspecific learning by GP faster. Next, a trial to improve the generalization ability for evolution of learning by genetic programming is explained.

2 Genetic Programming for Intraspecific Learning

We describe an approach using co-evolving organisms in cooperation to generate the emergence of job separation among the organisms. Genetic programming generates a program which specifies the behavior of an organism as the form of tree, and evolves the shape of the tree. We have three nonterminal symbols and three terminal symbols in order to make nodes and leaves of a tree.

The nonterminal symbols mean three kinds of functions :
a) If organism sees a food in front of him then perform first argument, else
   perform second argument[IfFoodAhead],
b) Perform two arguments in order[Prog2],
c) Perform three arguments in order[Prog3].

The terminal symbols mean three kinds of action :
a) Turn right at an angle of 90 degrees,
b) Turn left at an angle of 90 degrees,
c) Move forward a grid per unit time.

Genetic programming uses three operators that change the structure of a tree : mutation, inversion, and crossover. These are natural extension from GA operators which deal with only the bit strings. The paradigm of genetic programming is also nearly the same as the reproduction system of GA.

The overview of a simulation of genetic programming is as follows.
(1) Generate an initial population of random compositions of the functions and terminals of the problem(computer programs).

(2) Execute each program in the population and assign it a fitness value according to how well it solves the problem.

(3) Create a new population of computer programs by applying the following three primary operation :

a) Copy existing computer programs to the new population,

b) Create new computer programs by genetically recombining randomly chosen parts of two existing programs,

c) Create a new computer programs from an existing program by randomly mutating a node in the program.

(4) If terminate conditions are satisfied then END, else increase the number of generations and go to (2).

(5) The best computer program(the best individual) that appeared in any generation may be a solution to the problem.

We extended "sgpc1.1" and the Ant system written by Iba H[6] to simulate behavior of multiple organisms in various environments. Table 1 shows the values of GP parameters. Here we use homogeneous breeding strategy : All organisms in a team use the same program or algorithm evolved by GP. But, since each organism is always situated in a different environment, for example, his place, direction, or relative location to a same food are different, his movement is always vary from organism to organism even though they are controlled by a same tree.

Table 1. GP Parameters for sgpc1.1.

```
+-------------------------------+-----------------------------------+
| Number of populations:  1     | grow_method = RAMPED              |
| Number of generations:  500   | selection_method = FITNESSPROP    |
| population_size = 100/250/500  | tournament_K = 6                 |
| steady_state = 0              | deme_search_radius_sigma = 1.000  |
| max_depth_for_new_trees = 6   | crossover_func_pt_fraction = 0.2  |
| max_depth_after_crossover = 17| crossover_any_pt_fraction = 0.20  |
| max_mutant_depth = 4          | fitness_prop_repro_fraction =0.1  |
| parsimony-factor = 0.0/0.2/0.5/variable according to fitness      |
|                    and complexity                                 |
+-------------------------------+-----------------------------------+
```

3 Feeding Simulation with No Interaction

Our experimental domain consists of a square divided into 32 x 32 grids. On the first experiment the nest site of organisms is at the upper left and foods are scattered in the domain on a trail. Organisms can search for their foods as long as their energy last. An organism can change his direction to east, west, north, and south, but can not change the direction diagonally. He moves forward a grid a unit time, and can look for a food in front of him with his sensor. The organism spends a unit of energy whenever he takes an action: move forward, right, or left. Organisms must search foods efficiently because their energy is limited and they are killed when they spend all energy. Since fitness is evaluated as the number of foods organisms could not find, the smaller value of fitness means the better feeding by organisms. Organisms are like any social insect that live in a group, for example, ants, bees, or a kind of cockroaches.

Each organism starts his action in turn after an unit movement of previous organism has completed. The start point of all organisms is the nest site, and the initial moving direction of each organism are all same or different from previous organism at an angle of 90 degrees according to the objectives of simulations. This domain is not a torus and the east end is not connected to the west end, nor the north end connects to south end. The organism which has reached to the east end must turn to other directions. Figure 1(right) shows the trails made by a simulation with four organisms at 22nd generations with the value of fitness 36. There are trajectories of four organisms, where the newer trajectory(bigger number) overwrote the older trajectories partially. Here we used the "Santa Fe Trail" for the initial foods pattern.

3.1 Improvement and saturation of efficiency by co-operation

Figure 2 shows the fitness value vs. generation. The values of fitness are the lower the better. We can see if the number of organisms were more than or equal to 8 the fitness value converged to less than 20 after nearly 200th generation. On the contrary, the fitness values of one-organism and 4-organisms never converge and stay near the initial fitness values even at the 200-th generation. Although one or four organisms could not evolve to raise fitness in this environment eight or more organisms could evolve by co-operative behavior. But the important point here is that more than 16 organisms could not raise the efficiency of their work as a whole no matter how they devised plans. If there were too many organisms some extra organisms had no choice but to be out of work and ramble without finding foods. This is the reason why extra organisms must invent a new job and may be the origin of the job separation.

Figure 3 shows the return plots of the value of fitness f(g) and f(g+1) from one-organism(upper left) and 32-organisms(lower right), respectively. Return plot shows the relation between the value of fitness of g-th generation and (g+1)-th generation in the time series of fitness. We can see positive correlations and the regression coefficient is nearly "1.0" and the constant term is nearly zero on all charts of return plot, regardless of the increase of the number of organisms. But the mean value and the variance becomes smaller as the number of organisms increases.

3.2 Emergence of the job separation

We could see two kinds of emergence of job separation or cooperative behavior. The first one is division of worker organism and guard organism. The second is emergence of territories between worker organisms. In the following experiments we use the foods scattered in a nearly round shape. Figure 1 (left) shows the initial foods pattern and the place of nest site for simulations of emergence of

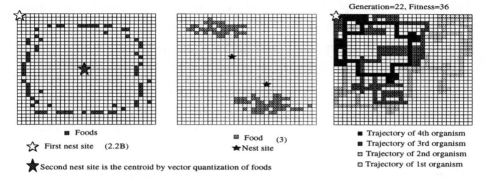

Fig. 1. left : Two foods pattern and nest site(left,middle); Trajectories of four organism(right) :

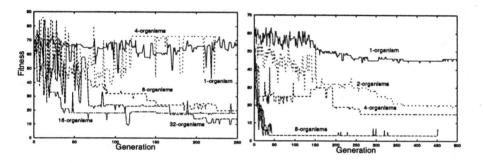

Fig. 2. Fitness value vs. Generation; if the number of organisms were more than or equal to 8 the fitness value converged to less than 20 after nearly 200th generation(Santa Fe trail;left); for round food pattern and center nest(fig 1,left)

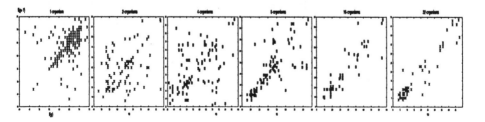

Fig. 3. Return plot of the value of fitness f(g) and f(g+1) ; 1-organism to 32-organisms.

the job separation. We tried two cases : that the nest site is set on left upper corner in the domain and at the center of foods.

Division of worker organism and guard organism

Figure 4 shows the emergence of division between worker organisms and guard organisms. The nest site is set on left upper corner(Figure 1, left). The trajectories from first to 13th organism are deleted here to make it easy to understand. The 16th organism traveled the world diagonally to find foods until generation=24, but he began to ramble around the nest site after generation=85. On the contrary, the 14th organism rambled around the nest site until generation=24, but evolved to travel around the world diagonally after generation=85. We could see the 16th organism and 14th organism exchanged their jobs on about generation=50. This experiment is an example of emergence of division of work and exchange of the work.

Here we have no concept of an enemy and a friend. But, here we would be able to regard and call the organisms who do not look for foods and ramble around the nest as guards. A professional guard usually does nothing except rambling and inspecting his territory, and he rarely meets a thief or an enemy. And so it is no problem to call "Guard organism" although we have no enemy in our simulations.

Emergence of territory between worker organisms

Figure 5 gives the emergence of territories between worker organisms. The nest site is at the center of the foods (Figure 1, left). Only four trajectories(from 29th to 32nd) are illustrated here and the others are deleted. We can see the 29th organism to 32nd organisms traveled the world directly to find foods in the shape of propeller on generation=2. They evolved to divide the world into four parts in the shape of four squares exclusively for searching foods on generation=21. Here we can see a spontaneous occurrence of territories between organisms.

We can see convergences of fitness on any number of organisms in this simulation with round food pattern and center nest site(figure 1,left). The speed of convergence becomes faster and the limit value becomes lower as the number of organisms increases. The graphs of return plot(figure 6) have stronger positive correlations than figure 3 whose food pattern is "Santa Fe.trail" and the net site is upper left. It is clear that the speed of convergence and the limit value of fitness are dependent on the shape of foods and the place of the nest site.

4   The more interaction the faster convergence

Here, in this simple world, the environment consisted of a square divided into 32 × 32 grids that is surrounded by walls on four sides, and foods that are scattered in various shape. The only interaction between an organism and the environment is made by the organism's sensor, which can only sense food directly in front of it and change next behavior according to the result of sensing. The computer program processes this interaction with the conditional branching operator IfFoodAhead in the function set, as explained in the above section. If

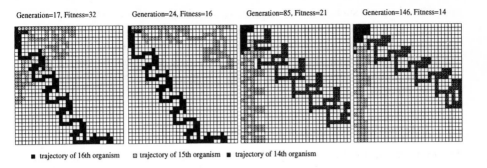

Fig. 4. Trajectory of 16-organisms (nest site is upper left), Emergence of Worker and Guard

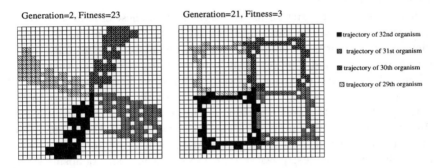

Fig. 5. Trajectory of 32-organisms (nest site is center), Emergence of Territories

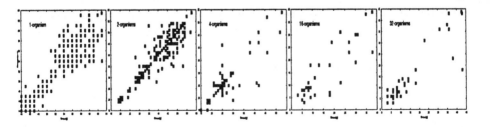

Fig. 6. Return plots for round food pattern and center nest site with various number of organisms

organisms had no such sensor or the programs had no IfFoodAhead functions, the organisms could not have a method to learn to find foods in their environment at all. Then an organism could not be called a living thing no longer. We are not concerned with the vertical and horizontal positions of the organisms or the directions the organisms are facing, but primarily concerned with finding foods by all organisms.

4.1 Interaction between organisms

The first interaction between organisms is a restriction that two or more organisms cannot enter in a grid at the same time. If an organism has already been in the grid to which another organism are going, the latter must turn his direction at an angle of +90 degrees. If there have been also a different organism he turns his direction at an angle of +90 degrees again. If three directions, forward, right and left were all surrounded by three organisms he could not move any directions until any empty grid appear around him except backward, since the organism can not move backward.

This is a natural restriction for the organisms in these simulations. But we had no such restriction in the above "no interaction" simulations in previous section, where two or more organisms could enter into a grid but only the first organism could eat food if he sensed it in front of him and moved there.

Figure 7 gives a comparison between generations and fitness values with various number of organisms. On the center figure all organisms start to same direction(east) from the nest site, but on the left figure each organism starts to different direction from previous organism at an angle of 90 degrees. We can see that the convergencies of feeding simulations where each organism start to different direction are clearly faster than the one where all organisms start to same direction.

Next comparison can be seen in the center and right figure in Figure 7. The difference between center and right is that the right figure has the interaction between organisms : one organism in a grid. All organisms start to same direction(east) from the nest site on both center and right figures. Here it is also clear that the interaction between organisms : one organism in a grid, makes the convergencies of feeding simulation of multi-organisms faster except 1-organism.

4.2 Interaction between an Organism and Environment

The second is an interactions between organism and environment : if an organism finds and eats a food the energy of the organism increases by one. And so the organisms that found and ate much foods could live longer time. This interaction between an organism and a food seems to be very natural law in nature.

Figure 8 gives an comparison between generation and fitness values with various number of organisms. The difference between left and center is all organisms start to same direction(east) from the nest site on left figure, but the organisms start to different directions on right figure. All other conditions are same; if an

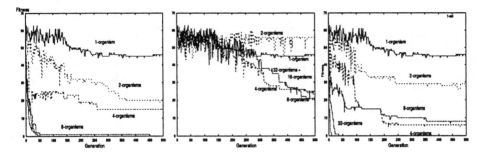

Fig. 7. Fitness vs generation for round food pattern and center nest site; no interaction and four initial directions(left), no interaction and same initial directions(center), with interaction and same initial directions(right)

organism eat a food the energy of the organism increases by one, and they have the restriction that one organism in a grid. Here we can see the convergencies of various number of organisms are fastest in our simulations this time, since there are most interactions between organisms and between an organism and food on the right figure. The curves of convergence became so quick and smooth as the number of organisms increased that they seem to be exponential curves turned at an angle of -90 degrees.

Figure 9 shows the return plots of the value of fitness(g) and fitness(g+1) from one-organism(left) and 16-organisms(right), respectively. This corresponds to figure 8(right). We can see positive correlations and the regression coefficient is nearly "1.0" and the constant term is nearly zero on all charts of return plot, regardless of the increase of the number of organisms. But the mean value and the variance becomes smaller as the number of organisms increases. The speed of becoming smaller values for the mean value and the variance is increased compared with the results of no interaction: figure 3 and 6.

5   Bayesian Methods for Evolution of Learning

Bayesian probability theory provides a unifying frame work for evaluation of learning systems. Exhaustion of exploration ability caused by an excess of the size of tree by genetic programming could be seen on some cases. Minimum Description Length principle(MDL) can not only decrease the size of the resulting evolved algorithm or tree, but also increases their generality and the effectiveness of the evolution process. We give a method to apply Bayesian Occam's factor and MDL to the evolution of learning by genetic programming, and give an application with multiple-organisms learning by genetic programming.

The principle of Occam's Razor says that the simpler model should be preferred to complex models if all other things are nearly equal. And the MDL

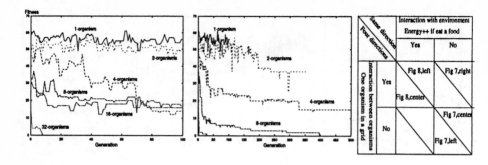

Fig. 8. Fitness vs generation for round food pattern and center nest site, energy++ if an organism eats a food, with interaction; Same initial direction(left), Four initial directions(center), Combinations of interactions(right)

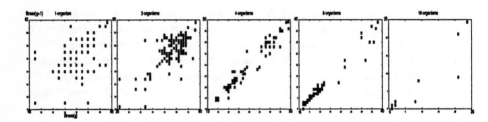

Fig. 9. Return plots for round food pattern and center nest site, energy++ if an organism eat a food

principle states that one should prefer models which can communicate the data in the smallest number of bits[10]. This means the size of the resulting evolved algorithm or tree by genetic programming should be smaller in order to increases their generality and the effectiveness of the evolution process.

First, we explain methods of generalization technique for adaptive fitness evaluation to deal with the parsimony problem using MDL. Parsimonious method could obtained the balance between the complexity of the model and the error range without losing the population diversity to achieve the efficiency of evolution. Next, an example that exhaustion of exploration ability caused by an excess of the size of tree is rescued by MDL-based fitness evaluation is explained.

5.1 Probability Theory and Occam's Razor

Two level of inference could be distinguished in the task of data modeling. First inference is to fit that model to the data assuming that a particular model is true. A model includes free parameters and the fitting the model to the data involves inferring what values those parameters should probably take, given the data. Second inference is the task of model comparison. This is comparing the models in light of the data, and assign some sort of preference or ranking to the alternatives.

Bayes' rule states the posterior probability as:

$$P(B_i|A) = \frac{P(A|B_i)P(B_i)}{P(A)} \tag{1}$$

where $P(B_i)$ is the prior probability and

$$P(A) = \sum_k P(A|B_k)P(B_k). \tag{2}$$

We study the probability of the data given a model i to explain how the Bayesian Occam's razor works for model comparison: which model is most plausible given the data. The probability of the data given a model i is the normalizing constant for eq.(1). For many problems, it is common for the posterior $P(B_i|A) \propto P(A|B_i)P(B_i)$ to have a strong peak at the most probable parameters. Then the probability of the data given a model i can be approximated by the height of the peak of the $P(A|B_i)P(B_i)$ times its width $\delta_{B_i|A}$, in one dimensional case:

$$P(A) \simeq P(A|B_i)' P(B_i)' \times \delta_{B_i|A}. \tag{3}$$

where, $P(A|B_i)'$ and $P(B_i)'$ mean the probabilities at the most probable parameters. The eq.(3) means that the probability of the data given a model i is found by taking the best-fit likelihood that the model can achieve and multiplying it by an Occam factor:

$$P(A) \simeq Best\ fit\ likelihood \times Occam\ factor. \tag{4}$$

The quantity $\delta_{B_i|A}$ is the posterior uncertainty in the parameter i. The Occam factor is equal to the ratio of the posterior accessible volume of model i's parameter space to the prior accessible volume, or the factor by which the model i's hypothesis space collapse when the data arrive. The logarithm of the Occam factor can be said as the amount of information gained about the model when the data arrive. The Occam factor is a measure of complexity of the model and relates to the complexity of the predictions that the model makes in data space.

5.2 MDL-Based Fitness evaluation

A complementary view of Bayesian model comparison is obtained by replacing probabilities of events by the length in bits of message length that communicate the event without loss to a receiver. Message length $L(x)$ corresponding to a probabilistic model over events x via the relations:

$$P(x) = 2^{-L(x)}, \qquad L(x) = -log_2 P(x). \tag{5}$$

This gives the code length $L(x)$ if the probability $P(x)$ is given. In order to increase the probability $P(B_i|A)$ we must decrease the code length:

$$L(B_i|A) = -log\frac{P(A|B_i)P(B_i)}{P(A)} = L(A|B_i) + L(B_i) - L(A). \qquad (6)$$

here $L(A|B_i)$ is the code length of the data when encoded using the model B_i as a predictor for the data A, and $L(B_i)$ is the length of the model itself. $L(A)$ is the length of the data A, which is called the normalizing constant and commonly ignored since it is irrelevant to the choice of parameters for B_i.

Next chart explains an intuitive model comparison by minimum description length. Note that $L(B_i) \leq L(B_{i+1})$ and $L(A|B_i) \geq L(A|B_{i+1})$.

```
Model1: <--L(B1)--><--------L(A|B1)--------------->
Model2: <----L(B2)----><--------L(A|B2)--------->
Model3: <-------L(B3)------><--------L(A|B3)------->
```

Model with a small number of parameters have only a short parameter block $L(B_i)$ but do not fit the data well, and so the data message $L(A|B_i)$ is long. As the number of parameters increases, the parameter block lengthens, and the data message becomes shorter. Model 2 is an optimum model complexity for which the sum $L(A|B_i) + L(B_i)$ is minimized on the above chart.

However, obtaining the exact formula for the fitness function is difficult since the true underlying probability distribution is unknown. Zhang BT proposed a measuring method for fitness of a tree B_i, given a training set A[11]:

$$F(B_i|A) = F_A + F_{B_i} = \beta E(A|B_i) + \alpha C(B_i). \qquad (7)$$

where $E(A|B_i)$ is fitting error and $C(B_i)$ is complexity of the model B_i. The problem is how to balance parameters α and β to be optimum for trade-off between the fitting error and the complexity.

Our basic approach used here is to change the complexity factor α adaptively with respect to the fitting error, and to fix the error factor $\beta = 1$. The training set A is usually constant and can be regarded fixed during evolutionary process. And so the first term $\beta E(A|B_i)$ in the right side of the eq.(5) is replaced with $E_j(g)$, which means the error of the best individual j up to the g-th generation. The complexity term consists of the number of nodes and leaves in a tree evolved by genetic programming. Then the fitness is calculated as follows:

$$F_j(g) = E_j(g) + \alpha(g)C_j(g), \quad \alpha(g) = k\frac{C_{best}(g-1)}{E_{best}(g-1)}. \qquad (8)$$

where, $C_{best}(g)$ is the complexity of the best individual at g-th generation, $E_{best}(g)$ is the error of the best individual at g-th generation, and k is the positive constant.

Since $C_{best}(g)$ is generally small in the early stage of evolution it does not prevent the size of the best tree from growing. This gives a counter effect to Occam's Razor to ensure growth of tree if necessary. But if the size of tree grow too much it encourages fast reduction of the size of tree to obtain parsimonious solutions, because the Occam's factor $\alpha(g)$ increases in proportion to the size of tree.

On the other hand, $E_{best}(g)$ is generally large in the early stage of evolution. And so this also does not prevent the size of the best tree from growing in cooperation with smaller value of $C_{best}(g)$. But if the value of error become small it encourages fast reduction of the size of tree to obtain parsimonious solutions, because the Occam's factor $\alpha(g)$ decreases in reverse proportion to the value of error.

Although the above method is very simple compared with Zhang's method, our trial had a quite good results as followed in the next chapter.

5.3 Experiments with intraspcific learning simulation by GP

Here we give an example that MDL-based fitness evaluation rescues from over-fitting on multi-organisms learning by GP.

Constant Occam factor

Figure 10 shows the fitness values vs generation both on Occam factor=0.0 (left) and =0.5(right). This is the case that Occam factor in eq.(6) is an constant. Here we used two nest sites and two clusters of foods in the experimental domain(fig 1,middle). As the number of organisms increases, the convergence value does not get smaller on figure 10(left). This result seems clearly erroneous since fitness should be proportional to number of organisms as we have confirmed on many experiments[4]. On the contrary, we can see figure 10(right) with the value of Occam factor=0.5 shows the seemingly normal result; fitness generally decreases as the number of organisms increases. The fitness values of 8 or 32 organisms are converged to lower values, which seems to be a right result, however the fitness values of 2 or 4 organisms show chaotic vibration.

Table 2 shows program(tree) sizes and fitness values on 100-th generation with three kinds of numbers of organisms and values of Occam factor. We can see, on the case of 8-organisms, the program size was decreased from 480 steps (Occam factor=0.0 ; abnormally big tree) to 19 steps(Occam factor=0.5 ; small tree) on 100-th generation. This small program is shown under Table 2. On this experiment all the fitness values had chaotic vibration when the value of Occam factor is greater than 1.0. We can see that the program size decreases abruptly as the value of parsimony-factor increases.

Variable Occam factor

This corresponds to the case that Occam factor in eq.(6) changes adaptively with the fitting error and the complexity of the tree . Figure 11(right) gives the result for $k = 0.3$ in eq.(8), which shows the better generalization ability than constant Occam factor on figure 10(right). Figure 11(left) shows the result for $k = 0.5$ in eq.(8), which shows chaotic vibrations. These are examples that exhaustion of exploration ability caused by an excess of the size of tree is corrected by minimum description length principle.

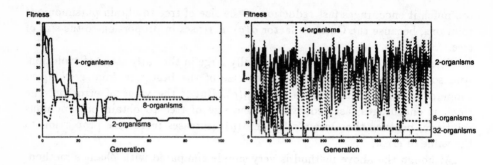

Fig. 10. Result by MDL-based fitness evaluation : Occam factor=0.0 (left),=0.5(right)

Table 1. Program(tree) size and Fitness values at 100-th generation

		Number of organisms		
		2-organisms	4-organisms	8-organisms
Occam factor=0.0	Program size	446	465	480
	Fitness	4	17	16
Occam factor=0.5	Program size	35	48	19
	Fitness	46	15	6
Occam factor=1.0	Program size	6	5	31
	Fitness	62	67	26

(PROG2 (PROG2 FORWARD (PROG2 (IfFoodAhead FORWARD RIGHT) (PROG2 FORWARD (IfFoodAhead FORWARD (PROG# RIGHT (PROG# FORWARD FORWARD) LEF T))))) FORWARD)

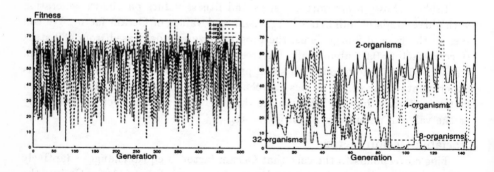

Fig. 11. Result by MDL-based fitness evaluation : variable Occam factor k=0.6(left),=0.3(right)

6 Conclusion

(1) Efficiency of finding foods gets better as the the number of organisms increases until a limit value.

(2) Even in an environment where only one or few organisms could not evolve to raise the fitness over long generations, many organisms could evolve. This is the emergence of co-operation.

(3) When the number of organisms reaches the upper limit extra organisms start to ramble near the nest site. They became the guard instead of searching foods. This is an example of the origin of emergence of the job separation among multiple organisms.

(4) The more interaction between organisms and between an organism and the environment make the convergence of multiple-organisms learning faster.

(5) Some cases where exhaustion of exploration ability caused by an excess of the size of tree by genetic programming is corrected by MDL-based fitness evaluation.

Appendix: Model for Spatial Segregation

Here we study a one dimensional model by Shigesada N briefly. Our intraspcific simulations by GP corresponds to some results of asymptotic behavior and stability of semilinear diffusion equations with constant flux, since the population in GP simulations is constant.

$n(x,t)$ denotes the population density at time t and position x. The flux of density of organism $J(x,t)$ is given by:

$$J(x,t) = -\frac{\partial}{\partial x}((\alpha + \beta n)n) - \frac{dU}{dx}n. \tag{9}$$

The first term represents the sum of flow associated with random movements of individual and the dispersal that is proportional to the density n. If $\beta = 0$ this means the usual diffusion process. As the density of organisms increases the interaction between them is forced to get bigger, and each organism has a tendency to disperse. This is well known in the population ecology and called "population pressure effect". The second term means the flow due to directed movements of individuals toward favorable environment. Shigesada calls $U(x)$ as the "environmental potential".

The equation of continuity(conservation law) can be used when the population is constant:

$$\frac{\partial}{\partial t}n(x,t) = -\frac{\partial}{\partial x}J(x,t). \tag{10}$$

In the case of steady distribution of organisms the flux is zero. And there is no population flow through the boundaries, so this model is subject to the zero flow and boundary conditions:

$$\frac{\partial}{\partial x}J(x,t) = 0, \quad J(0,t) = J(L,t) = 0, \quad x \in [0,L]. \tag{11}$$

The solution of the equation $J(x, t) = 0$ gives a stable distribution $n_s(x)$:

$$2\beta n_s(x) + \alpha log n_s(x) + U(x) = C. \tag{12}$$

The stable distribution $n_s(x)$ has following properties :

a. If there is no population pressure effect: $\beta = 0$, the distribution is proportional to $e^{-U(x)}$ and this is unrelated to population size N.

b. When population pressure effect works: $\beta \neq 0$, the distribution gets flatter as N and β increase.

The moves of organisms in our simulations is equivalent to the diffusion term in eq.(9), the interactions between organisms match to the distribution term in eq.(9), and the environmental potential $U(x)$ may be foods in our simulations, although the foods should be also the function of time t.

Our result: the more interactions the faster convergence, could be explained with the increase of the value of β in the eq.(9). As "N" and "β" get larger the distribution of organisms becomes flatter. If individuals spread out rapidly in all directions they could find foods easily.

The above results show that two or three dimensional domain has some stable equilibrium solutions. Emergence of territory may be one of some stable solutions.

References

1. Koza JR, Genetic Programming, On the Programming of Computers by means of Natural Selection,MIT Press,1992
2. Koza JR, Genetic Programming II, Automatic Discovery of Reusable Subprograms,MIT Press,1994
3. Shigesada N, On the Spatial Distributions of Dispersing Animals, in "Lectures in Applied Mathematics and Informatics", Manchester University Press,1990
4. Kauffman SA, The Origins of Order, Oxford, 1993
5. Kimura M, The natural theory of molecular evolution, Cambridge University Press 1983
6. Iba H, Genetic Programming, Tokyo Electric University Press(Japanese),1996
7. Iba H, Emergent Cooperation for Multiple Agents using Genetic Programming, Parallel Problem Solving from Nature, Springer,1996
8. Yoshida A, A Model of Mutual Associative Memory for Simulations of Evolution and Learning,SEAL'98, LNAI 1585 p.446, Springer,1999
9. Yoshida A, Multiple-Organisms Learning and Evolution by Genetic Programming, AJWIES'99
10. Rissanen J, Stochastic complexity and modeling. The Annals of Statistics,14,1080,1986
11. Zhang BT, Evolutionary Neural Trees for Modeling and Predicting Complex Systems, Artificial Intelligence. Vol.10,No.5,p.473 Elsevier Science,1997

An Evolutionary Approach to Multiperiod Asset Allocation

Stefania Baglioni[1], Célia da Costa Pereira[3], Dario Sorbello[1], and Andrea G. B. Tettamanzi[2]

[1] Fideuram Capital S.p.A., Research Department
Via San Paolo 10, I-20121 Milano, Italy
E-mail: {sbaglioni,dsorbello}@ifam.it
[2] Università degli Studi di Milano
Dipartimento di Scienze dell'Informazione, Sezione di Crema
Via Bramante 65, I-26013 Crema (CR), Italy
E-mail: tettaman@dsi.unimi.it
[3] Université Paul Sabatier
118 Route de Narbonne, 31077 Toulouse, France
E-mail:costa@irit.fr

Abstract. Portfolio construction can become a very complicated problem, as regulatory constraints, individual investor's requirements, nontrivial indices of risk and subjective quality measures are taken into account, together with multiple investment horizons and cash-flow planning. This problem is approached using a tree of possible scenarios for the future, and an evolutionary algorithm is used to optimize an investment plan against the desired criteria and the possible scenarios. An application to a real defined benefit pension fund case is discussed.

1 Introduction

The recent rise of interest in mutual funds and private pension plans in Italy has brought about a powerful drive to provide investors with more and more sophisticated, flexible and customized products.

Typically, a process of investing consists of two main steps.

- First, benchmarks for some asset categories (asset classes) are selected. These constitute what is called the *opportunity set*.
- In the second step, a mix of asset classes in the opportunity set is strategically chosen according to a specified criterion. This step is the main task of asset allocation. Following this kind of process, asset allocation policy decisions may have a relevant impact on portfolio performance, as shown by the seminal work of Brinson and colleagues [2].

Institutional investors have to manage a complex cash-flows structure. In addition, they wish to have the highest return as possible while, at the same time, guaranteeing their invested wealth to meet liabilities over time. In other words, their financial needs can be seen as a trade-off between the maximization of return and the chance of achievement of minimal targets.

The traditional asset allocation framework, based on the mean-variance approach [6], is not adequate to model the problem described above. The best way to solve the problem involves a dynamic approach, which takes into account the randomness of the time paths of the asset returns, and allocation becomes essentially a multi-period optimization problem. A sequence of portfolios is sought which maximizes the probability of achieving in time a given set of financial goals. The portfolio sequences are to be evaluated against a probabilistic representation of the foreseen evolution of a number of asset classes, the opportunity set for investment.

Evolutionary algorithms [1, 3, 4] are used to optimize the investment plan against the desired criteria and the possible scenarios. Some of the ideas presented below stem from a former study on the use of evolutionary algorithms for single-period portfolio optimization [5]. Another source of inspiration was an application of stochastic programming to dynamic asset allocation for individual investors, jointly developed by Banca Fideuram S.p.A. and Frank Russell Company (more details can be found in several papers in [8]).

2 The Problem

Objective data (expected conditions of financial markets) consist of the possible realizations of the return of the asset classes which are described by means of a scenario tree.

The multi-period optimization problem for institutional investors involves maximizing a function measuring the difference between the overall wealth and the *cost* of a failure to achieve a given overall wealth, taking into account the cash-flow structure.

Wealth (return) and the cost of falling short of achieving a predefined level (down side risk approach) are evaluated for each subperiod.

Along with objective data and criteria, which are relatively easy to quantify and account for, a viable approach to multi-period asset allocation should consider a number of subjective criteria. These comprise desired features for an investment plan, as seen from the stand point of the institutional investor and guidelines dictated by regulations or policies of the asset manager. The former are in fact criteria used to personalize the investment plan according to preferences of the customer, even at the expense of its potential return. The latter criteria may consist of specific product requirements or an *ad hoc* time-dependent risk policy.

In order to match subjective advices both from the institutional investors and the asset manager, constraints are implemented by means of the penalty functions technique (cf. Section 2.4). In this way, the relative importance of the various constraints can be fine-tuned by the asset manager to ensure a higher degree of customization for solutions. This approach makes it possible to treat advices as "soft" constraints, allowing trade-offs with other criteria. As the weight associated to a soft constraint increases, the relevant advice becomes more and more binding.

Overall, the criteria (objectives and constraints) of the multi-period portfolio optimization problem considered in this article can be grouped in five types:

1. Maximize the final wealth (EW_{Fin});
2. Minimize the risk of not achieving the intermediate and final wealth objectives (MR);
3. Minimize turn-over from one investment period to the next one, in order to keep commissions low (TC);
4. Respect absolute constraints on the minimum and maximum weight for any asset classes or groups thereof (WAG);
5. Respect relative constraints on the weight of asset classes or groups thereof (e.g.: European bonds shall be at least twice the weight of non-European bonds), (WRG);

Optimization is carried out on a portfolio tree corresponding to the scenario tree of returns, considered as a whole. A portfolio tree can be regarded as a very detailed investment plan, providing different responses in terms of allocation, to different scenarios. These responses are hierarchically structured, forming a tree whose root represents the initial allocation of assets. Each path from the root to a leaf represents a possible evolution in time of the initial asset allocation, in response to, or in anticipation of, changing market conditions. The objective function depends on all the branches of the scenario tree (path dependence).

2.1 Notation

The following conventions will hold in the rest of the article:

N number of asset classes;
T number of periods (also called *levels*);
Fin   set of leaf nodes;
Int   set of internal nodes;
$Pr(i)$ absolute probability associated with node i
$W(i)$ wealth at node i, *without* considering cash-flows
$W^*(i)$ wealth at node i, considering cash-flows
$\ell(i)$ period (level) of node i
$K(t)$ goals for downside risk calculation
$q(t)$ risk aversion coefficients
$ag(t)$ set of asset groups at period t
$\|X\|$ cardinality of set X

A group of asset classes (or asset group in short) $g \in ag(t)$ for some t can be thought of as a set of indices $g = \{i_0, \ldots, i_{m-1}\}$ referring to the asset classes that belong in it.

2.2   Constraints

For each asset class and each asset group, the minimum and maximum weight is given for portfolios of all periods. In addition, the maximum variation of each asset class with respect to the previous period is given. These constraints are defined by six matrices:

$W_{it}^{\min}$ minimum weight for asset class i in period t;
$W_{it}^{\max}$ maximum weight for asset class i in period t;
D_{it}^- maximum negative variation for asset class i in period t;
D_{it}^+ maximum positive variation for asset class i in period t;
$\mathrm{WG}_{gt}^{\min}$ minimum total weight of asset group $g \in \mathrm{ag}(t)$ in period t;
$\mathrm{WG}_{gt}^{\max}$ maximum total weight of asset group $g \in \mathrm{ag}(t)$ in period t.

Beside absolute constraints like the above, relative constraints on asset groups can be defined in the form

$$\begin{aligned} &\mathrm{wrelg}(\mathbf{w}) > 0, \text{with} \\ &\mathrm{wrelg}(\mathbf{w}) = k + k_0 * \mathrm{wag}(0) + \ldots + k_{m-1} * \mathrm{wag}(m-1), \end{aligned} \tag{1}$$

with $k, k_0, \ldots, k_{m-1}$ arbitrary constants and

$$\mathrm{wag}(g) = \sum_{i \in g} w_i$$

being the total weight of asset group g in portfolio $\mathbf{w}$; here m is the number of asset groups at the period of portfolio $\mathbf{w}$. For each period, any number of such constraints can be imposed, meaning that all portfolios at that period must satisfy them.

2.3   Wealth

The primary objective of an investor is to maximize wealth. Accordingly, the objective function, whose details are given in Section 2.4, is defined in terms of the wealth of single portfolios in the portfolio tree.

By wealth of a portfolio, we mean the quantity $W^*(\mathbf{s}, \mathbf{w})$, measuring the market value of portfolio $\mathbf{w}$ if sold in node $\mathbf{s}$ of the scenario tree. Wealth is calculated at each node while keeping into account the effect of cash-flows $F(t)$:

$$W_t^* = W_t + F(t). \tag{2}$$

2.4   Objective Function

The objective function z, to be minimized, is given by

$$z = -\mathrm{EW}_{\mathrm{Fin}} + \mathrm{MR} + \mathrm{TC} + \mathrm{WAG} + \mathrm{WRG}. \tag{3}$$

The first four terms in Equation 3 are the objectives. The first two terms ($\mathrm{EW}_{\mathrm{Fin}}$, the final wealth and MR, the mean cost of risk) are the fundamental objectives, i.e. optimization criteria, in that an investor wishes to maximize wealth while trying to minimize risk. The following term (TC, transaction cost) is an optional objective, depending on the valuations of the fund manager. All the other terms are penalties triggered by any violation of a number of constraints. Other constraints can be satisfied "by construction", as explained in Section 3.2, and therefore they need not be associated with a penalty term.

For each constraint, constant $\alpha_t^{\text{constraint}}$ is a non-negative weight allowing the fund manager to express how much the relevant penalty is to affect the search of an optimal portfolio.

The individual terms of the objective functions are discussed in the following paragraphs.

Expected Wealth The expected wealth at final nodes (the leaves of the scenario tree)

$$\text{EW}_{\text{Fin}} = \sum_{i \in \text{Fin}} \Pr(i) \cdot W^*(i).$$

Mean Risk This term is calculated using a penalty function of a portfolio *underperformance*, which is tightly linked with the concept of *downside risk*. The motivations for this choice are illustrated and discussed in a previous work [5].

The mean risk is the mean cost of downside risk,

$$\text{MR} = \sum_{i:W^*(i)<\theta_i} \alpha_{\ell(i)}^{\text{MR}} \Pr(i)[k(\theta_i - W^*(i))]^{q(\ell(i))},$$

where $\Pr(i)$ is the probability of scenario i, and $\theta_i = \max\{K(\ell(i)), W_\pi(i)\}$ is the threshold for wealth under which one can speak of an underperformance of the investment plan; furthermore,

- $\ell(i)$ is the period of node i,
- $K(t)$ is a parameter provided by the investor for each period t, indicating the (intermediate or final) accumulation goal, also called *strike level* at period t,
- $W_\pi(i)$ is the value at node i of the benchmark portfolio π,
- k is a scale factor for underperformance,
- $q(t) \geq 0$ characterizes the investor's aversion to risk in period t.

Transaction Cost This takes into account the effect of transaction cost, calculated as

$$\text{TC} = \sum_{i \in \text{Int}} \alpha_{\ell(i)}^{\text{TC}} \Pr(i) \sum_{0 \leq j < N} \Delta_{ij} \left([\Delta_{ij} > 0]c_j^{\text{buy}} - [\Delta_{ij} < 0]c_j^{\text{sell}}\right),$$

where $\Delta = w_{ij} - \hat{w}_{\text{parent}(i),j}$ and the c_j^{buy} and c_j^{sell} are parameters specified by the fund manager, providing the percent cost respectively for the purchase and for the sale of assets in class j. Here, $\hat{w}_{\text{parent}(i),j}$ stands for the weights of the parent portfolio $w_{\text{parent}(i),j}$, modified by the price variations of the asset classes since the last period. Of course, for $i = 0$, $\Delta_{ij} = 0$ for all j provided that the investor does not have an initial portfolio to re-optimize, in which case this is assumed to be the parent portfolio for the calculation of Δ_{ij}.

Absolute Asset Group Constraints The degree of violation of absolute constraints on asset groups

$$\text{WAG} = \sum_{0 \leq t < T-1} \frac{\alpha_t^{\text{WAG}}}{\|\text{nal}(t)\|} \sum_{g \in \text{ag}(t)} \frac{1}{\|\text{ag}(t)\|} \sum_{i \in \text{nal}(t)} d(g, i),$$

where $\text{nal}(t)$ is the set of nodes at level t, $d(g, i)$ is the distance of the weight of asset group g from the interval $[\text{WG}_{gt}^{\min}, \text{WG}_{gt}^{\max}]$ for portfolio $\mathbf{w}_i$.

Relative Asset Group Constraints The degree of violation of relative constraints on asset groups

$$\text{WRG} = \frac{1}{\|\text{Int}\|} \sum_{i \in \text{Int}} -\alpha_{\ell(i)}^{\text{WRG}} \sum_{j=1}^{J_{\ell(i)}} \min(0, \text{wrelg}_{\ell(i),j}(\mathbf{w}_i))$$

where all wrelg_{tj} are defined according to Equation 1, and J_t represents the number of relative constraints on asset groups at period t. All wrelg_{tj} represent constraints on asset groups at period t, the constraint being satisfied if its value is non-negative: that is why the objective function is penalized for each negative value of wrelg.

3 The Optimization Engine

The optimization engine of the Galapagos system employs evolutionary algorithms to solve a multi-period optimization problem.

3.1 The Algorithm

The optimization algorithm, which operates on an array *individual[popSize]* of individuals (i.e. the population) is a standard, generational replacement, elitist evolutionary algorithm.

The various elements of the algorithm are illustrated in the following subsections.

3.2 Encoding

A solution to the problem, that is a portfolio tree, is encoded as a string of $\|\text{Int}\|N$ bytes. Each portfolio in the tree is encoded by a substring of N bytes, starting from the root and visiting all the nodes of the tree breadth-first. While the encoding of the root portfolio is direct, the encoding of all the children portfolios is *differential*. This means that only the changes with respect to the parent portfolio are encoded. The differential encoding makes it easy to enforce the D_{it}^- and D_{it}^+ constraints, as explained below. When these constraints are absent, the whole encoding turns out to be absolute.

The decoding of the genotype is performed in such a way that the $W_{it}^{\min}$, $W_{it}^{\max}$, D_{it}^- and D_{it}^+ constraints are satisfied by construction for all asset classes i and for all periods t.

The decoding of the genotype of a portfolio tree starts by decoding the root portfolio, as follows.

Let $\mathbf{g} = (g_1, \ldots g_N)$ be the genotype substring for the root portfolio; let $\mathbf{w} = (w_1, \ldots w_N)$ and $\bar{\mathbf{w}} = (\bar{w}_1, \ldots \bar{w}_N)$ be respectively the non-normalized weight vector and the corresponding normalized vector for the same portfolio.

It is assumed

$$w_i = W_{i0}^{\min} + \left(W_{i0}^{\max} - W_{i0}^{\min}\right) \frac{g_i}{\sum_{k=1}^N g_k}.$$

Therefore, vector $\mathbf{w}$ satisfies the $W^{\min}$ and $W^{\max}$ constraints, but not necessarily $\sum_{i=1}^N w_i = 1$. The normalized vector $\bar{\mathbf{w}}$ is obtained from $\mathbf{w}$ as follows:

Let $\Delta = 1 - \sum_{i=1}^N w_i$, and

$$r_i = \begin{cases} W_i^{\max} - w_i & \text{if } \Delta \geq 0, \\ W_i^{\min} - w_i & \text{if } \Delta < 0. \end{cases}$$

The normalized weights are given by

$$\bar{w}_i = w_i + r_i \frac{\Delta}{\sum_{k=1}^N r_k}.$$

This procedure is illustrated in Figure 1 by means of a three asset class example.

The decoding then continues with the children portfolios, in an analogous way.

3.3 Initialization

The initial population is seeded with random genotypes. This means that every byte is assigned a value at random with uniform probability over $\{0, \ldots, 255\}$.

It is important to notice that, because of the complicated decoding procedure illustrated in Section 3.2, the corresponding portfolio trees will not be uniformly distributed over the space of all feasible portfolio trees.

3.4 Crossover

Uniform balanced crossover, as illustrated in [5], was adopted. Let γ and κ be substrings of length N in the two parent chromosomes corresponding to the same portfolio. For each gene i in the offspring, it is decided with equal probability whether it should be inherited from one parent or the other. Suppose that the ith gene in γ, γ_i, has been chosen to be passed on to a child genotype substring ζ. Then, the value of the ith gene in ζ is

$$\zeta_i = \min\left(255, \gamma_i \frac{\sum_{j=1}^N \gamma_j + \sum_{j=1}^N \kappa_j}{2 \sum_{j=1}^N \gamma_j}\right).$$

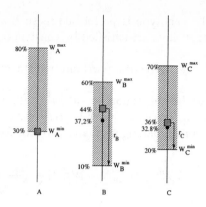

Fig. 1. Example of the derivation of a root portfolio for a problem with three asset classes, A, B and C and constraints $W_A^{\min} = 30\%$, $W_A^{\max} = 80\%$, $W_B^{\min} = 10\%$, $W_B^{\max} = 60\%$, $W_C^{\min} = 20\%$ and $W_C^{\max} = 70\%$. The feasible weight region for each asset class is indicated by the dashed texture. Assuming that the genotype be the integer vector $\mathbf{g} = (g_A, g_B, g_C) = (0, 15, 7)$, the non-normalized weights $w_A = 30\%$, $w_B = 44\%$ and $w_C = 36\%$, represented as grey squares, are calculated. Since their sum is 110%, $\Delta = -10\%$, while $r_A = 0$, $r_B = -34\%$ and $r_C = -16\%$. The normalized weights, represented as solid bullets, are therefore calculated as $\bar{w}_A = 30\%$, $\bar{w}_B = 37.2\%$ and $\bar{w}_C = 32.8\%$.

The main motivation behind this operation is to preserve the *relative* meaning of genes, which depends on their context. Indeed, genes have a meaning with respect to the other genes of the same portfolio, since the corresponding weight is obtained by normalization. Therefore, crossing the two substrings $(1, 0, 0)$ and $(0, 0, 10)$ to obtain $(1, 0, 10)$ would not correctly interpret the fact that the 1 in the first substring means "put everything on the first asset".

3.5 Mutation

Mutation perturbs genotypes by randomly changing each gene (consisting of one byte, corresponding to the weight of an asset class in one of the portfolios in the portfolio tree) with independent and identical probability p_{mut}.

There is a difference in how this random change is carried out between the first N bytes, encoding for the root portfolio, and the rest of the genotype, reflecting the fact that while the former are a direct encoding of a portfolio, the latter encode for the variation in weight from the parent portfolio.

Therefore, the first N bytes are increased or decreased by one with equal probability, while the remaining bytes are completely overwritten by a new random value from $\{0, \ldots, 255\}$. In other words, the first N bytes mutate very gradually, and the others can mutate much widely.

3.6 Fitness

The fitness of an individual (i.e. of a portfolio tree) is a positive real number f obtained from the objective function z via the transformation

$$f = \begin{cases} \frac{1}{z+1}, & \text{if } z > 0; \\ 1 - z, & \text{if } z \leq 0. \end{cases}$$

3.7 Selection

Elitist fitness proportionate selection, using the roulette-wheel algorithm, was implemented, using a simple fitness scaling whereby the scaled fitness $\hat{f}$ is

$$\hat{f} = f - f_{\text{worst}}, \tag{4}$$

where f_{worst} is the fitness of the worst individual in the current population.

Overall, the algorithm is elitist, in the sense that the best individual in the population is always passed on unchanged to the next generation, without undergoing crossover or mutation.

4 Experimental Results

4.1 A Case Study

We present a proposal-study made by Fideuram Capital to a defined benefit pension fund of the Italian subsidiary of a large multinational company at the end of December 1998.

The data set, provided by the pension fund, concerned its demographic structure and the wealth level at the moment of proposal.

Furthermore, the fund board of directors, as mandated by Italian regulations, gave some investment directives: maintain the equity exposure of the allocation plan preferably under 30%.

To begin with, we processed the demographic features of the fund members, in order to infer the pattern of cash-out flows. Furthermore, we carried out an actuarial analysis of the liabilities, to calculate their modified duration. The synopsis of cash flows and duration is given in Table 1.

Year	1	2	3	4	5	...
Cash-flow (in millions)	3593.979	3559.601	3522.412	3482.310	3439.196	...
Duration	10.02					

Table 1. Cash flows for the first five years and duration of liabilities for all years.

4.2 Problem Formulation

In this subsection we show how to approach the problem illustrated above using our model. The generation of the scenario tree was based on the data provided by the econometric model for forecasting, jointly developed by the Research Department of Fideuram Capital and Frank Russell Company. The holding period of the investment plan is five years.

It is common sense that matching the duration of asset portfolio with the liability duration minimizes future funding uncertainty [7]. This widespread methodology leads to portfolios having a return approximately equal to the yield of a bond with the same maturity as the duration. We took this return as the minimal return for an optimal allocation plan. Likewise, given the attitude toward risk of the institutional investor, we try to perform better than an alternative suitable risk-free investment. Because of the defined benefit nature of the pension fund, an optimal plan is supposed to outperform the "duration maturity" bond *each* year, once again to reduce uncertainty. In order to translate this methodology in the language of the Galapagos system, we calculated the strike levels (target returns) in terms of wealth, assuming that the initial wealth is one.

The preferences communicated by the fund board of directors were implemented in terms of "soft" constraints, by associating a penalty to plans with equity exposure above the suggested threshold. However, we still admit investment plans with a higher equity exposure, provided that violation is compensated for by a much higher performance.

For this particular application, transaction costs (TC) were not taken into account. Parameter q, modeling the investor's aversion to risk was set to 2, reflecting the fact that a typical pension fund has a conservative investment profile.

The values for the other parameters were dictated by sensitivity analysis studies previously carried out by the asset allocation bureau at Fideuram Capital. The parameter setting is summarized in Table 2.

4.3 Results

The evolutionary algorithm, with a population size of 100 and a scenario tree of 14808 nodes, was run several times for the problem presented in Section 4.1 on a Pentium 233 MHz PC with the Microsoft Windows NT operating system and 128 Mbytes of RAM. All runs converged to the same optimal investment plan, whose details are given in Tables 3, 4 and 5 for the first five years of the overall holding periods. Average running time before convergence was little more than 8 hours.

The solution abtained was judged completely satisfactory by experts, because it fulfilled all requirements with a high chance of achievement and, in addition, provided a fair level of expected annual returns, as an inspection of Table 3 can confirm.

Year t	0	1	2	3	4	5
cash flow $F(t)$	-1	0.07037	0.06969	0.06896	0.06817	0.06733
recursivity	1	1	1	1	1	1
strike level $K(t)$	n/a	0.9556	0.9108	0.8649	0.8192	0.7732
aversion $q(t)$	n/a	2	2	2	2	2
α_t^{MR}	n/a	4	4	4	4	4
$WG_{EQUITY,t}^{max}$	n/a	30%	30%	30%	30%	30%
α_t^{WAG}	n/a	10	10	10	10	10
tree branching	50	6	4	3	2	0
D_{it}^+ ($\forall i$)	n/a	40%	40%	40%	40%	40%
D_{it}^- ($\forall i$)	n/a	40%	40%	40%	40%	40%

Asset groups: {CASH, BOND, EQUITY}

Table 2. Summary of parameter values.

Date	expected wealth	stdev of wealth	cash flows	expected return	stdev of return	chance of achievement
01-Jan-99	51,075,957,450	n/a	n/a	n/a	n/a	n/a
01-Jan-00	51,539,522,840	3,593,979,000	6.23%	7.94%	6.23%	85.33%
01-Jan-01	52,162,291,989	3,559,601,000	10.85%	8.07%	9.76%	86.28%
01-Jan-02	52,232,981,114	3,522,412,000	16.71%	6.96%	12.31%	87.36%
01-Jan-03	52,712,022,519	3,482,310,000	23.02%	7.70%	14.38%	88.41%
01-Jan-04	54,052,204,567	3,439,196,000	30.46%	9.16%	16.94%	90.83%

Table 3. Expected Wealth and Wealth Objectives for the optimal investment plan. Wealth is expressed in December 1998 real ITL.

Date	01-Jan-99	01-Jan-00	-Jan-01	-Jan-02	-Jan-03
JPM_EMU_6M	20.97%	16.54%	15.08%	8.55%	9.91%
SSB_EMU_S	3.71%	10.64%	0.00%	14.57%	8.67%
SSB_EMU_M	3.73%	19.59%	18.36%	8.55%	0.39%
SSB_EMU_L	0.00%	0.00%	8.31%	3.18%	5.84%
SSB_US	15.08%	8.33%	0.87%	14.71%	8.35%
SSB_JP	22.28%	7.40%	7.15%	5.50%	13.08%
SSB_UK	0.00%	0.48%	19.81%	5.24%	10.74%
COMIT	3.41%	0.85%	2.71%	3.18%	6.51%
MSUK	3.94%	11.37%	2.71%	9.41%	6.34%
MSEUXUK	16.39%	15.99%	13.33%	12.59%	11.96%
MSUS	3.61%	7.50%	1.06%	3.98%	8.29%
MSJP	6.88%	1.31%	10.63%	10.54%	9.91%

Table 4. Asset Class Weights for the sequence of portfolios corresponding to the mean-return scenarios, i.e. those scenarios in which the returns for each asset class correspond to the mean of the probability distribution.

5 Conclusions

This paper proposes a new approach to solve a multiperiod asset allocation problem. If we take into account factors like regulatory constraints, individual investors requirements, non-trivial indices of risk and subjective quality measures, portfolio construction becomes a very complicated highly non-linear problem. As the complexity of the problem increases, traditional optimization techniques are no longer useful tools for finding satisfactory solutions. This is so because their performance does not scale nicely and they typically lack flexibility, whereby accommodating new criteria for optimization along with old ones almost always requires recoding large parts of the algorithms.

Evolutionary algorithms are particularly attractive for their flexibility because they do not make any assumptions on the objective function. The user who decides about criteria and constraints can consider them in any form, i.e. he or she is not obliged to choose them linear or derivable because all kind of functions are allowed.

Date	CASH	BOND	EQUITY
01-Jan-99	24.68%	41.10%	34.22%
01-Jan-00	27.18%	35.80%	37.02%
01-Jan-01	15.08%	54.49%	30.43%
01-Jan-02	23.12%	37.18%	39.70%
01-Jan-03	18.58%	38.40%	43.02%

Table 5. Weights of the asset groups in the portfolio sequence of Table 4.

References

1. T. Bäck. *Evolutionary algorithms in theory and practice.* Oxford University Press, Oxford, 1996.
2. G. P. Brinson, L. R. Hood, and G. P. Beebower. Determinants of portfolio performance. *Financial Analysts Journal*, pages 39–44, July-August 1986.
3. J. H. Holland. *Adaptation in Natural and Artificial Systems.* The University of Michigan Press, Ann Arbor, 1975.
4. J. R. Koza. *Genetic Programming: on the programming of computers by means of natural selection.* The MIT Press, Cambridge, Massachussets, 1993.
5. A. Loraschi and A. Tettamanzi. An evolutionary algorithm for portfolio selection within a downside risk framework. In C. Dunis, editor, *Forecasting Financial Markets*, Series in Financial Economics and Quantitative Analysis, pages 275–285. John Wiley & Sons, Chichester, 1996.
6. H. M. Markowitz. *Portfolio Selection.* John Wiley & Sons, New York, 1959.
7. S. A. Zenios. *Financial Optimization.* Cambridge University Press, Cambridge, 1993.
8. W. T. Ziemba and J. M. Mulvey. *Worlwide Asset and Liability Modeling.* Cambridge University Press, Cambridge, 1998.

Acquiring Textual Relations Automatically on the Web Using Genetic Programming

Agneta Bergström[1], Patricija Jaksetic[2], and Peter Nordin[3]

[1] Interverbum AB, Localization, P.O. Box 6016,
SE-102 31 Stockholm, Sweden
agneta.bergstrom@interverbum.se
[2] PLAY: Applied research on art and technology, Viktoria Institute, Box 620,
SE-405 30 Gothenburg, Sweden
patricia@viktoria.informatics.gu.se
[3] Complex Systems, Chalmers University of Technology, CTH/GU,
SE-412 96 Gothenburg, Sweden
nordin@fy.chalmers.se

Abstract. The flood of electronic information is pouring over us, while the technology maintaining the information and making it available to us has not yet been able to catch up. One of the paradigms within information retrieval focuses on the use of thesauruses to analyze contextual/structural information. We have explored a method that automatically finds textual relations in electronic documents using genetic programming and semantic networks. Such textual relations can be used to extend and update thesauruses as well as semantic networks. The program is written in PROLOG and communicates with software for natural language parsing. The system is also an example of computationally expensive fitness function using a large database. The results from the experiment show feasibility for this type of automatic relation extraction.

1 Introduction

Today we talk about our society as the information society where an ever-increasing flood of information is pouring over us in both professional and private contexts. The amount of available electronic information is growing exponentially and this phenomenon is sometimes referred to as information overload (cf. [7]). Although the technology to produce, duplicate, spread and store information has moved forward rapidly, the capabilities for searching and structuring information have not made corresponding progress. Both Internet and company/organizational intranets are growing every day and new, often inexperienced, users are continuously joining the on-line community. This puts an extra strain on the use of powerful information retrieval methods.

The research within the information retrieval community has been focused mainly on three paradigms: statistics [10,11], semantics and context/structure [8]. The work presented here is addressing contextual/structural methods that use thesauruses and we will especially focus on how to find the information needed to extend, update and construct up-to-date thesauruses and semantic networks. In this case the information will have the form of textual relations automatically extracted from text masses by a genetic program [1].

2   Thesauruses and Semantic Networks

Thesauruses and semantic networks are sometimes used as synonyms since they have common features. However, the difference is in the number of connections between lexical entries.

In a semantic network every lexical entry is connected to a number of other lexical entries through various semantic relations. For instance, the hyponym relation is a subordinate relation used for organizing nouns into hierarchies, e.g. *dog* is hyponym of *mammal*. The relation is often seen in text as "a dog is a (kind of) mammal" or "a mammal such as a dog". Due to the transitivity of the hyponym relation, nouns can be arranged into hierarchies as in Fig. 1, where the noun *mammal* is superordinate (hyperonym) to its children, but subordinate (hyponym) to its parent. Examples of other semantic relations are synonyms and meronyms (a part-of relation: an arm is a part of a body).

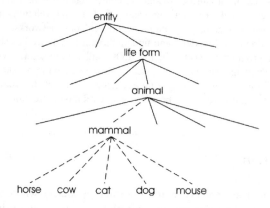

Fig. 1. Part of a noun hierarchy.

A thesaurus is kind of dictionary, but as well as organizing terms alphabetically in word forms (though for reference reasons), it is focusing on organizing the entries in word or concept meaning. The problem with thesauruses and semantic networks is that the number of look-ups are increasing, which is very time consuming for the user. This is of course the kind of mechanical work a computer rapidly and efficiently can perform.

2.1 WordNet.

Wordnet [2] is an electronic lexical database (http://www.cogsci.princeton.edu/~wn/) and the design of WordNet is influenced by the theories of psycholinguistics and human lexical memory resulting in the four lexical categories used - nouns, verbs, adjectives, and adverbs. Every lexical entry is organized into synonym sets, each representing one underlying lexical concept and therefore it is quite natural to think of WordNet's semantic relations as pointers between synonym sets. We use WordNet's prolog database for fitness calculation and the database contains 15 different relations derived from these synonym sets.

3 Related Work

There are varying degrees of automation that can be used for extracting textual relations. One option is to use various grammar formalisms (cf. [5]) and by syntactically parse text, extract relations from it. Unfortunately, grammars tend to be brittle since people often are ungrammatical or use words and concepts that disqualifies the use of full grammatical rules. This makes it very difficult and some would even argue that it is impossible to write complete grammars.

Probabilistic methods on the other hand offer a solution to this kind of problem by allowing frequent well-formed expressions to conquer ill-formed infrequent expressions. One problem though, is the "zero-frequency" assignment, where found ungrammatical n-grams[1] will be as probable as unseen grammatical n-grams.

Another way is to find short segments or patterns which often identifies a certain relation. For instance, the pattern "X is a part of Y" represents a meronym while "such X as Y" might indicate a hyponym. Hearst [6] has detected such patterns and phrases for hyponyms, by hand, and used them to extend the WordNet database [2]. However, to detect patterns by hand is extremely time consuming and also, it is hard to find good coverage of useful patterns and adapt to changes in vocabulary and phrasing.

In our approach we use a machine learning technique to automatically induce feasible patterns/segments for extraction of semantic relations, i.e. we are not only extracting new relations in unknown text but we extract, induce or learn patterns for the relation extraction itself. The machine learning technique genetic algorithms (GA) has previously been used for inducing natural language grammars [12] and the results indicate that GA is a robust and fault-tolerant method for inferring natural language grammars. We use genetic programming (GP) as machine learning technique, which, amongst other things, has been used for text classification [9] and datamining [3]. However, this kind of work in textual relation extraction is not very well explored.

---

[1] The n-gram language model is defined as $P(w_1...w_n) = \prod P(w_i \mid w_{i-n+1}...w_{i-1})$, where each word depends probabilistically on the n-1 preceding words. Each string is assigned the probability of that string in relation to all other strings of the same length, i.e. probabilities for strings of the same length sum to 1.

4   System Implementation

Today most genetic programming systems are written in different C languages for efficiency, but the original GP systems were written in LISP for its capabilities in string and meta processing. We have chosen PROLOG (SICStus prolog, http://www.sics.se) as implementation language since it is very well suited for natural language tasks. Also, WordNet offers an additional lexical database implemented in PROLOG, which is well suited for the kind of fitness calculation we want to perform.

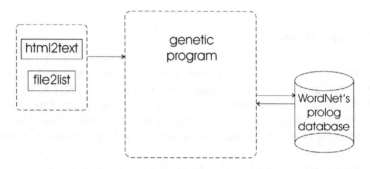

Fig. 2. Overview of the system

4.1   The Architecture of the system

The system consists of two separate modules and an extensive database (see Fig. 2). The first module is concerned with the electronic documents, and in particular, the transformation of them into the training set. The second module, which is our main focus in this paper, extract the textual relations based on a steady-state GP algorithm using the tournament selection model (see Fig. 2) and WordNet, which is an of the shelf semantic network product, for fitness evaluation.

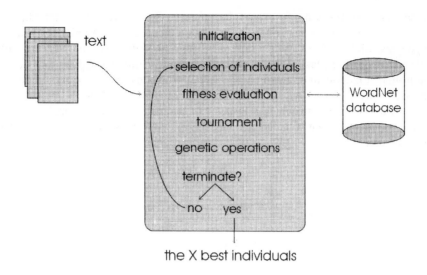

Fig. 3. Detailed overview of GP module

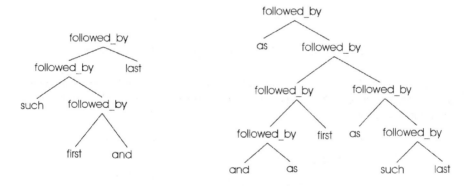

Fig. 4. An example of individuals initialized by the system.

The initial population is filled by random individuals created using grow method (see Fig. 4) and with a maximum depth of 5.

4.2 Fitness evaluation using WordNet.

The more interesting and crucial part of our system is the fitness function. Today, the system only handles the hyponym/hyperonym relation.

An individual in our system picks out pairs of words (see Fig.5) from the training set, i.e. text masses. Different text predicates, for instance predicates defining order of words, are part of the tree structures of the evolved patterns. In the initial population there are randomly constructed tree structures and any word pairs are likely to be output. The general principle behind the fitness evaluation is that an individual, who processes a lot of word-pairs of the right relation (according to WordNet), in this case hyponyms, while giving very few false picks (also according to WordNet) probably is rather fit to solve the task. When testing the system we are then willing to accept output from that individual on word-pairs where WordNet has no information. Each word pair that matches an instance of a hyponym relation in WordNet will be scored one point. If the word pair matches an instance of *another* relation in WordNet it will be scored minus one point. Finally, if no match is found in WordNet, the word pair is given zero points.

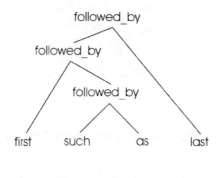

[..., phones, such, as, GA628, ...]

Fig. 5. An individual matching parts of the training set.

According to the rated word pairs, all the scores for one individual are added and used as fitness value in the competition. When the four individuals has been evaluated, the fight can begin.

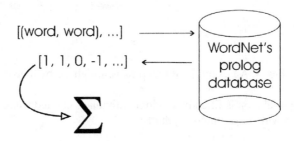

Fig. 6. Parts of the fitness function.

5 Experiment and initial results

The system was trained on preprocessed texts from the Web. The only manual part involved in the experiment was when the texts for the training set were selected. Since

WordNet decides whether a relation is valid or not, we had to make sure that the relations in the text existed in WordNet. By evolving patterns that could extract hyponyms from Web texts, we wanted to achieve a "proof-of concept" and initial results confirm the feasibility of the method. This is shown in Table 1 where the top 10 individuals, with duplicates removed, are the result from a population of 1000 individuals that has been evolved for 1500 steady state generations.

The results in Table 1 are from the second experiment and the training took about 89 hours to execute. The training set contained 41 hyponym relations and 20 of them was found by the two first individuals, although only 13 of the hyponym relations were found in WordNet. These two individuals had obviously already found the relations that the individuals that scored 1 point found, since their structure was more general. The rest of the relations' patterns where either to complex for any individual to match or could not be found in WordNet.

Table 1: The 10 fittest individuals from a population with 1000 individuals and their fitness after 6000 fitness evaluations.

Individuals	Fitness
[followed_by, [followed_by, fst, such], [followed_by, as, lst]]	13
[followed_by,[followed_by,lst,[followed_by,such,as]],fst]	13
[followed_by,[followed_by,[followed_by,and,lst],[followed_by,such, as]],fst]	1
[followed_by,[followed_by,fst,[followed_by,such,as]],[followed_by, lst,and]]	1
[followed_by,[followed_by,fst,such],[followed_by,as,[followed_by, lst,and]]]	1
[followed_by,[followed_by,and,lst],[followed_by,such,[followed_by, as,fst]]]	1
[followed_by,and,[followed_by,[followed_by,fst,such],[followed_by, as,lst]]]	1
[followed_by,[followed_by,[followed_by,[followed_by,lst,fst], [followed_by,such,fst]],lst],[followed_by,[followed_by,[followed_by, and,fst],[followed_by,and,as]],[followed_by,[followed_by,as,and], [followed_by,such,and]]]]	0
[followed_by,[followed_by,[followed_by,[followed_by,lst,and], [followed_by,lst,such]],[followed_by,[followed_by,lst,fst], [followed_by,such,such]]],[followed_by,[followed_by,[followed_by,lst, lst],[followed_by,and,lst]],such]]	0
[followed_by,[followed_by,[followed_by,[followed_by,fst, such], [followed_by,fst,fst]],[followed_by,lst,fst]],fst]	0

One of the two fittest individuals from this experiment is shown in Fig. 7. and it consists of the function *followed_by* and the terminals *first*, *such*, *as*, and *last*. The variable *first* matches anything that is followed by the word "such" and the variable *last* matches anything following the word "as". This sentence was taken from the training set:

"By hardwood, I mean any broadleaf tree such as maple, almond, ash, alder, hickory, cherry, etc."

The individual in Fig. 7 will find one match in this string and produce one word-pair (tree, maple). Fortunately, this word pair exists in WordNet and the individual receives one point for this particular match. This indicates that it is possible to evolve structures representing hyponyms, although this individual did not capture all the hyponyms present in the sentence.

[followed_by, [followed_by, first, such], [followed_by, as, last]]

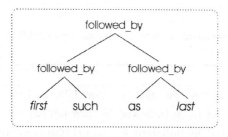

Fig. 7. An individual evolved by the system, matching one simple pattern representing hyponyms.

6 Discussion

This work is in the early stage of evolving hyponym relations and at the moment the system is only able to find such patterns that are already well know as hyponym patterns and that easily could be extracted from texts by simple pattern recognition. The idea is to extend the system to evolve patterns denoting hyponyms and other textual relations that we as humans have not been able to find or recognize as a pattern denoting some specific textual relation before.

7 Future Work

The first and most important direction for future work is to enable the system to construct more complex individuals in order to cover more complex structures such as the example sentence from the training set (Fig.7) by extending the function and terminal sets. Secondly comes the implementation of other textual relations and also developing functions for confidence grading of output. This system will also be compared to other methods of relation extraction and evaluated accordingly.

Due to the programming language the training time is extensive with several days of training. The experiment that resulted in the individuals shown in Table 1 took almost 90 hours to evolve. However, the training time could be reduced if the code were to be written in a C language, e.g. C++ instead of PROLOG. Another way to speed up the process is to use parallel computers.

8 Conclusions

The result from the training indicates that it is possible to automatically evolve patterns for relation extraction from Web text using genetic programming. If the system is used for extending, updating and/or constructing semantic networks, the technique could be an important countermeasure against information overload when used by contextual/structural methods within the area of information retrieval.

Although the training time is extensive, we do not consider it to be a bottleneck when extending, updating or constructing semantic networks. Once the system has properly evolved confident individuals, it will search your company intranet and update or extend the semantic network when needed. After a while, it might be necessary to additionally train the system, especially if it is supposed to cover a fast evolving, domain specific area, such as the IT-domain.

9 Acknowledgements

There are two persons at the Department of Linguistics at Göteborg University we would like to express our gratitude towards/for. First, Professor Robin Cooper for his guidance and supervision during our journey to finish our master's thesis. Secondly, our personal PROLOG guru, Mr. Robert Andersson, who luckily (for us) spends most of his awaken time at the department and is always more or less willingly to assist us in solving the trickiest PROLOG problems.

References

1. Banzhaf, W., Nordin, P. Keller, R. E., and Francone, F. D. Genetic Programming: An Introduction on the Automatic Evolution of Computer Programs and Its Applications. Morgan Kaufmann, Germany, 1997.
2. Fellbaum, C. WordNet - An Electronic Lexical Database. MIT Press, Cambridge MA, 1998.
3. Freitas, A. A. A genetic programming framework for two data mining tasks: Classification and generalized rule induction. In Genetic Programming 1997: Proceedings of the Second Annual Conference, pp. 96-101, Stanford University, CA, USA, 1997.
4. Gauch, S., and Smith, J. B. An Expert System for Searching Full Text. Information Processing and Management, 25(3), 1989.

5. Gazdar, G., and Mellish, C. Natural Language Processing in PROLOG: An Introduction to Computational Linguistics. Addison-Wesley Publishing Company, Wokingham, England, 1994.

6. Hearst, M. Automatic Acquistion of Hyponyms from Large Text Corpora. In Proceedings of the Fourteenth International Conference on Computational Linguistics, Nantes, France, 1992.

7. Nelson, M. R. We Have the Information You Want, But Getting It Will Cost You: Being Held Hostage by Information Overload. Crossroads 4-1: Fall 1997.

8. Marcus, R. S. Computer and Human Understanding in Intelligent Retrieval Assistance. In Proceedings of the 54th American Society for Information Science meeting. Vol. 28. 1991.

9. Masand, B. Optimizing Confidence of Text classification by Evolution of Symbolic Expressions. In Advances in Genetic Programming, MIT Press, USA, 1994.

10. Salton, G. och McGill, M. J. Introduction to Modern Information Retrieval. McGraw-Hill, 1983.

11. Salton, G. Automatic Text Processing: The Transformation, Analysis, and Retrieval of Information by Computer. Addison-Wesley Publishing, 1989.

12. Smith T. C., and Witten I. H. A genetic algorithm for the induction of natural language grammars. In Proceeding of IJCAI-95 Workshop on New Approaches to Learning for Natural Language Processing, Montreal, Canada, pp. 17-24, 1995.

Application of Genetic Programming to Induction of Linear Classification Trees

Martijn C.J. Bot[1] and William B. Langdon[2]

[1] Vrije Universiteit, De Boelelaan 1081, 1081 HV Amsterdam
mbot@cs.vu.nl
[2] CWI, Kruislaan 413, 1098 SJ Amsterdam
W.B.Langdon@cwi.nl

Abstract. A common problem in datamining is to find accurate classifiers for a dataset. For this purpose, genetic programming (GP) is applied to a set of benchmark classification problems. Using GP we are able to induce decision trees with a linear combination of variables in each function node. A new representation of decision trees using strong typing in GP is introduced. With this representation it is possible to let the GP classify into any number of classes. Results indicate that GP can be applied successfully to classification problems. Comparisons with current state-of-the-art algorithms in machine learning are presented and areas of future research are identified.

1 Introduction

Classification problems form an important area in datamining. For example, a bank may want to classify its clients in good and bad credit risks or a doctor may want to classify his patients as having diabetes or not. Classifiers may take the form of decision trees [11] (see Figure 1). In each node, a test is made in which one or more variables is used. Depending on the outcome of the test, the tree is traversed to the left or the right subtree (see Section 2.1). In our decision trees, the tests are linear combinations of some of the variables. This allows classification of continuous and integer valued datasets with an (unknown) inherent linear structure. An optimal tree is one which makes as few misclassifications as possible on the validation set.

Well known decision tree algorithms such as ID3, CART, OC1 and C4.5 are greedy local search algorithms which construct trees top-down [11]. Genetic programming (GP) [5] is used as a global stochastic search technique for finding accurate decision trees. Previous work on evolving decision trees with GP was done in [5] and [12]. The standard representation of GP was used in these experiments. Therefore, the trees look different from most Machine Learning decision trees, where nodes contain linear combinations of variables.

A new representation of decision trees in GP using Strong Typing is introduced. The classification accuracy of the GP is compared to that achieved by several other decision tree classification techniques, such as the OC1-algorithm, C5.0, and the M5' algorithm (Section 5.2).

In Section 2 the theoretical background behind the system is given. Section 3 explains the experimental setup for the experiments. The results are given in Section 4. In Section 5 an analysis is made of the performance of the GP-system and a comparison is made to other decision tree algorithms. Section 6 contains our conclusions.

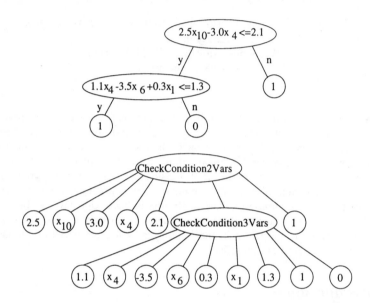

Fig. 1. Example decision tree and its representation in the GP. x_{10} means the tenth variable from the dataset. Each function node's first children are the weights and variables for the linear combination. The last two children are other function nodes or classifications. When evaluating the CheckCondition2Vars node on a certain case, if $2.5x_{10} + -3.0x_4 \leq 2.1$, the CheckCondition3Vars node is evaluated; otherwise the final classification is 1 and the evaluation of the decision tree on this particular case is finished.

2 Background

2.1 Decision Trees

Decision trees[11] are a well known technique in machine learning for representing the underlying structure of a dataset.

An **axis-parallel** decision tree is one in which each node contains only one variable. All hyperplanes (multi-dimensional planes) are parallel to the axes, hence the name. See Figure 2 for an example. A decision tree is **oblique** when the nodes contain one or more variables. Now the hyperplanes are not necessarily parallel to the axes, but can have any orientation in the attribute space. See

Figure 3 for an example. Note that axis-parallel trees are special cases of oblique trees. Clearly, oblique trees are more general than axis-parallel trees. There exist many domains in which oblique hyperplanes will form the best classification model. For example, any domain in which the (weighted) sum of two variables is crucial for correct classification needs an oblique hyperplane.

Because of this larger generality however, the search space is also much larger: there are many more possible oblique models than axis-parallel models.

Most decision tree algorithms create linear decision trees. Although often a linear tree can describe the data very well, there may be situations where non-linear trees are better. For example, in [1] hyper-ellipses were compared to hyper-rectangles. Neither hyper-ellipses nor hyper-rectangles were systematically better in size or accuracy of the solutions found. Our system focuses on linear trees.

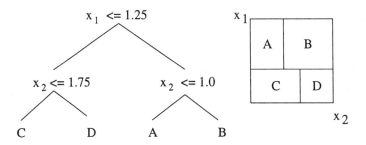

Fig. 2. The left side shows a simple axis-parallel decision tree. The right side shows the partitioning that this tree creates in the two-dimensional attribute space.

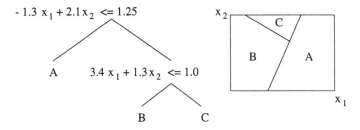

Fig. 3. The left side shows an oblique decision tree. The right side shows the partitioning that this tree creates in the two-dimensional attribute space.

When evaluating a decision tree on an individual case from the database, in each non-leaf node, a linear combination is made of some of the available variables. Each function node has ($\{c_i, x_i\}$, threshold, ifTrue, ifFalse) as its children with c_i and x_i the ith constant and the ith variable. The ifTrue and i fFalse branches are either direct classifications or other function nodes. Terminals are either constants (doubles), variables (integers) or classifications (integers). The node is evaluated as follows:

if $\sum_i c_i x_i$ <=threshold then return value of ifTrue branch
else   return value of ifFalse branch

The Classification terminal that is reached when evaluating a specific training of validation case is the classification the GP makes for that case.

This system could be extended with non-linear function nodes. For example, functions could be added which multiply two variables or which describe a hyper-ellipse. In [1] experiments were done with GA's with hyper-rectangles and with hyper-ellipses. The accuracies were not statistically different on most non-artificial databases.

2.2 Strong Typing

In order to ensure that only valid individuals are constructed in the GP, strong typing is used [9]. This is a technique that allows several datatypes to be used in one tree. The datatype of each function can be specified and the datatypes of its children. When generating a random tree, only nodes of the correct datatype are inserted at each child node. Our strong typing system contains three datatypes: Variable, Constant and Classification.

Variable

    Terminals A terminal of the Variable type is an integer which ranges between 0 and the number of variables in the database − 1. It represents the number of a variable in the database. When evaluated, it looks up the value in the database and returns it as a double.

    Functions There are no functions of this type.

Constant

    Terminals A terminal of the type Constant is a double within a certain range. In the experiments, the range is [-10, 10].

    Functions There are no functions of this type.

Classification

    Terminals A terminal of this type is an integer which ranges between 0 and the number of possible classifications − 1.

    Functions All functions have this return type.

2.3 Bloat

In GP, individuals tend to become larger over time [5, 2, 13, 7, 16, 15, 6]. This phenomenon is known as bloat. Disadvantages include overtraining, longer execution time for evaluating individuals and lower understandibility of the trees for humans. In our experiments, several measures are compared to avoid this problem. One way is to give penalties to larger individuals [15]. This may be done by lowering the fitness value by some factor times the number of nodes in the tree or its depth.

3 Experimental Setup

We used the GP system by Qureshi (GPSys) [1], which is written in Java. GPSys is a steady state, elitist system with tournament selection.

3.1 Parameter Settings

In Table 1, the standard settings for the experiments are given.

Table 1. Standard settings of the GP

Objective	Classify all training cases correctly
Terminal set	Variable, Constant, Classification
Function set	CheckCondition1Var, CheckCondition2Vars, CheckCondition3Vars
Fitness Cases	Various databases (see Section 3.2)
Selection	Tournament of size 7
Hits	not used
Wrapper	not used
Parameters	Population size = 250 No of runs = 30 No of generations=1000 Steady state, elitist 10-fold crossvalidation Mutation rate 50% Initial population creation RAMPED_HALF_AND_HALF Pareto sample size 50
Termination criterium	All cases correctly classified

The more generations allowed, the more accurate individuals are, so the number of generations was set quite high. Ten-fold crossvalidation [8] was used, with 3 runs in each split. After each run, the best individual is reported. The focus could be on accuracy, speed, generalization and simplicity of the trees. The best individual in our system is the most accurate one. Only after the runs are completed we look at speed, generalization and simplicity.

Some initial runs were performed to compare settings for the tournament size, mutation and crossover rates and bloating-penalties. In Section 4 the effects of different mutation rates are examined. We found that the exact amount of mutation and crossover was of no high importance, as long the mutation rate is larger than 0%. Thus, the mutation rate was set to 0.5 and the crossover rate also at 0.5. The GROW-population creation method was also used, but this made no significant difference to the RAMPED method.

[1] http://www.cs.ucl.ac.uk/staff/A.Qureshi/gpsys.html

Different bloating penalties were applied, on the number of nodes and on the depth of the tree. A penalty of 0.5 times the number of nodes or 2 times the depth proved to be the best penalties. Those are compared to the fitness sharing techniques.

3.2 Machine Learning Repository Databases

Four databases from the Machine Learning Repository [2] were used in the experiments:

- The Glass database contains 214 instances of 9 continuous variables each plus a classification (the type of glass). There are 7 classes (1...7), one of which (4) isn't used.
- The Ionosphere database contains 351 instances of 34 continuous variables plus a class attribute.
- The Pima database contains 768 instances of 8 continuous variables plus a class attribute. Classification is binary, either the Pima-Indians are positive of negative for diabetes.
- The Segmentation database comes in two parts. The training database consists of 210 and the validation database of 2100 cases. For crossvalidation, these were added together and crossvalidation took place on all 2310 cases. There are 19 attributes plus a class variable. There are 7 different classes.

4 Results

4.1 Mutation rate

In Figure 4 the performance of the GP on the Pima dataset with different mutation rates is shown. There's no significant difference between most mutation rates when a one-sided t-test is performed. This can also be seen in the figure, since most confidence intervals overlap. Only a rate of 0% is significantly worse than the other settings. Since at 50% there's a peak in validation accuracy, this setting will be used in the other experiments.

4.2 Bloating penalties

In Figure 5 the effects of different nodes penalties on the validation accuracy of the GP is shown. In Figure 6 depth penalties are examined. Again, most differences aren't significant. The nodes penalty that will be used in the other experiments is 0.5. The depth penalty is 2.0. Like the setting for mutation rate, these values are taken because there are peaks at those values.

---

[2] http://www.ics.uci.edu/~mlearn/MLRepository.html

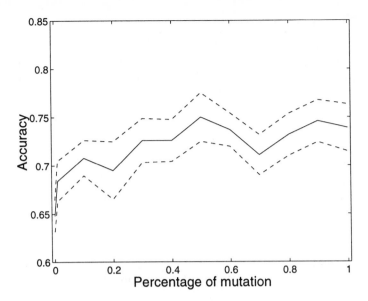

Fig. 4. Average validation accuracy and 95% confidence interval of the GP on the Pima dataset with different mutation rates

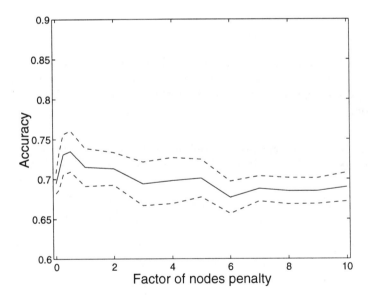

Fig. 5. Average validation accuracy and 95% confidence interval of the GP on the Pima dataset with different nodes penalties

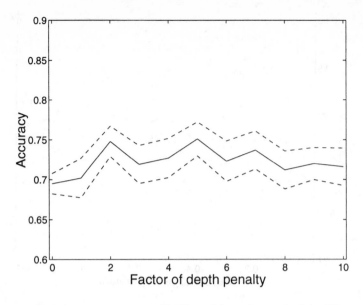

Fig. 6. Average validation accuracy and 95% confidence interval of the GP on the Pima dataset with different depth penalties

4.3 Results on all datasets

In Table 2 the performance of the GP with a nodes penalty of 0.5 and that of the GP with a depth penalty of 2.0 are compared. The mean validation accuracy and tree size of the best individuals from the 30 runs is reported, plus standard deviation.

Table 2. Comparison between nodes penalty = 0.5 and depth penalty = 2.0. Mean training and validation accuracy and tree size of the best individual of the runs plus standard deviation

Problem	Nodes penalty		Depth penalty	
	accuracy in %	tree size	accuracy in %	tree size
Glass	63.2 ± 3.1		66.5 ± 4.2	
Validation	64.1 ± 9.1	17.6 ± 4.5	63.0 ± 10.8	43.2 ±22.5
Ionosphere	93.7 ± 1.7		94.6 ± 1.3	
Validation	90.2 ± 5.5	19.7 ± 4.1	92.0 ± 5.9	40.5 ±21.2
Pima	75.3 ± 1.4		75.4 ± 2.0	
Validation	71.1 ± 5.0	14.7 ± 7.3	69.8 ± 5.8	43.8 ±23.6
Segmentation	73.1 ± 5.2		78.6 ± 5.8	
Validation	72.0 ± 6.4	53.8 ±14.3	78.2 ± 5.3	191.9 ±90.9

5 Analysis

5.1 Accuracy and Tree Size

In most runs, there is some overtraining, since training accuracy is higher than validation accuracy.

In Figures 7 and 8 plots are drawn of the average number of validation errors over time on the four datasets.

Clearly, the performance of the GP is different per database. On the ionosphere database, the GP improves for a few generations after which there's no more improvement. On the Pima and Glass database, the validation accuracy slowly improves for approximately 500 generations. After that, it stays more or less the same. The Segmentation keeps improving for 1000 generations and possibly goes on even after that. The trees that the GP creates for this database are much larger than those for the other three databases. This may mean when the optimal tree is large the GP needs more generations than when the optimal tree is smaller. The fact that OC1 produces larger trees for the Segmentation database than for the three other databases supports this hypothesis. (No exact figures are available because the OC1 system doesn't output the average tree size. However, the best tree of all crossvalidation runs is reported and the Segmentation tree is about three times larger than the other trees.) Runs on more databases are needed to substantiate this hypothesis. It is clear that the number of generations that is needed to find a good decision tree depends on the database.

The extra generations are neither beneficial nor harmful. They don't cause extra overtraining. Confidence intervals are very regular, but fairly wide (note the different scales in the figures). This suggests that the performance of the GP is stable in the different runs.

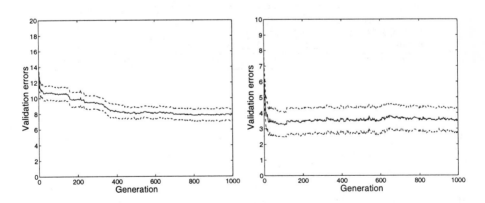

Fig. 7. Average validation errors and 95% confidence interval over time on the Glass dataset (left figure) and the Ionosphere dataset (right figure)

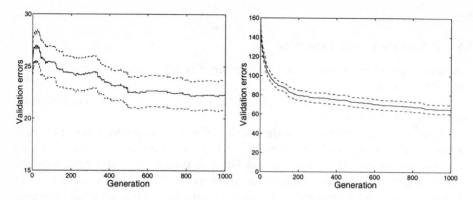

Fig. 8. Average validation errors and 95% confidence interval over time on the Pima dataset (left figure) and the Segmentation dataset (right figure)

5.2 Comparison to Machine Learning Algorithms

In Table 3 the performance of the GP is compared to that of three other decision tree classification algorithms, namely OC1 [10], C5.0 [14] and M5' [3]. Ten-fold crossvalidation and standard parameter settings are used in the other algorithms.

Table 3. Comparison between the GP and the OC1, C5.0 and M5' algorithm. The GP is with fitness sharing Pareto and LEF. Mean training and validation accuracy of the best individual of the runs plus standard deviation. Stars mark significantly better accuracy compared to the GP or better accuracy of the GP.

Problem	GP	OC1	C5.0	M5'
Glass	64.1 ± 9.1	62.3 ± 13.4	67.5 ± 2.6	70.5 ± 2.8 *
Ionosphere	92.0 ± 5.9 *	90.0 ± 5.8	88.9 ± 1.2	89.7 ± 1.2
Pima	71.1 ± 5.0	74.1 ± 6.1	74.5 ± 1.2	76.2 ± 0.8
Segmentation	78.2 ± 5.3	95.4 ± 1.5 *	96.8 ± 0.2 *	97.0 ± 0.2 *

The GP performs as well as or better than reported decision tree algorithms (OC1, C5.0 and M5') on two datasets (Ionosphere and Pima), but worse on Glass and Segmentation. Tree sizes of the GP are usually 5−10 times smaller than those from OC1 and C4.5 (from M5' no data was reported on tree sizes). However, if we let the GP run without any size restrictions, the produced trees are even larger than those from OC1 and C4.5 and overtraining increases much (training accuracy goes up and validation accuracy goes down). Determining characteristics of databases on which the GP does well or poorly is one of the subjects for future research.

The GP is slower than the other techniques. One run on a dataset of 700 cases on an Pentium III 450 takes approximately 5-8 minutes. A run of the other techniques (which are written in C or C++) typically takes about one or two minutes.

Future research will aim at improving execution speed and accuracy (for example by dynamic sampling of training cases (DSS [4]) or by seeding the population with trees constructed by standard decision tree algorithms such as C5.0).

6 Conclusions

Our representation for decision trees in GP can be used succesfully for inducing accurate decision trees. On some datasets, the GP accuracy is as well as or better than reported accuracies for decision tree algorithms, but on others, the GP does worse (see Table 3). A disadvantage of this approach is its long run time.

The GP usually overtrains: training accuracy is higher than validation accuracy (see Table 2). On different datasets, different numbers of generations are needed before validation accuracy stops improving. This ranges from 30 generations (Ionosphere dataset) to over 1000 generations (Segmentation dataset). Indications were found that the number of generations that are needed may correlate with the size of the optimal tree. Extra generations don't affect validation accuracy in a positive or negative way. There is no statistical difference in accuracy when a nodes penalty or a depth penalty is applied.

References

1. J. Aguilar, J.Riquelme, and M. Toro. Three geometric approaches for representing decision rules in a supervised system. In *Late Breaking Papers at the 1999 Genetic and Evolution Computation Conference*, pages 8–15, 1999.
2. Tobias Blickle and Lothar Thiele. Genetic programming and redundancy. In J. Hopf, editor, *Genetic Algorithms within the Framework of Evolutionary Computation (Workshop at KI-94, Saarbrücken)*, pages 33–38, Im Stadtwald, Building 44, D-66123 Saarbrücken, Germany, 1994. Max-Planck-Institut für Informatik (MPI-I-94-241).
3. E. Frank, Y. Wang, S. Inglis, G. Holmes, and I.H. Witten. Using model trees for classification. *Machine Learning*, 32:63–76, 1998.
4. Chris Gathercole. *An Investigation of Supervised Learning in Genetic Programming*. PhD thesis, University of Edinburgh, 1998.
5. J.R. Koza. *Genetic Programming: On the Programming of Computers by Means of Natural Selection*. MIT Press, Cambridge, MA, USA, 1992.
6. W.B. Langdon, T. Soule, R. Poli, and J.A. Foster. The evolution of size and shape. In L. Spector, W.B. Langdon, U. O'Reilly, and P.J. Angeline, editors, *Advances in Genetic Programming 3*, chapter 8, pages 163–190. MIT Press, Cambridge, MA, USA, May 1999. Forthcoming.
7. Nicholas Freitag McPhee and Justin Darwin Miller. Accurate replication in genetic programming. In L. Eshelman, editor, *Genetic Algorithms: Proceedings of the Sixth International Conference (ICGA95)*, pages 303–309, Pittsburgh, PA, USA, 15-19 July 1995. Morgan Kaufmann.
8. T. Mitchell. *Machine Learning*. WCB/McGraw-Hill, 1997.

9. D.J. Montana. Strongly typed genetic programming. *Evolutionary Computation*, 3(2):199–230, 1995.

10. S. Murthy, S. Kasif, S. Salzberg, and R. Beigel. Oc1: Randomized induction of oblique decision trees. In *Proceedings of the Eleventh National Conference on Artificial Intelligence*, pages 322–327. AAAI, MIT Press, 1993.

11. S. K. Murthy. Automatic construction of decision trees from data: a multi-disciplinary survey. In *Data Mining and Knowledge Discovery*, number 2, pages 345–389, 1998.

12. Nikolay I. Nikolaev and Vanio Slavov. Inductive genetic programming with decision trees. In *9th European Conference on Machine Learning*, Prague, Czech Republic, 3-26 April 1997.

13. Peter Nordin and Wolfgang Banzhaf. Complexity compression and evolution. In L. Eshelman, editor, *Genetic Algorithms: Proceedings of the Sixth International Conference (ICGA95)*, pages 310–317, Pittsburgh, PA, USA, 15-19 July 1995. Morgan Kaufmann.

14. J.R. Quinlan. *C4.5 Programs for Machine Learning*. Morgan Kaufmann, 1993.

15. Terence Soule. *Code Growth in Genetic Programming*. PhD thesis, University of Idaho, Moscow, Idaho, USA, 15 May 1998.

16. Terence Soule, James A. Foster, and John Dickinson. Code growth in genetic programming. In John R. Koza, David E. Goldberg, David B. Fogel, and Rick L. Riolo, editors, *Genetic Programming 1996: Proceedings of the First Annual Conference*, pages 215–223, Stanford University, CA, USA, 28–31 July 1996. MIT Press.

A Metric for Genetic Programs and Fitness Sharing

Anikó Ekárt and S. Z. Németh

Computer and Automation Research Institute
Hungarian Academy of Sciences
1518 Budapest, POB. 63, Hungary
E-mail: ekart@sztaki.hu, snemeth@oplab.sztaki.hu

Abstract. In the paper a metric for genetic programs is constructed. This metric reflects the structural difference of the genetic programs. It is used then for applying fitness sharing to genetic programs, in analogy with fitness sharing applied to genetic algorithms. The experimental results for several parameter settings are discussed. We observe that by applying fitness sharing the code growth of genetic programs could be limited.

1 Introduction

The search space of a search problem may contain more than one peak. In artificial genetic search, the population may converge to one of these peaks. In genetic algorithms there are several so-called *niching* techniques that prevent the convergence of the whole population to just one of the peaks [5, 4]. Niching proceeds by penalizing individuals that are "close" to other individuals in the population. It is based on the fact that the search space of genetic algorithms is a metric space (the individuals are represented as real-valued or boolean vectors).

Fitness sharing [4, 5] was introduced as a technique for maintaining population diversity in genetic algorithms. The population is distributed over the different peaks in the search space: in the neighborhood of each peak, a number of individuals proportionate to the height of that peak are allowed.

In this work, we extend the applicability of fitness sharing to tree-based genetic programming by defining a distance function for genetic programs that reflects the structural difference of trees. As a result of fitness sharing, we obtain smaller genetic programs as solutions.

Code growth is a general problem of genetic programming systems [8, 14] and several methods for limiting code growth have been proposed:

- introducing a maximum allowed depth [8];
- editing out the irrelevant parts [8]; and
- including program size in the fitness measure [7, 17, 14].

We show here that fitness sharing based on our metric induces a preference for smaller programs, and it could be used for limiting code growth in genetic programming systems.

2 State of the Art

For genetic algorithms several niching techniques have been developed [10]. *Crowding* is a restricted replacement method, where the offspring of crossover and mutation replaces the most similar individual from a sample of the population. Another approach is the *restricted competition* among dissimilar individuals during selection. Both algorithms restrict their selection method, but in a different way.

Fitness sharing [5, 4] treats fitness as a shared resource of the population, and thus requires that similar individuals share their fitness. The fitness of each individual f_i is worsened by its niche count m_i. The niche count estimates crowding in the neighborhood of individual i and is calculated over the individuals in the current population:

$$m_i = \sum_{j=1}^{N} S(d(i,j)),$$

N being the population size, $d(i,j)$ the distance of individuals i and j, and S the sharing function. S is a decreasing function, since sharing with a close individual should be greater than sharing with some farther individual. The typical sharing function has the form

$$S(d) = \begin{cases} 1 - (d/\sigma)^\alpha & \text{if } d \le \sigma, \\ 0 & \text{if } d > \sigma. \end{cases}$$

The niche radius σ is fixed by the user at some minimum distance between peaks, and usually $\alpha = 1$. By using fitness sharing, the population does not converge as a whole, but convergence takes place within the niches.

The main drawback of fitness sharing is that the computation of the shared fitness for the entire population in each generation could be very time-consuming.

By using tournament selection, it is possible to overcome this difficulty. Oei, Goldberg and Chang [12] propose a scheme for combining sharing with binary tournament selection. They calculate the shared fitness by continuously updating the fitness as the new population is filled.

Yin and Germay [16] propose the application of a clustering algorithm before sharing. Thus, the population is first divided into niches, and then the shared fitness of any individual is computed only with respect to the individuals that are in its niche.

Horn, Nafpliotis and Goldberg [6] propose a tournament selection scheme, where two individuals are picked randomly from the population, a comparison set is also picked randomly from the population and the two individuals are compared against the individuals from the comparison set. They implement sharing as counting the individuals from the comparison set that are in the niche of the two competing individuals. The individual with the smallest niche count is the best candidate.

The other difficulty of fitness sharing consists in determining the niche radius.

Beyond these difficulties, one has to make the appropriate choice of a genotypic or a phenotypic distance measure for the given problem.

In the case of tree-based genetic programming, a straightforward phenotypic distance measure is the Euclidean distance between the values of the two genetic programs on the fitness cases.

A genotypic distance measure should be a distance function on trees. Several distance-functions on trees have been proposed in different domains of artificial intelligence, such as inductive logic programming, bio-informatics, pattern recognition.

The *edit distance* between labeled trees was defined as the cost of shortest sequence of editing operations that transform one tree to the other [13, 15, 9]. An edit operation consists in inserting or deleting one node (subtree in [13]) or changing the label of one node. The cost of each edit operation could be specified separately [15] or the cost is unique for insertions, deletions, and changes, respectively [9]. The algorithms for computing the edit distance of trees are very time-consuming. In the case of ordered labeled trees (where the order of children of a node is significant) T_1 and T_2, the time complexity of the algorithm for the simplified edit distance [13] is $O(|T_1| * |T_2|)$, $|T|$ being the number of nodes in tree T. In the case of unordered labeled trees, the time complexity is exponential [18]. However, the edit distance has been used in pattern recognition [9] and instance-based learning [2]. But because of time complexity, the edit distance is not widely used in genetic programming [1].

Nienhuys-Cheng [11] defines a metric bounded by 1 that takes into consideration the depth of nodes in the trees:

$$d(p(s_1, s_2, \ldots, s_n), q(t_1, t_2, \ldots, t_m)) = \begin{cases} 1 & \text{if } p \neq q, \\ \frac{1}{2n} \sum_{i=1}^{n} d(s_i, t_i) & \text{if } p = q, \end{cases} \quad (1)$$

p and q being the root labels of the two trees. This metric is used then in inductive logic programming for measuring the quality of approximations to a correct program . This distance can be computed in $O(min(|T_1|, |T_2|))$.

3 The Proposed Metric

Our goal was to define a proper metric for genetic programs that reflects the structural difference of the genetic programs and can be computed efficiently.

We show the experimental results for the metric in the case of a symbolic regression problem.

The distance of two genetic programs is calculated in three steps:

1. The two genetic programs (trees) to be compared are brought to the same tree-structure (only the contents of the nodes remain different).
2. The distances between any two symbols situated at the same position in the two trees are computed.
3. The distances computed in the previous step are combined in a weighted sum to form the distance of the two trees.

We construct the distance function step by step, as follows.

We make the convention that before comparing two trees of different structure, we complete them by NULLs so that the resulting trees have the same structure, as it is shown in Figure 1. In the case of commutative operations (unordered trees) we do not know in advance which pairing of subtrees is better. Trying out each possible pairing makes the distance computation exponential in tree depth. When the trees have different arities, the missing subtree(s) could also be completed by NULLs. In this case, it would be difficult to choose where to insert the missing subtree. Therefore, in all of our experiments we use binary trees and in most of them we consider ordered trees.

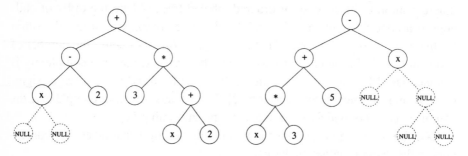

Fig. 1. The trees are completed by NULLs to have the same layout.

The trees that can be constructed by genetic programming contain functions and terminals (variables and constants). The functions may be grouped in several groups according to their similarity (E.g. + and -, * and / for trees encoding arithmetic expressions). The terminal set is initially differentiated into variables and constants, but one could further partition the variable set.

We partition the set of symbols into n sets $A_0, A_1, ..., A_n$; A_0 being the set containing NULL and A_1 the set containing the constants. We define the set A_0 for the single element NULL in order to treat the comparison between some node and NULL in the same way as a comparison between two elements of different sets.

Let $\delta : \{0, 1, \ldots, n\} \times \{0, 1, \ldots, n\} \to (0, +\infty)$ be a function with properties:

$$\delta(i, j) \leq \delta(i, k) + \delta(k, j), \tag{2}$$
$$\delta(i, j) = \delta(j, i), \tag{3}$$
$$\delta(i, i) < \delta(j, k), \quad j \neq k, \tag{4}$$

for all $i, j, k \in \{0, 1, \ldots, n\}$.

We define the distance between two different elements $x \in A_i$ and $y \in A_j$ as follows:

$$d(x, y) = \begin{cases} 0 & \text{if } x = y, \\ C * \dfrac{|x-y|}{\underset{v,w \in A_1}{Max} |v-w|} & \text{if } i = j = 1, \\ \delta(i, j) & \text{otherwise,} \end{cases}$$

C being a constant. By choosing $C < \delta(i, j), \forall i, j \in \{0, 1, \ldots, n\}$, the distance of two constants becomes less than the distance of two other elements. With this distance definition, we provide that the distance between two elements of the same set is less than the distance between two elements of different sets.

Now, after defining the distance between elements, we can define the distance between trees. We take into consideration the fact that a difference at some node closer to the root could be more significant than a difference at some node farther from the root (in a similar way to [11]). First we complete the trees by NULLs, so that both trees have the same structure. Let $T_1 = p(s_1, s_2, \ldots, s_m)$ be the tree with root p and subtrees $s_1, s_2, \ldots, s_m$, and $T_2 = q(t_1, t_2, \ldots, t_n)$ the tree with root q and subtrees $t_1, t_2, \ldots, t_m$, respectively. Then, we calculate the distance between trees T_1 and T_2 as:

$$dist(T_1, T_2) = \begin{cases} d(p, q) & \text{if neither } T_1 \text{ nor } T_2 \text{ have any children,} \\ d(p, q) + \frac{1}{K} * \sum_{l=1}^{m} dist(s_l, t_l) & \text{otherwise,} \end{cases} \tag{5}$$

where $K \geq 1$ is a constant, signifying that a difference at any depth r in the compared trees is K times more important than a difference at depth $r + 1$. By choosing different values for K, we could define different shapes for the neighborhoods (see Section 4).

It can be easily proved that all conditions for $dist$ to be a **metric** [3] are met:

$dist(T_1, T_2) \geq 0,$

$dist(T_1, T_2) = 0 \Leftrightarrow T_1 = T_2,$

$dist(T_1, T_2) = dist(T_2, T_1)$ (symmetry),

$dist(T_1, T_2) \leq dist(T_1, T_3) + dist(T_3, T_2)$ (triangle inequality).

Our metric can be seen as a generalization of metric (1):

- instead of fixing the distance of any two different symbols at 1, we partition the set of symbols into subsets of "similar" symbols and we set the values of distances on the basis of this partitioning;
- we introduce constant K in (5), and thus we can experiment with different shapes of niches.

If $K > m$, our $dist$ is bounded by

$$M = \frac{\underset{i,j\in\{0,1,...,n\}}{Max}\ \delta(i,j)}{1 - m/K}$$

(generalization of Theorem 5 of [11]). Then, if two trees are the same to at least depth r,

$$dist(T_1, T_2) \leq \frac{m * M}{K^{r+1}}$$

(generalization of Lemma 6 of [11]). If two trees are not the same to depth r,

$$dist(T_1, T_2) \geq \frac{\underset{i,j\in\{0,1,...,n\}}{Min}\ \delta(i,j)}{K^r}$$

(generalization of Lemma 7 of [11]).

4   Empirical Validation

We used the metric in a symbolic regression problem. We applied fitness sharing with different parameter settings and compared the results. The goal was to evolve the programs that approximate the function $f(x) = 1.5 + 24.3x^2 - 15x^3 + 3.2x^4$ in 100 data points that were randomly selected in the $[0, 1]$ interval. The general parameter setting is shown in Table 1. The presented results are the average of 50 runs in each parameter setting.

We define the distance function for two genetic programs in the case of symbolic regression of a function of one variable, with function set $F = \{+, -, *, /\}$. All of the functions take two arguments, thus the resulting trees are binary. We mainly consider ordered trees. We partition the set of symbols as follows:

$$A_0 = \{NULL\},$$
$$A_1 = \{\text{constants in the range -100 ... 100}\},$$
$$A_2 = F,$$
$$A_3 = \{x\}.$$

Table 1. The genetic programming parameter setting

Objective	Evolve a function that fits the data points of the fitness cases
Terminal set	x, real numbers $\in [-100, 100]$
Function set	$+, *, /$
Fitness cases	$N = 100$ randomly selected data points (x_i, y_i), $y_i = 1.5 + 24.3x_i^2 - 15x_i^3 + 3.2x_i^4$
Raw fitness and also standardized fitness	$\sqrt{\frac{1}{N}\sum_{i=1}^{N}(gp(x_i) - y_i)^2}$, $gp(x_i)$ being the output of the genetic program for input x_i
Population size	100
Crossover probability	90%
Probability of Point Mutation	10%
Selection method	Tournament selection, size 10 (as in [6])
Termination criterion	none
Maximum number of generations	200
Maximum depth of tree after crossover	20
Initialization method	Grow

Then we choose function δ, such that it satisfies requirements (2),(3),(4), for example

$$\delta(0, i) = \delta(i, 0) = 5, \quad i \in \{1, 2, 3\},$$
$$\delta(1, 2) = \delta(2, 1) = 2,$$
$$\delta(2, 2) = 1,$$
$$\delta(3, i) = \delta(i, 3) = 3, \quad i \in \{1, 2\},$$

where the other values of the function need not be specified.

4.1 Application of the Metric for Fitness Sharing

We experimented fitness sharing with $K = 1$, $K = 2$ and $K = 10$ in the expression of the distance function of (5). We set the distance of two symbols $d(x, y) \in [0, 5]$, so that the distance of two nodes at the same position in the trees counts for the final distance with a value depending on this position, as it is shown in Table 2.

Thus, if the distance of two trees is a value less than 0.005 in the case of $K = 10$, they only differ at some node at depth 3(or more) or they differ just at the values of constants.

Table 2. The difference of two nodes counted for the distance of the trees containing them

Position(depth)	Range of added value		
	K=10	K=2	K=1
0 (root)	0-5	0-5	0-5
1	0-0.5	0-2.5	0-5
2	0-0.05	0-1.25	0-5
3	0-0.005	0-0.625	0-5

Table 3. The ranges for niche size

K	Niche size
10	$[5*10^{-4}, 0.5]$
2	$[5*10^{-3}, 0.5]$
1	$[1, 10]$

We experimented with niche sizes corresponding to the values of K, as shown in Table 3. By selecting different values for K, we can experiment different niche shapes. For $K = 1$, for a certain tree, any other tree that differs from it at just one internal node, is in its neighborhood of size $\sigma \geq 1$. A niche size $\sigma < 1$ in this case would restrict only to differences in the values of constants. In the mean time, for $K = 10$, if $\sigma = 1$, for a certain tree, any other tree with the same root is in its neighborhood. But for $K = 10$ and $\sigma = 5*10^{-3}$, the neighborhood of a tree contains only the trees with differences at depth 3 or more (and slight differences in the values of constants at depth 1 or 2).

We experimented with our metric (1) on trees with + and * counted as commutative and (2) on ordered trees. In the first case, the final solution of fitness sharing was slightly more accurate than the solution of original genetic programming. Since the computational cost was very high (the distance computation was exponential in tree depth) and there was little improvement in fitness, we decided to conduct the rest of the experiments on ordered trees. In the case of ordered trees, the final solution of fitness sharing was slightly worse than the solution of original genetic programming. The error of the final solution was 0.42 (averaged over 50 runs) for simple genetic programming and 0.45 for fitness sharing with $K = 10$ and niche radius $\sigma = 0.05$. For bigger niche radii, the error increases as an effect of fitness sharing with too many neighbors (E.g., for $K = 10$ and $\sigma = 0.5$, sharing is performed among individuals that can differ at one node at the first depth or any number of nodes at depth more than 1). But the scope of fitness sharing is to maintain diversity in the population.

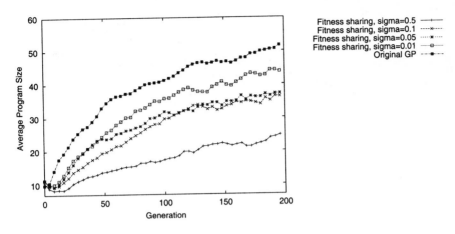

Fig. 2. The average program size for $K = 10$.

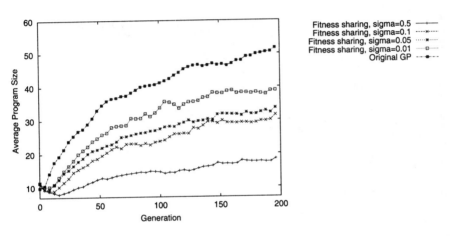

Fig. 3. The average program size for $K = 2$.

If we look at the average program size (Figures 2, 3, 4), we can immediately notice, that the size of programs in the case of fitness sharing does not grow as fast as the size of programs in the case of original genetic programming. There is a niche size depending on K ($K = 10 \Rightarrow \sigma \in [0.05, 0.1]$, $K = 2 \Rightarrow \sigma \in [0.01, 0.1]$, $K = 1 \Rightarrow \sigma \in [1, 2]$), for which code growth is much slower and fitness is not significantly worsened.

In order to see whether the difference in program size between original genetic programming and fitness sharing is significant, we computed the 99% confidence interval for this difference over generations. The evolution of two confidence intervals for $K = 10$ is shown in Figure 5. Since at generation 200 the difference is in the range $[8.18, 20.95]$ with probability 99% when $\sigma = 0.05$, we can conclude that the difference is significant.

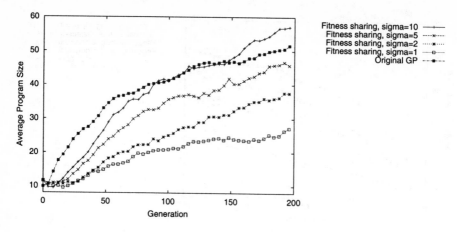

Fig. 4. The average program size for $K = 1$.

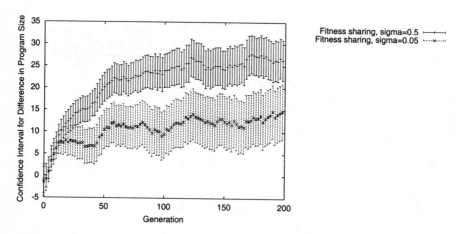

Fig. 5. The 99% confidence interval of the difference in program size between original genetic programming and fitness sharing.

The moderated code growth is explained by the following:

1. By the application of fitness sharing, fitness is shared among not highly fit individuals containing the same nonfunctional code. Their shared fitness is even worse, and they have less chance of being selected.
2. By using tournament selection and fitness sharing (see [6]), if two competing individuals are of similar (unshared) fitness, the smaller one is selected. The individual with less neighbors or farther ones is preferred to the individual with more or closer neighbors. In other words, the individual that has less common parts with the individuals from the comparison set is se-

lected. *But a smaller individual has less common parts with other random individuals, than a bigger one.*
3. The probability of selecting a crossover point closer to the root is higher in small trees than in big trees. A change near the root moves the offspring farther from its parents. Thus, *the offspring of smaller parents are more likely to find new niches.*

4.2 Comparison to Other Metrics

We also conducted experiments with two other metrics: the phenotypic distance and the edit distance [13]. In the case of phenotypic sharing, the results were very similar to original genetic programming. In the case of editing distance, the accuracy of best program was slightly better than in original genetic programming, but the computational cost was high.

5 Conclusions

In the present paper, we introduced a metric for genetic programs that reflects their tree-structure and can be efficiently computed.

We used the metric for applying fitness sharing to genetic programming. By the application of fitness sharing, we could obtain smaller genetic programs of similar (unshared) fitness. Fitness sharing is usually applied for maintaining population diversity. Thus, for interpreting its result, a proper measure of population diversity in genetic programming is needed [1]. Our metric could be used for defining such a measure of population diversity, since it is based on the structural difference of the programs.

Our fitness sharing did not improve accuracy, but the existing literature on the beneficial effects of fitness sharing in *genetic algorithms* suggests that further work might give results in this direction in *genetic programming*, too.

Acknowledgements

This work has been supported by grants No. T023305, T029572 and T25471 of the National Research Foundation of Hungary. The second author was supported by the Bolyai János research fellowship. The authors are grateful to A. Márkus and J. Váncza for many helpful discussions.

References

1. Wolfgang Banzhaf, Peter Nordin, Robert E. Keller, and Frank D. Francone, *Genetic Programming: An Introduction*, Morgan Kaufmann, 1998.

2. Uta Bohnebeck, Tamás Horváth, and Stefan Wrobel, 'Term comparisons in first-order similarity measures', in *Proceedings of the 8th International Workshop on Inductive Logic Programming*, ed., D. Page, volume 1446 of *LNAI*, pp. 65–79. Springer-Verlag, (1998).

3. J. R. Giles, *Introduction to the Analysis of Metric Spaces*, Australian Mathematical Society Lecture Series, 1987.

4. David E. Goldberg, *Genetic Algorithms in Search, Optimization, and Machine Learning*, Addison-Wesley, 1989.

5. John H. Holland, *Adaptation in Natural and Artificial Systems*, Ann Arbor: The University of Michigan Press, 1975.

6. Jeffrey Horn, Nicholas Nafpliotis, and David E. Goldberg, 'A niched Pareto genetic algorithm for multiobjective optimization', in *Proceedings of the First IEEE Conference on Evolutionary Computation*, pp. 82–87, (1994).

7. Hitoshi Iba, Hugo de Garis, and Taisuke Sato, 'Genetic programming using a minimum description length principle', in *Advances in Genetic Programming*, ed., Kenneth E. Kinnear, 265–284, MIT Press, (1994).

8. John R. Koza, *Genetic Programming: On the Programming of Computers by Means of Natural Selection*, MIT Press, 1992.

9. Shin-Yee Lu, 'The tree-to-tree distance and its application to cluster analysis', *IEEE Transactions on PAMI*, **1**(2), 219–224, (1979).

10. Samir Mahfoud, 'Niching methods for genetic algorithms', Illigal report 95001, University of Illinois at Urbana-Champaign, (1995).

11. Shan-Hwei Nienhuys-Cheng, 'Distance between Herbrand interpretations: a measure for approximations to a target concept', in *Proceedings of the 7th International Workshop on Inductive Logic Programming*, eds., N. Lavraĉ and S. Džeroski, volume 1297 of *LNAI*, pp. 213–226. Springer-Verlag, (1997).

12. C. K. Oei, David E. Goldberg, and S. J. Chang, 'Tournament selection, niching and the preservation of diversity', Illigal report 91011, University of Illinois at Urbana-Champaign, (1991).

13. Stanley M. Selkow, 'The tree-to-tree editing problem', *Information Processing Letters*, **6**(6), 184–186, (1977).

14. Terence Soule, James A. Foster, and John Dickinson, 'Code growth in genetic programming', in *Genetic Programming 1996: Proceedings of the First Annual Conference*, eds., John R. Koza, David E. Goldberg, David B. Fogel, and Rick L. Riolo, pp. 215–223, (1996).

15. Kuo-Chung Tai, 'The tree-to-tree correction problem', *Journal of the ACM*, **26**(3), 422–433, (1979).

16. Xiaodong Yin and Noël Germay, 'A fast genetic algorithm with sharing scheme using cluster analysis methods in multimodal function optimization', in *Artificial Neural Nets and Genetic Algorithms*, eds., Rudolf F. Albrecht, Nigel C. Steele, and Colin R. Reeves, pp. 450–457, (1993).

17. Byoung-Tak Zhang and Heinz Mühlenbein, 'Balancing accuracy and parsimony in genetic programming', *Evolutionary Computation*, **3**(1), 17–38, (1995).

18. Kaizhong Zhang, Rick Statman, and Dennis Shasha, 'On the editing distance between unordered labeled trees', *Information Processing Letters*, **42**, 133–139, (1992).

Using Factorial Experiments to Evaluate the Effect of Genetic Programming Parameters

Robert Feldt[1] and Peter Nordin[2]

[1]Dept. of Computer Eng., [2]Complex Systems, Chalmers University of Technology,
SE-412 96 Gothenburg, Sweden

Abstract. Statistical techniques for designing and analyzing experiments are used to evaluate the individual and combined effects of genetic programming parameters. Three binary classification problems are investigated in a total of seven experiments consisting of 1108 runs of a machine code genetic programming system. The parameters having the largest effect in these experiments are the population size and the number of generations. A large number of parameters have negligible effects. The experiments indicate that the investigated genetic programming system is robust to parameter variations, with the exception of a few important parameters.

1   Introduction

The Genetic Programming (GP) method might be the first instance of real *automatic programming* [6]. In an even more general sense, GP could be the first technique to tell the computer *what* to do without having to specify *how* to do it. However, in order for that to be true the user must be able to run the GP system using only a minimal set of natural parameters. In an ideal case there should be no parameters or only parameters that make immediate sense to the user's requirements such as *maximal search time* etc. This is far from true with present genetic programming systems. A modern GP system with additions such as Automatically Defined Functions (ADFs), Demes, and Dynamic Subset Selection have a very large number of parameters and settings creating a combinatorial explosion for the complete parameter space. This enormous parameter search space makes the search for an optimal or near optimal parameter setting difficult for the user.

What is even more severe is the theoretical implication of numerous parameters and settings. Each time we set a parameter we supply information to the search algorithm. If we set too many specific parameters, we might *"point out"* the solution with the parameters and we will not get *more out of the system than we put in*. We are supplying more information than the system is giving us back or in other words we are spending more effort and intelligence on the search for the right combination of parameters than the system does for the right solution.

The standard defence against this argument is that GP is very robust and accepts a wide range setting with little degradation in performance. This is usually only a hunch from GP researchers since there has been no large, systematic investigation of parameter effect using genetic programming. Such an investigation would have the additional benefit of enhancing experiments by providing close to optimal parameter settings. The only broad directions in the literature are experience-based, *rule-of-thumb-type* parameter recommendations [1].

In this works we describe the first series of experiments that address parameter influence in a broad and systematic way.

The questions that we are addressing are:

- Is GP robust toward different parameter settings or do settings have an effect on performance?
- If there is an effect on fitness, which parameters have the largest effect?
- Is the parameter effect dependent on single parameter settings or are combinations of parameters important?
- Can some parameters be ignored and can general guidelines be devised for the most important ones?

This paper address these questions using statistically sound experimental methods for parameter screening based on *fractional factorial designs* [2]. These methods reduce the number of runs needed and increase the amount of knowledge that can be gained.

2 Method

To overcome the combinatorial explosion in the number of parameter combinations that need to be considered we use experimental design methods studied in mathematical statistics. Statistical Design of Experiments (DoE) provides a framework to design and analyze comparative experiments, ie. experiments with the purpose to determine the quantitative effects of inputs on some output [6] [3]. In this context the inputs are called *factors* and the output is called the *response*. The major advantage of using DoE designs is that experimentation becomes more efficient: both the effects of individual factors and their interaction can be investigated with limited experimental effort. This is achieved by changing more than one factor at a time.

A basic DoE experimental design is the factorial design where each factor has a discrete number of levels. An example of a two-level factor in GP is whether a certain function should be included in the function set or not. Continuous factors, such as for example the population size, can be used in factorial experiments if two discrete levels are chosen from their valid range. In a full two-level factorial design all combination of factor levels are included, resulting in 2^k different parameter settings, where k is the number of factors. Even for relatively small k:s the number of combinations needed is impractical. To overcome this fractional factorials are used. They utilize the fact that higher-order interactions between factors, i.e. that two or more factors have a combined effect different from each one of them in isolation, often have negligible effects. By letting lower-order effects, such as the main effects of the parameters and their two-factor interactions, be confounded with each other only a fraction of the full factorial design needs to be run.

The amount of confounding between effects in a design is determined by the design resolution. Design resolution refers to the amount of detail, separate identification of factor effects and interactions, that a design supports. For example, in a design of resolution five the main effects are confounded with four-factor interactions while two-factor interactions are confounded with three-factor interactions. The confounding pattern can be calculated from the design generators that define how the design is to be constructed. For more information on factorial designs see [3].

A typical strategy for experimentation using DoE is to make sequential use of designs with increasing resolution [3]. In the first experiment a large number of factors are included since we do not yet know which of them may have large effects on the response. A heavily fractionalized design with low resolution is often used to *screen out* a majority of the factors. The remaining factors are studied in more detail in later experiments. Later experiments typically have higher resolution to permit separation of main and two-factor effects.

Tradtional DoE have been developed for physical and medical sciences and its development has been biased by the typical applications in these fields. For example, when an experiment is conducted in the real world it is often impractical to control more than 15 factors. Kleijnen [8] points out that a number of things are different when conducting experiments on a computer simulation: there are often more factors to be studied, we can practically control many more factors, and we do not need to randomize the run order of the experiments to get results that are robust to uncontrolled, and possibly even unknown, factors. These issues apply, in a similar way to genetic programming experiments.

3 Experiments

A total of 1108 GP runs were performed in seven different experiments on three different problems. In these runs, a total of about 2.5 billion individuals have been evaluated. Below we describe the problems, GP system, factors and response variable used in the experiments. We also describe the design of the experiments.

3.1 Problems

We believe in the importance of evaluating machine learning algorithms over *several* problems. In this work we have used three different binary classification problems. However, we plan to expand the number and types of evaluated problems significantly in future work, see section 6. The problems used are all standard machine learning problems: Ionosphere, Gaussian, and Pima Indians Diabetes Database:

Ionosphere Problem: This real-world radar echo classification problem has been donated by Vincent Sigillito of the Space Physics Group at John Hopkins University in the US. It is taken from the UCI Machine Learning repository [11]. There are 200 instances in the training set and 151 instances in the validation set. The problem has thirty-four attributes and a binary-valued response indicating whether the echoes have detected any structure in the ionosphere.

Gaussian Problem: The gaussian classification problem is an artificial problem for heavily overlapping distributions with non-linear separability. The class 0 is represented by a multivariate normal distribution with zero mean and standard deviation equal to 1 in all dimensions, and the class 1 by a normal distribution with zero mean and standard deviation equal to 2 in all dimensions. There are 1000 patterns, 500 in each class. We have used a variant of the standard eight-dimensional version, where there are 16 additional false (random) inputs in addition to the eight true inputs. Theoretical maximal classification for the pure 8-D problem is 91%. The problem is probably *not* easier with the false inputs added.

Pima Indians Diabetes Problem: This real-world medical classification problem has been donated by National Institute of Diabetes, Digestive and Kidney Diseases in the US. It is taken from the UCI Machine Learning repository [11]. The diagnostic, binary-valued response variable indicates whether the patient shows signs of diabetes according to World Health Organisation criteria (i.e., if the 2 hour post-load plasma glucose was at least 200 mg/dl at any survey examination or if found during routine medical care). The population lives near Phoenix, Arizona, USA. There are 576 instances in the training set and 192 in the validation set. The problem has eight attributes and a binary-valued response value.

3.2 Genetic programming system, its parameters and their values

For our experiments we used the Discipulus™ system, a commercial implementation of machine code GP [9]. Discipulus™ is based on the AIM-GP approach, a very efficient method for genetic programming formerly knows as CGPS [7]. The system uses a linear representation of individuals and a substring-exchanging crossover. In this survey we have used most of the parameters in Discipulus™. These parameters are used as factors in the experiments described below. Their factor identifier (A to Q) and their value at the low and high level used for the experiments are given in table 1. The levels of the continous parameters were chosen to represent qualitatively distinct levels based on our previous experience with the GP system in use. The parameters are briefly described below:

A. PopSize: The number of individuals in the population. At the low level the population size is 50 and at the high level it is 2000.
B. Generations: The system uses steady-state tournament selection so the generation parameter is the number of *generation equivalents* computed from *number of tournaments*. At the low level 50 generations are used and at the high level 250 are used.
C. MutationsFreq: Mutation frequency is the probability that an offspring will be subject to mutation. At the low level the mutation frequency is 10% and at the high level it is 90%.
D. CrossoverFreq: Crossover frequency is the probability that an offspring will be subject to crossover. At the low level the crossover frequency is 10% and at the high level it is 90%.
E. Demes: Determines whether the population is subdivided into subpopulations. In each experiment with demes we used 5 subpopulations, a crossover rate between

demes of 3% and a migration rate of 3%. At the low level demes are not used and at the high level they are used.

F. ErrorMeasurement: The error measurement determines whether fitness is the sum of *absolute values* of errors (parameter at low level) *or* the sum of *squared* errors (parameter at high level).

G. DynamicSubsetSelection: Dynamic Subset Selection (DSS) is a method that only uses a subset of all the fitness cases in each evaluation. The selection of fitness cases was based on their individual difficulty (40%), the time since they were last used in fitness calculation (40%) and randomly (20%) [5]. At the low level DSS is not used and at the high level it is used.

H. MissClassificationPenalty: The classification problems are mapped to symbolic regression problems; each class is given a unique number. The fitness value is either the absolute distance or the squared distance between the actual and desired value. This parameter governs the amount of extra penalty that is added to the fitness for incorrect (miss) classifications. At the low level it is 0.0 and at the high level it is 0.25.

I. FunDiv: Determines whether division instructions are in (high level) or not in (low level) the function set.

J. FunCondit: Determines whether conditional instructions such as comparison, conditional loads and jumps are in (high level) or not in (low level) the function set.

K. FunTrig: Determines whether trigonometric functions are in (high level) or not in (low level) the function set.

L. FunMisc: Determines whether other non- trigonometric, non-arithmetic and non-conditional functions are in (high level) or not in (low level) the function set.

M. InitSize: The maximal initial size of the individuals, measured in number of instructions. At the low level it is 50 and at the high level it is 100.

N. MaxSize: The maximal allowed size of an individual, in number of instructions. At the low level it is 128 and at the high level it is 1024.

O. Constants: Determines the number of constants used in each individual. At the low level it is 1 and at the high level it is 10.

P. MutationDistr: When an instruction block is mutated it can be done on several different levels; the block level, instruction level or sub-instruction level (RML 1998). At the low level the distribution between them is 80%, 10%, and 10% respectively, and at the high level it is 10%, 10% and 80%.

Q. HomologousCrossover: Determines the percentage of crossovers that are performed as homologous crossover [8]. At the low level it is 5% and at the high level it is 95%.

3.3 Response variable

We have used the maximum validation hit rate as the response variable for all problems and runs. This value was obtained by extracting the best individual on *training data* and running it on the validation set. It is reported as the percentage of correctly classified instances in the validation set.

Our choice of response variable defines the unit for the effects from the analysis of the experimental data. If, for example, an effect is calculated to be 5 this means that the average effect that can be expected when changing the factor from its low to its high level will be 5 percentage units (*not* 5%). Thus if the average response is 65% we would expect 70% on average with the factor at its high level.

3.4 Experimental designs

We have used three different experimental designs in a sequential fashion, each one based on the results from the previous one. The first two designs have been used on all three problems with the settings of factor levels described above. The third design uses different levels for the factors and has only been used on the gaussian problem. The purpose of the first experiment is to screen the large number of factors down to a more manageable set. Later experiments study the effects of the remaining factors in more detail.

To reduce the number of runs in the screening experiment we have employed a saturated design first described by Ehlich [3] [10]. This design allows the estimation of the main effects of seventeen factors in eighteen runs. The confounding patterns for this design is very complicated; main effects are confounded with several two- and higher-order effects.

The factors that had the largest effect in the screening experiments are varied in the second round of experiments. The rest of the factors are held constant at intermediate levels ($N = 256$, $P = (40, 40, 20)$, $Q = 50$) or at the level indicated by the sign of its effect from the screening experiment (I, K, L, M, O at their low level and J at its high). We employ a fractional factorial experiment of resolution four. In this design the main effects are confounded with three-factor interactions which are assumed to be negligible. This allows the estimation of all main effects. Two-factor effects can be estimated but are confounded with each other. The actual design used is a 2^{8-4} fractional factorial with 16 runs [3]. The generators for this design are D=ABC, E=BCH, F=ACH and G=ABH where a low level is represented by -1 and a high level by 1.

In order to estimate all two-factor interactions individually we need a design of resolution five. This is illustrated for the gaussian problem in the third experiment, which uses a 2^{5-1} fractional factorial with 16 runs [3]. The generator for this design is H = ABCD.

In this third experiment we study the five factors that had the largest effect in experiment 2 on the gaussian problem. We alter the levels of these factors to gain more knowledge of their effect. The population size and number of generations had a significant effect and by increasing them (A to (500, 5000) for low and high level respectively and B to (100, 500)) we want to investigate if this effect holds also for higher levels. By increasing the low level of the mutation and crossover probabilities to 50% and keeping the high level at 95%, we can investigate if the level of 95% was extreme. By altering the values of both the low (to 0.05) and high levels (to 0.5) of the miss-classification penalty we can investigate if it is only important to have this penalty regardless of level or if the level in itself is important.

4   Results

Below we document the results for the seven experiments conducted. All values reported for the effect of factors and for confidence intervals is in the same unit as the response variable, see section 3.3. We have conducted a sensitivity analysis to evaluate how sensitive our results are to the number of replicates used for each parameters setting. This analysis is briefly described below.

4.1 Results of the screening experiment on IONOSPHERE

For each of the eighteen factor settings ten (10) replicates were run on the ionosphere problem. The standard error calculated from these 180 runs was 2.04 giving a 95% confidence interval of 4.63. The effects that were statistically significant at this confidence level are (in order of decreasing effect): A, B, G, C, H, D, E, F. The effect of A was about 45% larger than the effect of F.

4.2 Results of the screening experiment on Gaussian

For each of the eighteen factor settings eight (8) replicates were run on the gaussian problem. The standard error calculated from these 144 runs was 0.74 giving a 95% confidence interval of 1.94. The effects that were statistically significant at this confidence level are (in order of decreasing effect): A, B, C, H, E, D, G, F, J*, O*, P*, L*. However, note that the four factors marked with an asterisk had much smaller effect than the previous eight. For example the effect of F is more than four times higher than the effect of J.

4.3 Results of the screening experiment on PIMA-diabetes

For each of the eighteen factor settings, eight (8) replicates were run on the pima-diabetes problem. The standard error calculated from these 144 runs was 0.33 giving a 95% confidence interval of 0.96. The effects that were statistically significant at this confidence level are (in order of decreasing effect): A, C, G, B, F, E, H, D, L*, N*, P*, M*. However, note that the four factors marked with an asterisk had much smaller effect than the previous eight. For example the effect of D is more than eight times the effect of L.

4.4 Result of Second experiment on ionosphere

For each of the sixteen factor settings ten (10) replicates were run on the ionosphere problem. The standard error calculated from these 160 runs was 0.66 giving a 95% confidence interval of 1.50. The effects that are statistically significant at this confidence level are as follows: Population size (A) effect *4.9*. Number of generations

(B) *2.23*. The combination of AD+BC+EH+FG gives *2.16*. The combination of AG + BH + CE + DF gives *1.89*. The combination of AH + BG + CF + DE gives *1.75*. For all experiments, a 95% confidence interval corresponds to +/- 1.50 units in *effect*.

The population size (A) has the largest effect while the number of generations (B) and three different two-factor-interaction combinations have similar effects. The values reported in the table should be interpreted in the following way: if we change the level of factor A from its low to its high level we can expect an average increase in the validation hit rate by 4.19 units with a 95% confidence interval from 2.69 to 5.69 units. The same type of interpretation can be made for all effects reported in this paper.

The average validation hit rate was 92.1%, with a maximum of 98.7% and a minimum of 66.9%. The maximum average for a particular setting of the factors was 98.2% and the minimum 85.8%. These results can be compared with the maximum reported result from the UCI database describing the ionosphere problem: an average of 96% obtained by a backprop NN and 96.7% obtained with the IB3 algorithm [11]. However, we measure generalisation in a slightly different way: In the GP community it is common to look for the best generalizer in the population at reporting intervals in contrast to noting generalization capabilities among the best performing solution candidate on the training set. This difference applies for all experiments in this paper.

4.5 Result of second experiment on gaussian

For each of the sixteen factor settings ten (10) replicates were run on the gaussian problem. The standard error calculated from these 160 runs was 0.84 giving a 95% confidence interval of *1.94*. The effects that are statistically significant at this confidence level are as follows; A with effect 11.51, C with effect 5.21, B with effect 5.14, the combination of AD + BC + EH+ FG with effect 3.39, D with effect 2.82, the combination of AH + BG + CF + DE with effect 2.76 and the combination of AF + BE + CH + DG with effect 2.35. For all experiments, a 95% confidence interval corresponds to +/- 1.94 units in *effect*.

The population size (A) clearly has the largest effect with the mutation probability (C) and number of generations (B) having about half the effect of A. Three different two-factor-interaction combinations and the crossover probability (D) have smaller effects.

The average validation hit rate was 63.8%, with a maximum of 88.9% and a minimum of 48.6%. The maximum average for a particular factor setting was 83.7% and the minimum 52.3%. This can be compared to the theoretical limit for this problem with a dimensionality of eight: 91%. However, note that this limit does not take the eight false inputs into account.

4.6 Result of second experiment on pima-diabetes

For each of the sixteen factor settings ten (10) replicates were run on the pima-diabetes problem. The standard error calculated from these 160 runs was 0.72 giving a 95% confidence interval of 1.63. The effects that are statistically significant at this confidence level are as follows: A with effect 5.72, B with effect 2.12 and G with effect 2.02, all with +/- 1.63 corresponding to 95% confidence.

The population size (A) have the largest effect while the number of generations (B) and the dynamic subset selection (G) have smaller effects.

The average validation hit rate was 65.46%, with a maximum of 77.6% and a minimum of 61.5%. The maximum average for a particular setting of the factors was 72.8% and the minimum 61.5%. These results can be compared with the maximum reported result from the UCI database describing the pima-diabetes problem: 76% using the ADAP learning algorithm [11].

4.7 Result of third experiment on gaussian

For each of the sixteen factor settings ten (10) replicates were run on the gaussian problem. The standard error calculated from these 160 runs was 0.89 giving a 95% confidence interval of 2.02. The effects that are statistically significant at this confidence level are as follws: B effect 9.39, A effect 7.53, H effect 3.71 and the combination of AD has effect -2.30, given 95% confidence at +/-2.02. Note that the levels used for the factors in this experiment are not the same as for the previous experiments. Hence, the actual effects are not comparable between experiments 2a and 3.

The number of generations (B) and the population size (A) have the largest effects. The positive effect of the increased miss-classification is smaller but still significant. The same is true for the interaction between the population size (A) and the crossover probability (D). Note that since this design has resolution five this two-factor interaction is not confounded with any other two-factor interaction, as was the case in previous experiments. The somewhat surprising negative effect of this interaction means that some caution is called for when using large population sizes; increasing the crossover probability might have a detrimental effect.

The average validation hit rate was 76.4%, with a maximum of 88.9% and a minimum of 61.6%. The maximum average for a particular factor setting was 85.8% and the minimum 65.1%.

5. Discussion

We have presented our first results in a larger project attempting to investigate the effect of GP parameters. Even though these results stem from a limited number of problems and experimental designs we believe that some interesting conclusions can be drawn. However, we are far from settling the questions raised in the introduction, but we can identify interesting patterns.

In all three screening experiments the same eight parameters had the largest effects with the remaining nine factors having small or statistically insignificant[1] effects. Among these nine factors that were consistently screened *out*, we can find the factors determining the function set, the initial and maximal size of the individuals, the number of constants, the distribution of different mutation operators and the amount of crossovers that are homologous. It will be interesting to see if this result is valid for other problems and in other ranges of the continuos parameters.

Consistently, on all three problems, the population size and the number of generations are the most significant parameters. The population size comes out on top in the second experiments on all three problems with the number of generations a close second or third. However, note that the effect of the population size is numerically much larger than the other effects; this indicates that having a large population is important to get good results with GP. *Effort* has not been individually targeted in this survey, but it is interesting to note that choosing a large population size sometimes is more important than a large number of generations. In other words: a large population size running for very small number of generations could be better than a small population size running for a "normal" number of generations. More investigation is needed on this.

It is interesting to note that the mutation and crossover probabilities have rather large effects on the gaussian problem. This somewhat contradicts the notion that mutation probability should be low. However, these factors did not have a statistically significant effect on the two real-world problems.

Dynamic subset selection can have a positive effect on the performance [4]. The fact that it, in addition, decreases the execution time of a run considerably would further speak for a more widespread use.

On both the gaussian and the ionosphere problem there are significant two-factor interactions. Since the design for experiment number two had a resolution of four we cannot separate the effect of different two-factor interactions. If we would like to do so we could add further runs to the existing designs or use a design of resolution five. Note that it can often be wise to use a design with lower resolution first and then add runs to separate between two-factor interactions of interest. In general, this will reduce the total number of runs needed. For example, to separate the four two-factor interactions having a combined effect of 3.39 on the gaussian problem in table 5 would require 3 extra experiments. Using a design of resolution five would require 64 runs; 48 more runs than for the design used herein.

The third experiment on the gaussian problem was included to show an example of a design of resolution five. Furthermore, the levels of the factors studied were changed to see their effect in other ranges of values. It is notable that the population size and number of generations are still the dominant factors. Note, however that the population size is no longer dominating; this could indicate that there is a limit to what can be gained from increasing the population size. The positive effect of the miss-classification penalty factor indicates that not only is it good to have such a penalty, but a relatively large penalty is better than a smaller one.

---

[1] If an effect is not statistically significant it can not be separated from the natural variation in the response, ie noise.

Our results partly support the notion that GP systems are robust to different parameter settings, as long as we choose the right values for the most important ones: population size and number of generations. On some problems the crossover and mutation probability can give good results with large levels. However, the negative interaction between population size and crossover probability in experiment 3 indicates that some caution must be taken.

The methodology used in this work can be used to optimize the results from a GP system. For example, note that the average response on the third experiment on gaussian is higher than the average response on the second experiment on the same problem. This is because the levels used in the third experiment were chosen based on the results from the second experiment. Thus, in addition to giving researchers a way to map out the effect of different parameters, DoE techniques may be used to optimize the response on a particular problem.

A drawback with the kind of DoE techniques used in this work is that they assume that higher-order interactions between factors are negligible. The empirical evidence for making this assumption are abundant; experimental investigations frequently show that the effect can be explained by a few important factors [5]. However, we can never be fully sure and it will probably be wise to conduct full factorial experiments on some problems to validate this assumption. We have also noted that the responses in our experiments are often not normally distributed but grouped into clusters. In theory this makes statistical analysis of effects difficult since it violates the assumption of normally distributed responses. In practice, most statistical techniques have shown to be robust against deviations from normality [3].

It is worth noting that the GP system consistently performed very well compared to the previously reported best results on the test problems but with the caveat that generalization is measured differently. In future work we plan to change generalisation measurements to comply with the methods used in the UCI-database.

6. Conclusions

The Design of Experiments (DoE) techniques, from mathematical statistics, have been introduced as a solid methodology for evaluating the effect of genetic programming parameters. These techniques can also be used to increase the performance of a GP system, by guiding the user in choosing 'good' parameter combinations.

Our experiments show that, on three binary classification problems, the most important parameter was the population size followed by the number of generations. On one problem, large mutation and crossover probabilities had a positive effect. Furthermore, on all three problems, the same and large number of factors could be screened out because their effect could not be distinguished from noise. The result supports the notion that GP systems are robust against parameter settings but highlights the fact that there are a few parameters that are crucial.

This work reports the first results from a larger project attempting to investigate the effect of GP parameters. Much more work, involving more detailed designs as well as more varied test problems, is needed before we can address the questions as to the role and effect of GP parameters. We believe that such findings can be of great importance to the applicability of genetic programming in both industry and academia.

Acknowledgements

The authors wish to acknowledge Martin Hiller and Bill Langdon, whose comments increased the quality of this paper. Peter Nordin gratefully acknowledges support from the Swedish Research Council for Engineering Sciences.

References

1. Banzhaf, W., Nordin, P. Keller, R. E., and Francone, F. D. (1998). Genetic Programming - An Introduction. On the automatic evolution of computer programs and its applications. Morgan Kaufmann, Germany
2. Box, G. E., Hunter, W. G., Hunter, J. S. (1978). Statistics for Experimenters – an Introduction to Design, Data Analysis and Model Building. Wiley & Sons, New York, USA.
3. Ehlich, H. (1964). Determinantenabschatzungen fur binare Matrizen. Math. Z. 83, 123-132.
4. Gathercole C. and Ross P. (1994) Dynamic Training Subset Selection for Supervised Learning in Genetic Programming, Chris. In proceedings of the 3rd conference on Parallel Problem Solving from Nature (PPSN III), Springer-Verlag, Berlin, Germany.
5. Kleijnen, J. P. C. (1998). Experimental Design for Sensitivity Analysis, Optimization, and Validation of Simulation Models. In Handbook of Simulation, (ed.) Banks, Wiley & Sons, New York, USA.
6. Koza J.R, Andre D., Bennett F.H., Keane M. A. (1999) Genetic Programming III: Darwinian Invention and Problem Solving. Academic Press/Morgan Kaufmann.
7. Nordin, J.P. (1997), Evolutionary Program Induction of Binary Machine Code and its Application. Krehl Verlag, Muenster, Germany.
8. Nordin J. P., Banzhaf W., and Francone F. (1999) Efficient Evolution of Machine Code for CISC Architectures using Blocks and Homologous Crossover. To appear in *Advances in Genetic Programming III*, (eds.) Langdon, O'Reilly, Angeline, Spector, MIT-Press, USA
9. RML (1999) Register Machine Learning Incorporated. http://www.aimlearning.com
10. StatLib (1999), Online Statistical resources library at the Department of Statistics, Carneige Mellon University, USA, http://lib.stat.cmu.edu/.
11. UCI ML repository (1999). Files for the Pima-diabetes and Ionosphere problems from the Machine Learning repository at University of California, Irvine describing the Ionosphere problem. http:// www.ics.uci.edu/~mlearn.

Experimental Study of Multipopulation Parallel Genetic Programming

Fernández F.[1], Tomassini M.[2], Punch III, W. F.,[3], Sánchez J.M.[1],

[1] Departamento de Informática. Escuela Politécnica. Universidad de Extremadura. Cáceres.
{fcofdez, sanperez} @unex.es
[2] Institut d'Informatique. Universite de Lausanne.
mtomassi@iissun4.unil.ch
[3] Computer Science and Engineering Department. Michigan State University.
punch@cse.msu.edu

Abstract. The parallel execution of several populations in Evolutionary Algorithms has usually given good results. Nevertheless, researchers have to date drawn conflicting conclusions when using some of the Parallel Genetic Programming models. One aspect of the conflict is population size, since published GP works do not agree about whether to use large or small populations. This paper presents an experimental study of a number of common GP test problems. Via our experiments, we discovered that an optimal range of values exists. This assists us in our choice of population size and in the selection of an appropriate Parallel Genetic Programming model. Finding efficient parameters helps us to speed up our search for solutions. At the same time, it allows us to locate features that are common to Parallel Genetic Programming and the classic Genetic Programming Technique.

1 Introduction

Many researchers have supported the use of several populations when working with Evolutionary Algorithm (EA) techniques. Different experimental and theoretical studies have reported the efficiency of parallel Genetic Algorithms and have studied the relationship between the classic model and the Island model [1][2]

But in the Genetic Programming (GP) domain, things are less clear. Some researchers talk about the usefulness of working with multiple large populations [3], while others question both the size of populations [4] and the efficiency of multiple populations in GP [5], at least for the few problems that have been intensively studied.

Nevertheless, there is no doubt about the nature of the Parallel Genetic Programming (PGP) algorithm. All authors agree that tuning parallel GP parameters, (e.g. the number of populations, the size of each population, the topological connections between populations) is important for gaining maximal performance.

Our goal is to study the parameters that affect parallel performance and study their interactions in common problems. By so doing we hope to develop a more robust model of how parameters might be set so as to maximize the performance of parallel GP on many different types of problems.

This article is structured as follows. The next section deals with the Island Model for parallel processing and the design of our experiments. Afterwards, we look at the tool we used that facilitated the development of these experiments. Next, we present the problems addressed, followed by a detailed description of results and regularities found in all the problems. Finally, we offer our conclusions, which lead us to the suggestion of the existence of an optimal parameter range of values in both GP and PGP.

2 Methodology

2.1 The Island Model

Many researchers have investigated how to parallelize classic evolutionary computation algorithms. Several ideas have been proposed and employed: parallelization using multiple interacting populations, parallelization of fitness evaluation etc.

In GP, parallelization using multiple interacting populations has been used in a number of interesting experiments [3][6][7]. This model introduces some new parameters into the algorithm and so we decided to use it to compare results.

While multipopulation models can be classified in a number of ways [8], we are most interested in coarse-grain parallelisation. In coarse grained models, also called island parallel models, the population of individuals is divided into several autonomous subpopulations, usually called demes. Demes are allowed to exchange individuals at a certain rate, called the migration rate, according to a usually predefined communication topology. The main reason for using this approach is its ability to avoid premature convergence by injecting new individuals, -these promote diversity- while at the same time exploring different portions of the search space within each deme.

Within each deme, a standard sequential evolutionary algorithm is executed between migration phases. Several migration policies have been described [8]. The most common one replaces the worst k individuals of a deme with the same number of individuals coming from other different populations, usually copies of the best individuals.

The topology of communication is also important. The most common topologies are ring structures, 2-d and 3-d meshes. Hypercubes and random graphs have also been employed. Here we introduce a dynamical topology, described in the following section, in which the exchange pattern changes during the run.

At present, there is no commonly accepted way of deciding the best set of parameters (number of demes, size of each demes, frequency of exchange, size of exchange, quality of exchange, topolgy, etc.). Moreover, changes on the part of an algorithm sometimes modify the behavior of the rest. These new parameters not only affect the "parallel" behavior of the system, but also affect the other aspects of all the algorithms; it is thus difficult to decide how parameters affect results.

Therefore, due to the high number of parameters and their interaction, we cannot simultaneously study all of them, and instead must select those that a priori seem to be the most important. Thus, in our work, we decided to study how the number of subpopulations, the number of individuals, and the migration rate affect the results. Furthermore, we wanted to see how a dynamically changing topology, which will be explained below, may help to solve several typical problems.

2.2 The Software Tool

We chose to modify a standard GP tool [9] to allow us to work with several demes over an heterogeneous network. Our tool [10] uses the Parallel Virtual Machine [11] communication primitives to connect processes, which in this case are subpopulations.

This tool allows us not only to decide classic parameters like the number of individuals per population or mutations and crossover probabilities, but also to choose the number of populations involved in the experiment, the migration rate, the number of migrating individuals and the communication topology.

The tool works by means of the client/server model, in which each client is a subpopulation and the server is a process that takes charge of the input/output buffers and of establishing the communication topology (Fig 1). Thus the server can either work with a predefined topology or dynamically change it during the run.

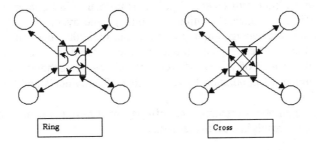

Fig. 1. Client/Server Architecture. The Server decides the topology.

The exchange method we explored most deeply was a random exchange topology. In fact, this means that no set topology was employed. When a population was ready for an exchange, the central server decided randomly from which population the new values would be drawn. This topology proved very effective on the test problems (see section 3.2) Executions and Effort

When using a parallel algorithm, one must carefully judge "improved performance" as these improvements could be due to two factors:

– The time saved by a simultaneous execution of code.
– The possible improvement due to the parallel nature of the algorithm.

While the first factor is useful for shortening the time required to find solutions in GP, which is often a slow process, the second is interesting because it shows us new features of the algorithms compared to the sequential version.

Although the first point certainly helped us to gather a lot of results in an affordable time, we designed the experiments to explore the second factor. We were interested in finding out how different parameters affect results.

Thus, in order to compare results, we decided to analyze data by means of the computational effort, calculated as the number of nodes that are evaluated in a GP tree. Let's suppose that we have executed the same experiment N times; we can compute the average size of individuals per generation in each of the executions (the number of nodes per individual solution tree), and then calculate the average of those N values. What we really have in each generation is the average number of nodes per individual over N execution in a particular problem. Once this average has been computed, the required effort in a particular generation will be: I*N*AVG_LENGTH, where I is the number of individuals in the population, and AVG_LENGTH is the average number of nodes previously calculated.

The computed effort is not necessarily useful for comparing results between very different problems, because different functions, nodes, etc, may require different time values. Nevertheless, it is a helpful measure when comparing different results from the same problem.

Since we are working with several populations instead of just one, it is also easy to compute the effort. In each execution we compute the average number of nodes, taking into account the average number of nodes per individual per generation in each of the populations. Then we can proceed as we would in a classic model, by averaging all the values of the N different executions per generation. Finally we multiply the number of population by the number of individuals and by the average number of nodes.

If we add up a number of generations' efforts, we'll have the computational effort required to obtain a result in a particular generation.

2.3. The Problems Addressed

GA and GP have proven to be powerful tools when solving problems in many different fields. Results often have improved on others previously obtained by other kinds of machine learning methods.

In our work we are not particularly interested in solving specific problems but in extracting some basic features that could be useful in many problems; thus we decided to use some classic GP test problems. We chose "the even parity 5 problem" and "the symbolic regression problem".

2.3.1 The Even Parity 5 (Evenp 5) problem
The goal here is to decide the parity of a set of 5 bits. The Boolean *even-k-parity function* of k Boolean arguments returns T if an even number of its Boolean arguments are T, and otherwise returns NIL. If $k=5$, 32 different possible combinations are available, so 32 fitness cases must be checked to evaluate the accuracy of a particular program in the population. The fitness can be computed as the number of mistakes over the 32 cases.

Every problem to be solved by means of GP needs a set of functions and terminals. In the case of the Evenp-5 problem, the set of functions we have employed is the following: F={NAND,NOR}, smaller than that described in the original version of the problem [12].

2.3.2 The Symbolic Regression Problem

The goal here is to find an individual i.e. a program which matches a given equation. For each of the values in the input set, the program must be able to compute the output obtained by means of the equation.   We employ the classic polynomial equation:

$$f(x) = x^4 + x^3 + x^2 + x \qquad\qquad (1)$$

And the input set is composed of the values 1 to 1000.

For this problem, the set of functions is the following: F={*,//,+,-} where // is like / but returns 0 instead of ERROR when the divisor is equal to 0, thus allowing syntactic closure.

3 Results: Optimal Parameter Range of Values

Once the problems were defined, the different parameters of the experiments had to be chosen. We decided to deal only with the number of populations, the migration rate, the number of individuals in each deme.

Different test cases were tried. First we used only one population to solve the problem, with different number of individuals each time. Afterwards we tried to solve the problems with several populations and different numbers of individuals.. Each of the graphs we show is an average over 60 executions, so the results are statistically significant. Note that overall we performed approximately 4000 runs to gather these statistics.

3.1 Even Parity 5

Figure 2 compares different sizes of populations when solving the Evenp-5 problem. We can see that there is an optimum number of individuals which produces the best results. This seems to agree with other researchers results [1][4]
The best possible result in this problem is the value 16, when a given effort is applied.

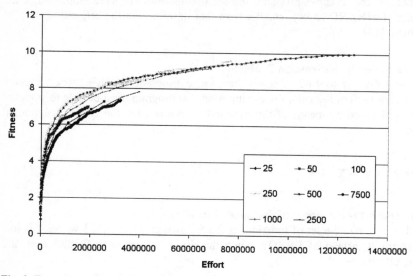

Fig. 2. Evenp-5 problem solved by means of Classic sequential GP.

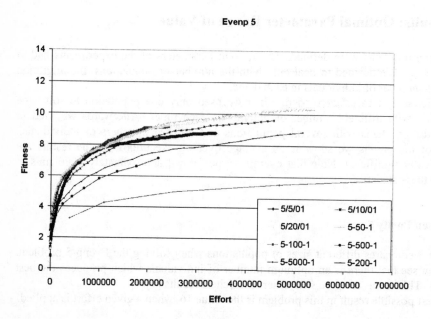

Fig. 3. The Evenp-5 problem solved by means of 5 population with a random communication Topology, where each population sends the best individual after 1 generation.

Figure 3 also compares different population sizes, but we are now working with the island model PGP. Again we find an optimum population size. In these experiments we used 5 populations communicating by means of random topology.

Finally, figure 4 depicts the same results as figure 3 but now using 2 populations. In both experiments, each population sends its best individual after each generation, according to random communication topology.

None of the three graphs finds solutions in the end, because we did not wait for them to be reached. However, some important information can be retrieved from the results.

EVENP 5

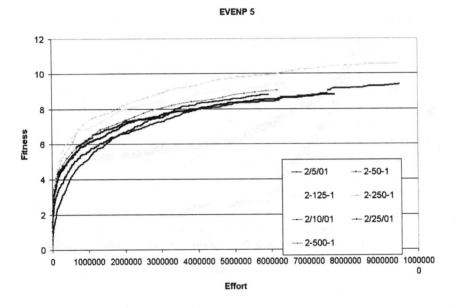

Fig. 4. The Evenp-5 problem solved by means of 2 population with a random communication Topology, in which each population sends the best individual after 1 generation.

We can observe an interesting point: the best choice in the classic problem corresponds with the best choice in PGP case: we just have to multiply the number of populations by the number of individuals. Although we cannot exactly say the optimum number of individuals in each of the cases, we can establish an optimal range limited by two numbers which indicate the best option each time. Let's imagine that the numbers are [A, B]. If we decide to use NP populations, in the PGP model, then this range tells us that the number of individuals per population should be chosen from between [A/NP, B/NP]. If we make NP equal to 1, then we are just talking about a number of individuals. For this problem the range seems to be [200,500].

But one should not generalize when talking about individuals. We think it's more accurate to talk about a pathway delimited by NUMBER_POPULATIONS * NUMBER_INDIVIDUALS.

3.2 Symbolic Regression

We present the results obtained from this problem when conducting the same experiments as with the evenp-5 problem.

In these diagrams, the best fitness value is 0. Therefore the best curves are at the bottom of each diagram.

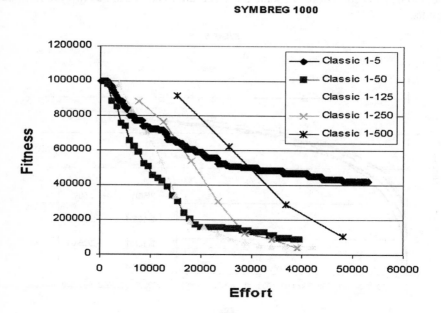

Fig. 5. The Symbolic Regression problem solved by means of Classic GP.

Figure 5 now compares different population sizes when solving the problem with 1000 checkpoints. Again, an optimum number of individuals can be observed, although it is different from that obtained in the evenp-5 problem (Figure 2).

Figure 6 is equivalent to figure 3 but the results are for the problem at hand and using 10 populations instead of 5. Figure 7 is also analogous to figure 4 employing 2 populations.

If we apply the same reasoning as in Evenp-5 problem, we observe that a different range appears again. Now, the limits could be [100,200]. So, when using one population we have a range of individuals [100/1, 200/1]. When using X, we have [100/X,200/X] which agree with the previous results: if we put 1 instead of X.

SYMBREG 1000

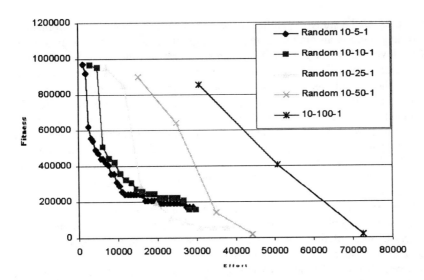

Fig. 6. The Symbolic Regression problem solved by means of 10 population with a random communication topology in which each population sends the best individual after 1 generation.

SYMBREG 1000

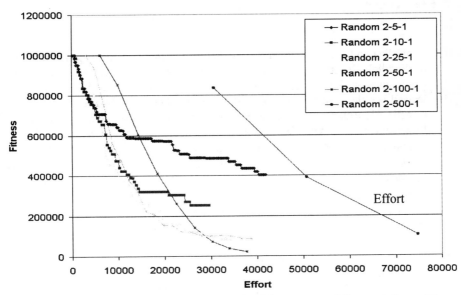

Fig. 7. The symbolic regression problem solved by means of 2 population with a random communication topology, in which each population sends the best individual after 1 generation.

4 Conclusions and Future Work

We have presented a study of parallel GP as applied to two standard GP problems. It indicates that there is an optimal parameter range of values that allows us to reduce the effort involved in solving these problems. Moreover, the range is different for each problem. Thus, we should try to classify problems by means fo this range. The validity of these results is being checked via investigation into other problems. To date, the results for these new problems have supported our previous conclusions concerning the existence of the pathway. These experiment's results help us to understand the working of GP and PGP and the relationship between several parameters, particularly between the number of populations and the number of individuals. We think that only by means of a series of experiments like that which we presented above will we be able to understand perfectly the dynamics of evolutionary techniques.

On the other hand, we must bear in mind that more experiments, and more kinds of problems should be addressed. We are planning to compare different communication topologies in order to find out which is the best, or at least to classify problems according to the results of the experiments.

Moreover, theoretical models of the multiple populations' dynamics should be built in order to obtain more explanations for the empirically observed phenomena.

In future papers we will present new comparisons between other interesting parameters when using PGP.

5 References

1. Cantú Paz and David Goldberg: "Predicting Speedups of Ideal Bounding Cases of Parallel Genetic Algorithms". Proceedings of the Seventh International Conference on Genetic Algorithms. Morgan Kaufmann.
2. Darrell Whitley, Soraya Rana, and Robert B. Heckendorn. "Island Model Genetic Algorithms and Linearly Sparable Problems". Evolutionary Computing: Proceedings of the AISB Workshop, Lecture notes in computer science, vol. 305 D. Corne and J. L. Shapiro (Eds), Springer-Verlag, Berlin, 109-125, 1997.
3. David Andre and John Koza: "Parallel Genetic Programming: A Scalabel Implementation Using the Transputer Network Architecture". Advances in Genetic Programming 2. The Mit Press.
4. Matthias Fuchs: "Large Populations are not always the best choice in Genetic Programming". Proceedings of the Genetic and Evolutionary Computation Conference GECCO 1999.
5. William F. Punch: "How effective are multiple populations in Genetic Programming". Genetic Programming 1998: Proceedings of the Third Annual Conference, J. R. Koza, W. Banzhaf, K. Chellapilla, K. Deb, M. Dorigo, D. B. Fogela, M. Garzon, D. Goldberg, H. Iba and R. L. Riolo (Eds),Morgan Kaufmann, San Francisco, CA, 308-313, 1998.

6. S. Arnone, , M. Dell'Orto, A. Tettamanzi, M. Tomassini,: "Highly Parallel Evolutionary Algorithms for Global Optimization, Symbolic Inference and Non-Linear Regression". Proceedings of the 6th Joint EPS-APS International Conference on Physics Computing. 1994.

7. M. Oussaidéne, B. Chopard, O. V. Pictet, , M. Tomassini: "Parallel Genetic Programming and its application to trading model induction". Parallel Computing, 23, pp. 1183-1198, 1997.

8. S-C Lin, W.F. Punch and E.D. Goodman, "Coarse-grain Genetic Algorithms, Categorization and New Approaches" Sixth IEEE Parallel and Distributed Processing Oct 94, pg. 28-37.

9. Weinbrenner, T.;"Genetic Programming Kernel Version 0.5.2 C++ Class Library". http://thor.emk.e-technik.th-darmstadt.de/~thomasw/gpkernel1.html

10. Francisco Fernández, Juan M. Sánchez, Marco Tomassini and Juan A. Gómez: " A parallel Genetic Programming Tool Based on PVM". Jack Dongarra, Emilio Luque, Tomás Margalef (Eds) Recent Advances in Parallel Virtual Machine and Message Passing Interface. Springer.

11. V.S. Sunderman "PVM: A Framework for Parallel Distributed Computing", Journal of Concurrency: Practice and Experience" pp. 315-339, Dec 1990.

12. J. Koza "Genetic Programming II". The MIT Press.

Genetic Programming and Simulated Annealing: A Hybrid Method to Evolve Decision Trees

Gianluigi Folino, Clara Pizzuti and Giandomenico Spezzano

ISI-CNR, c/o DEIS
Univ. della Calabria
87036 Rende (CS), Italy
{folino,pizzuti,spezzano}@si.deis.unical.it

Abstract. A method for the data mining task of data classification, suitable to be implemented on massively parallel architectures, is proposed. The method combines genetic programming and simulated annealing to evolve a population of decision trees. A cellular automaton is used to realise a fine-grained parallel implementation of genetic programming through the diffusion model and the annealing schedule to decide the acceptance of a new solution. Preliminary experimental results, obtained by simulating the behaviour of the cellular automaton on a sequential machine, show significant better performances with respect to C4.5.

1 Introduction

Data mining consists in the extraction of implicit, previously unknown and interesting knowledge from real-world databases [16, 3]. Data mining techniques have been originally developed from related research studies in machine learning, statistics and database systems. These research fields, however, did not address the problem of effectively mine information from large databases. One of the major data mining requirements is the applicability of the developed methods to huge amounts of data in databases. Thus the knowledge discovery algorithms must be efficient and scalable to large databases [1]. Several data mining tasks have been defined based on different kinds of available knowledge. Among them is *data classification*, which consists in identifying common characteristics in a set of objects contained in a database and categorising them into different groups (*classes*). To build a classification, a sample of the tuples (also called examples) of the database is considered as the *training set*. Each tuple is composed of the same set of attributes, or features, which are used to distinguish them, and an additional class attribute, which identifies the class that the tuple belongs to. The task of classification is to build a description or a model for each class by using the features available in the training data. The models of each class are then applied to determine the class of the remaining data (*test set*) in the database. *Decision trees* [17] currently represent one of the most highly developed techniques for the classification of databases. A decision tree is a tree where the leaf nodes are labelled with the classes, while the non leaf nodes (decision nodes) are labelled with the attributes of the training set. The branches living a node represent a test on the values of the attribute. The path from the root to a leaf represents a set of conditions attribute-value (a rule) which describes the class labelling that leaf. There is a rule for every leaf node, thus a class is modelled by a set of rules. Decision trees are evaluated with respect to two parameters: *accuracy* and *size*. Accuracy measures the rate of misclassification. A totally accurate tree should correctly predict the class of any example from the database. The size regards the number of nodes of the tree. The simpler is the tree, the more concise is the class description and the information described can be easily understood. C4.5 [17] is the most famous decision tree based classification method.

In this paper a hybrid method that couples *genetic programming* (GP) and *simulated annealing* (SA) for the data mining task of classification is proposed. The method is based on a cellular genetic programming environment like that proposed in [4] to evolve decision trees but enriched with the simulated annealing strategy for the selection of new individuals. A cellular genetic programming environment [4] uses a *cellular automaton* to assign a spatial location on a low-dimensional grid to each GP individual. Every GP individual encodes a decision tree and it is allowed to interact only within a small neighbourhood. The new proposed method, called CGA/SA, adopts the local selection strategy of *simulated annealing* (SA) to evolve decision trees. The current GP individual is occasionally substituted with a new generated decision tree even if the latter has a misclassification rate (in this case represented by the fitness) worst than the former. The substitution is done under the guidance of a control parameter called the *temperature*. The combination of a cellular automaton with genetic programming and a simulated annealing based selection strategy of new individuals allows for a number of advantages. In fact, on one hand, the utilisation of a cellular automaton to map a population of trees on a two dimensional grid enables for a direct parallelisation of genetic programming through the diffusion model [15]. On the other hand, simulated annealing avoids the problem of premature convergence inherent to genetic programming by allowing *uphill moves* to solutions of worse fitness. A preliminary sequential implementation of the method, which simulates the cellular automata framework, shows a very good behaviour in both the complexity (that is the number of nodes) of the generated decision trees and the ability to generalise unknown examples. The paper is organised as follows. In section 2 the standard approach to data classification through genetic programming is shown. In section 3 the simulated annealing method is presented. In section 4 the combination of genetic programming and simulated annealing is discussed. In section 5 the hybrid method to evolve decision trees, which combines cellular genetic programming and simulated annealing, is presented. In section 6, finally, we give the results of the method obtained by a sequential implementation.

2 Genetic Programming and Data Classification

Genetic programming is a variation of *genetic algorithms* [6] in which the evolving individuals are themselves computer programs instead of fixed length strings from a limited alphabet of symbols [9]. Programs are represented as trees with ordered branches in which the internal nodes are *functions* and the leaves are so-called *terminals* of the problem. The *GP* approach evolves a population of trees by using the genetic operators of *reproduction, recombination* and *mutation*. Each tree represents a candidate solution to a given problem and it is associated with a *fitness value* that reflects how good it is, with respect to the other solutions in the population. The reproduction operator copies individual trees of the current population into the next generation with a probability proportionate to their fitness (this strategy is also called roulette wheel selection scheme). The recombination operator generates two new individuals by crossing two trees at randomly chosen nodes and exchanging the subtrees. The two individuals participating in the crossover operation are again selected proportionate to fitness. The mutation operator replaces one of the nodes with a new randomly generated subtree.

Genetic programming can be used to inductively generate decision trees for the task of data classification. Decision trees can be interpreted as composition of functions where the function set is the set of attribute tests and the terminal set are the classes. The function set can be obtained by converting each attribute into an attribute-test function. Thus there are as many functions as there are attributes. For each attribute A, if $A_1, \ldots A_n$ are the possible values A can assume, the corresponding attribute-test function f_A has arity n and if the value of A is A_i then $f_A(A_1, \ldots A_n) = A_i$. When a tuple has to be evaluated, the function at the root of the tree tests the corresponding attribute and then executes the argument outcoming from the test. If the argument is a terminal, then the class name for that tuple is returned, otherwise the new function is executed. The fitness is the number of training examples classified in the correct class. Both crossover and mutation must generate syntactically correct decision trees. This means that an attribute can not be repeated more than once in any path from the root to a leaf node. In order to balance the accuracy against the size of the tree, the fitness is augmented with an optional parameter, the *parsimony*, which measures the complexity of the individuals [9]. Higher is the parsimony, simpler is the tree, but accuracy diminishes.

Several methods for data classification based on genetic programming have recently been proposed [11, 12, 18, 5, 4].

Interesting results, however, have been obtained when such methods are applied to problems that evolve small decision trees. If the database contains a high number of examples with many features, large decision trees are requested to accurately classify them. In data mining applications, databases with several millions of examples are common. A decision tree generator based on genetic programming should then cope with a population of large sized trees. Furthermore, it has already been pointed out [18] that, in order to obtain the same classification accuracy of a decision tree generated by C4.5, small population size is inadequate. Processing large populations of trees that contain many nodes considerably degrades the execution time and requires an enormous amount of memory. The utilisation of parallel strategies used to increase the performances of genetic programming and to realise a really scalable data classification package for data mining applications, seems to be the only choice.

3   Simulated Annealing

Simulated annealing [7] is a randomised technique for finding a near-optimal approximate solution of difficult combinatorial optimisation problems. A SA algorithm starts with a randomly generated candidate solution. Then, it repeatedly attempts to find a better solution by moving to a neighbour with higher fitness, until it reaches a solution where none of its neighbours have a higher fitness. Such a solution is called *locally optimal*. In order to avoid getting trapped in poor local optima, simulated annealing strategy occasionally allows for *uphill moves* to solutions of lower fitness by using a *temperature parameter* to control the acceptance of the moves. At the beginning the temperature has a high value and then a cooling schedule reduces its value. The new solution is kept if it has a better fitness than the previous solution, otherwise it is accepted with a probability depending on the current temperature. As the temperature becomes cooler, it is less likely that bad solutions are accepted and that good solutions are discarded. In this way it should be possible to avoid getting trapped into local minima early in the execution and to explore the search space in its entirety.

4   Parallel Genetic Programming and Simulated Annealing

Genetic programming is well suited to be implemented on parallel architectures because the population can be distributed across the nodes of the system. One of the main problems in parallelising GP comes from the global selection of individuals, proportionate to their fitness, both in the reproduction and recombination steps. This kind of selection forces the sharing of the new solutions until the new population can be chosen. Two main approaches to parallel implementations of GP have been proposed to avoid this bottleneck. The *island* model [13] and the *diffusion* model [15].

The island model divides the population into smaller subpopulations. A standard genetic programming algorithm works on each partition and is responsible for initialising, evaluating and evolving its own subpopulation. The standard GP algorithm is augmented with a migration operator that periodically exchanges individuals among the subpopulations.

In the diffusion model each individual is associated with a spatial location on a low-dimensional grid. The population is considered as a system of active individuals that interact only with their direct neighbours. Different neighbourhoods can be defined for the cells. The most common neighbourhoods in the two-dimensional case are the 4-neighbour (*von Neumann neighbourhood*) consisting of the North, South, East, West neighbours and 8-neighbour (*Moore neighbourhood*) consisting of the same neighbours augmented with the diagonal neighbours. Fitness evaluation is done simultaneously for all the individuals and selection, reproduction and mating take place locally within the neighbourhood. Information slowly diffuses across the grid thus clusters of solutions are formed around different optima.

Another shortcoming of genetic programming due to the roulette wheel selection scheme is that those candidates having the better fitness are allowed to assume more places in the new population. In this way, individuals having higher fitness rapidly spread through the population and low fitness individuals are gradually lost. After a few number of generations the population presents a high degree of homogeneity and the power of recombination is considerably weakened. As a consequence, GP is not able to improve the solution getting trapped into a local optimum.

Simulated annealing, on the contrary, always accepts the new solution if it has a better fitness than the previous one and it accepts an inferior solution with a probability depending on the current temperature. As the temperature becomes cooler, it is less likely that bad solutions are accepted and that good solutions are discarded. This strategy, as already pointed out, guarantees the convergence property of the method. The combination of the two methods can thus take advantage of the suitability of genetic programming to be parallelised and of the capability of simulated annealing to avoid poor local solutions and to maintain a good diversity in the population. In the next section we propose a cellular genetic programming method enriched with the simulated annealing strategy, called CGP/SA, for classification of databases.

5   The CGP/SA Method

A new method for the task of data classification, suitable to be implemented on massively parallel architectures, is proposed. The method combines genetic programming and simulated annealing to evolve a population of decision trees. A *cellular automaton* (CA) [21] is used to realise a fine-grained parallel implementation of GP through the diffusion model and the annealing schedule is applied to establish the acceptance of a new solution. Preliminary experimental results, obtained from a sequential implementation of the approach that simulates the behaviour of the cellular automaton, show significant better performances with respect to C4.5 and comparable performances with respect to CGP [4].

The method follows other recent hybrid methods [8, 2], that incorporate simulated annealing into genetic algorithms, and the Cellular Genetic Algorithm proposed in [22]. Our approach, however, is the first proposal that couples cellular genetic programming and simulated annealing for classifying databases.

A cellular automaton is composed of a set of cells in a regular spatial lattice, either one-dimensional or multidimensional. Each cell can have a finite number of states. The states of all the cells are updated synchronously according to a local rule, called a transition function. The state of a cell at a given time depends only on its own state at the previous time step and the states of its "nearby" neighbours (however defined) at that previous step. Thus the state of the entire automaton advances in discrete time steps. The global behaviour of the system is determined by the evolution of the states of all the cells as a result of multiple interactions. Thus the recombination mechanism, that is responsible to generate the offspring, can be done by choosing the mate of the current individual in the local neighbourhood.

A *Cellular Genetic Programming algorithm coupled with Simulated Annealing technique*, called CGP/SA, can be designed by associating with each cell of a CA two substates: one contains an individual (tree) and the other its fitness. At the beginning a population of individuals is randomly generated and the fitness is evaluated. Then, at each generation, every tree undergoes one of the genetic operators (reproduction, crossover, mutation) depending on the probability test. If crossover is applied, the mate of the current individual is selected as the neighbour, in the Moore's neighbourhood, having the best fitness and the offspring is generated. The current tree is then replaced by one of the two offspring, the one having the best fitness, if fitness increase is less than or equal to the current temperature, defined by an annealing schedule. The algorithm on a 2-dimensional toroidal grid can be described by the pseudo-code shown in figure 1. The initial temperature is different for each cell and it is randomly set between 2 and 6 percent of the number of tuples. A parameter α, which has a value between 0.95 and 1.0, is chosen to reduce the temperature at each generation and such that the temperature assumes the final value when $MaxNumberOfGeneration$ steps have been executed. $select(t_i, t_0, t_1, temperature)$ first chooses between t_0 and t_1 the one having the best fitness. Suppose it is t_0. Then, t_i is replaced by t_0 only if $fitness(t_0) - fitness(t_i) \leq temperature$. This deterministic criterion [2] has been shown to be less expensive and to perform equivalently to the random technique.

6 Implementation and Experimental Results

In this section we present the experiments and results obtained by a preliminary implementation of the method on a sequential machine. The CGP/SA classifier has been implemented in C by modifying the *sgpc1.1* standard tool for genetic programming [20] to meet the requirements of our classification method. A procedure that does not allow for the generation of trees with repeated attributes on the branches, after the application of crossover and mutation operators, has been added. In order to simulate the cellular automata framework, the population has been mapped into a two-dimensional array of fixed dimensions 20×20. CGP/SA accepts discretised data sets (training and test set) as input. The environment runs on a Sun Ultraspark workstation with two 200-Mhz processors and 256 Mbytes of memory.

Experiments have been executed on standard databases contained in the UCI Machine Learning Repository [14]. Table 1 contains the description of these

```
begin
  Let  p_c, p_m, be the crossover and mutation
    probability
  temperature= initial_temperature()
  for each cell i in CA do in parallel
    generate a random individual t_i
    evaluate the fitness of t_i
  end parallel for
  while not MaxNumberOfGeneration do
  for each cell i in CA do in parallel
    generate a random probability p
    if  (p < p_c) then
    select the cell j, in the neighbourhood of i,
      such that t_j has the best fitness
    (t_0, t_1)= crossover(t_i, t_j)
    t_i = select(t_i, t_0, t_1, temperature)
    else
    if  (p < p_m + p_c) then
    mutate the individual
    else
    copy the current individual in the new population
    end if
    end if
  end parallel for
    temperature = temperature × α
  end while
end
```

Fig. 1. Pseudo-code of the CGP/SA algorithm.

databases. They present different characteristics in the number and type (numerical and nominal) of attributes, two-classes versus multiple classes and number of examples. A population of 400 elements has been used with a probability of 0.095 for reproduction, 0.890 for crossover and 0.01 for mutation. The maximum depth of the new generated subtrees is 4 for the step of population initialisation, 5 for crossover and 2 for mutation. The algorithm stops after 200 generations. In table 2 the results generated by C4.5 with pruning compared with those of CGP/SA are shown. The results of CGP/SA have been obtained by running the algorithm 10 times. In the table the best result, with respect to the misclassification error on the test set, is shown along with the average result in parenthesis. It is clear from the table that the trees generated by the CGP/SA algorithm with respect to C4.5 are smaller, for almost all the dataset, they have a misclassification error on the training set comparable, but, more important, they generalise better than C4.5. In particular, for the cancer, monk1 and monk3 datasets the results are very good. The tree obtained for the cancer dataset contains 9 nodes with respect to 41 of C4.5 and the test error is 18.95 instead of 30.6. The tree generated for monk1, with a size of 37, is able to correctly classify both the training and the test sets. The decision tree for monk3, although has a tree size of 17 nodes with respect to 12 of C4.5, it has an error of 4.92 on the training set, with respect to 6.6 of C4.5 and correctly generalise to the test set. For these two last datasets, C4.5 is not able to find the correct tree.

In table 3 the results of the CGP/SA and the CGA methods are presented on a bigger set of examples of those reported in [4]. Both algorithms stops after 200 generations. The behaviour of CGP/SA is almost always better than CGP. For example, for the Australian and German databases CGA obtains a size of 69 and 66 respectively, while CGA/SA obtains 30 and 44. Thus it seems that the method takes advantages of the introduction of simulated annealing strategy which is essentially based on allowing the substitution of the current string with a high probability when the temperature is high, and with a decreasing probability as long as the temperature diminishes. A better tuning of the parameters and the effect of population size could improve the performance of the method.

7   Conclusions and Future Work

A new approach to data classification based on a cellular genetic programming framework, augmented with simulated annealing technique, has been presented. The introduction of SA strategy to decide the acceptance of a new individual proved to be profitable. The approach showed to outperform Quinlan's C4.5 method by generating both smaller and more accurate trees on standard machine learning problems. The sequential implementation of the cellular genetic programming algorithm, however, needs running times that, as expected, are not competitive with respect to C4.5. This behaviour is obvious since CGP/SA manages a population of trees, while C4.5 works with only one tree at a time. On the other hands C4.5 performances notably degrades as the size of the tree increases, thus it is not able to deal with real data mining applications, having

Table 1. Databases description

DATABASE	ATTRIBUTES	TUPLES
Australian	14	690
Cancer	9	286
Crx	15	690
German	20	1000
Heart	13	270
Hypo	29	3772
Iris	4	150
Monk1	6	124
Monk2	6	169
Monk3	6	122
Pima	8	768
Vote	16	435

Table 2. Results generated by C4.5 and CGP/SA

DATABASE	C4.5			CGA/SA		
	Size	Training set	Test set	Size	Training set	Test set
Australian	60	4.6	12.1	30 (35.0)	10.43 (11.46)	10.00 (11.74)
Cancer	41	19.9	30.6	9 (63.9)	25.67 (22.44)	18.95 (21.25)
Crx	44	5.9	11.7	25 (22.7)	10.41 (12.16)	14.00 (16.15)
German	127	10.2	23.8	44 (46.6)	24.92 (27.44)	22.75 (24.10)
Heart	27	7.2	17.6	30 (18.7)	14.44 (19.22)	12.22 (14.77)
Hypo	21	0.2	0.9	39 (28)	0.60 (0.95)	0.87 (1.25)
Iris	7	1.0	6.3	14 (10.4)	2.00 (2.60)	2.00 (3.80)
Monk1	18	16.1	24.3	37 (40.1)	0 (0)	0 (0)
Monk2	31	23.7	35.0	65 (54)	14.75 (19.10)	30.32 (33.10)
Monk3	12	6.6	2.8	17 (16.8)	4.92 (5.58)	0 (1.90)
Pima	99	6.2	18.3	77 (45.4)	19.14 (24.57)	21.48 (22.34)
Vote	7	4.3	6.9	16 (19)	3.33 (3.77)	2.22 (2.37)

Table 3. Results generated by CGA and CGP/SA

DATABASE	CGA			CGA/SA		
	Size	Training set	Test set	Size	Training set	Test set
Australian	69 (30.1)	10.65 (13.00)	9.56 (11.48)	30 (35.0)	10.43 (11.46)	10.00 (11.74)
Cancer	9 (51.2)	16.23 (22.25)	17.90 (21.05)	9 (63.9)	25.67 (22.44)	18.95 (21.25)
Crx	38 (26.4)	9.18 (11.96)	14.50 (16.15)	25 (22.7)	10.41 (12.16)	14.00 (16.15)
German	66 (61.4)	24.62 (24.56)	21.86 (23.92)	44 (46.6)	24.92 (27.44)	22.75 (24.10)
Heart	31 (32.9)	16.67 (17.50)	12.22 (14.87)	30 (18.7)	14.44 (19.22)	12.22 (14.77)
Hypo	30 (25.9)	0.48 (1.23)	0.87 (1.59)	39 (28)	0.60 (0.95)	0.87 (1.25)
Iris	15 (9.9)	2.00 (2.30)	2.00 (4.00)	14 (10.4)	2.00 (2.60)	2.00 (3.80)
Monk1	37 (39.4)	0 (0)	0 (0)	37 (40.1)	0 (0)	0 (0)
Monk2	21 (33.4)	27.87 (25.16)	32.18 (34.07)	21 (32.9)	27.87 (25.98)	32.17 (33.56)
Monk3	17 (15.4)	4.92 (5.82)	0.0 (1.76)	17 (16.8)	4.92 (5.58)	0 (1.90)
Pima	69 (43.2)	22.65 (24.08)	20.70 (22.15)	77 (45.4)	19.14 (24.57)	21.48 (22.34)
Vote	13 (27.40)	4.67 (4.53)	2.22 (2.58)	16 (19)	3.33 (3.77)	2.22 (2.37)

hundreds of attributes and thousands of tuples, in reasonable times. Parallel implementation of our method, which is in progress, should successfully cope with big sized databases.

References

1. M. Chen, J. Han and P.S. Yu (1996). Data Mining: an Overview from Database Perspective. *IEEE Transaction on Knowledge and Data Engineering*, 8(6), pp.866-883.
2. H.Chen, N.S.Flann and D.W.Watson (1998). Parallel Genetic Simulated Annealing: A Massively Parallel SIMD Algorithm. In *IEEE Transaction on Parallel and Distributed Systems*, vol. 9, No.2.

3. U.M. Fayyad, G. Piatesky-Shapiro and P. Smith (1996). From Data Mining to Knowledge Discovery: an overview. In U.M. Fayyad & al. (Eds) *Advances in Knowledge Discovery and Data Mining*, pp.1-34, AAAI/MIT Press.

4. G. Folino, C. Pizzuti and G. Spezzano (1999). A Cellular Genetic Programming Approach to Classification. *Proc. Of the Genetic and Evolutionary Computation Conference GECCO99*, Morgan Kaufmann, pp. 1015-1020, Orlando, Florida.

5. A.A. Freitas (1997). A Genetic Programming Framework for two Data Mining Tasks: Classification and Generalised Rule Induction. *GP'97: Proc. 2nd Annual Conference*, pp.96-101, Stanford University, CA, USA.

6. D.E. Goldberg (1989). *Genetic Algorithms in Search, Optimization and Machine Learning*. Addison Welsey.

7. S. Kirkpatrick, C. D. Gellant and M. P. Vecchi (1983). Optimization by Simulated Annealing. *Science* 220, pp. 671-680.

8. F. T. Lin, C. Y. Kao and C. C. Hsu (1991). Incorporating Genetic Algorithms into Simulated Annealing. *Proc. Fourth Int. Symposium Artificial Intelligence*, pp. 290-297.

9. J. R. Koza (1992). *Genetic Programming: On Programming Computers by Means of Natural Selection and Genetics*, MIT Press.

10. J. R. Koza and D.Andre (1995) Parallel genetic programming on a network of transputers. *Technical Report CS-TR-95-1542, Computer Science Department*, Stanford University.

11. N.I. Nikolaev and V. Slavov (1997). Inductive Genetic Programming with Decision Trees. *Proceedings of the 9th International Conference on Machine Learning*, Prague, Czech Republic, April 1997.

12. R.E. Marmelstein and G.B. Lamont (1998). Pattern Classification using a Hybrid Genetic Program - Decision Tree approach. *Proceedings of the Third Annual Conference on Genetic Programming*, Morgan Kaufmann.

13. W.N. Martin, J. Lienig and J. P. Cohoon (1997), Island (migration) models: evolutionary algorithms based on punctuated equilibria, in T. Back, D.B. Fogel, Z. Michalewicz (eds.), *Handbook of evolutionary Computation*. IOP Publishing and Oxford University Press.

14. C.J. Merz and P.M. Murphy (1996). UCI repository of Machine Learning. *http://www.ics.uci/mlearn/MLRepository.html*.

15. C. C. Pettey (1997), Diffusion (cellular) models, in T. Back, D.B. Fogel, Z. Michalewicz (eds.), *Handbook of evolutionary Computation*. IOP Publishing and Oxford University Press.

16. G. Piatesky-Shapiro and W. J. Frawley (1991). Knowledge Discovery in Databases. AAAI/MIT Press.

17. J. Ross Quinlan (1993). *C4.5 Programs for Machine Learning*. San Mateo, Calif.: Morgan Kaufmann.

18. M.D. Ryan and V.J. Rayward-Smith (1998). The Evolution of Decision Trees. *Proceedings of the Third Annual Conference on Genetic Programming*, Morgan Kaufmann.

19. W.A. Tackett (1993). Genetic Programming for feature discovery and image discrimination. *Proceedings of the Fifth International Conference on Genetic Algorithms*.

20. W.A. Tackett and A. Carmi. Simple Genetic Programming in C. Available through the genetic programming archive at *ftp://ftp.io.com/pub/genetic-programming/code/sgpc1.tar.Z*.

21. T. Toffoli and N. Margolus (1986). *Cellular Automata Machines A New Environment for Modeling*. The MIT Press, Cambridge, Massachusetts.

22. D.Whitley (1993). Cellular Genetic Algorithms. *Proceedings of the Fifth International Conference on Genetic Algorithms*, Morgan Kaufmann.

Seeding Genetic Programming Populations

W.B. Langdon[1] and J.P. Nordin[2]

[1] Centrum voor Wiskunde en Informatica, Kruislaan 413, NL-1098 SJ, Amsterdam
bill@cwi.nl http://www.cwi.nl/~bill
Tel: +31 20 592 4093, Fax: +31 20 592 4199

[2] Chalmers University of Technology, S-41296, Göteborg, Sweden
nordin@fy.chalmers.se http://fy.chalmers.se/~nordin
Tel: +49 31 772 3159, Fax: +46 31 772 3150

Abstract. We show genetic programming (GP) populations can evolve under the influence of a Pareto multi-objective fitness and program size selection scheme, from "perfect" programs which match the training material to general solutions. The technique is demonstrated with programmatic image compression, two machine learning benchmark problems (Pima Diabetes and Wisconsin Breast Cancer) and an insurance customer profiling task (Benelearn99 data mining).

1 Introduction

In every day speech we regard learning and remembering as somewhat similar. However *machine learning* means something very different from memorising. In machine learning we are not interesting in letting the computer memorise a number of facts, we know this is something it does very well. Instead we want our algorithm to find patterns in data, particularly patterns that enables generalisation on so far *unseen* data. We would like the computer find the "essence" of the information that we present to it.

Here we try and turn genetic programming (GP) on its head. Instead of asking GP to find a function which matches some training data and then seeing how well the evolved function generalises, we construct such a function before hand and give it to GP. We then run GP and see if it can evolve the function to be more general. That is GP starts from a solution rather than a random starting point. This is done by *seeding* the initial population with perfect individuals that can already solve the fitness cases. We thus skip the memorisation part and go right to the generalisation part.

We use simple deterministic algorithms to produce perfect individuals from the fitness cases used for training. There are many possible ways of doing this. In this paper (rather than creating initial programs at random) we assemble them from a large number of *if-then-else clauses* that either find a combination of input variables for each training case or we produce something like a decision tree where the desired output is narrowed down through interval tests on all of its input variables. The evolution is started with a *Pareto fitness function* [11] where multiple objectives are sought concurrently. In our case we want to optimise (keep) a good performance while reducing the size of the individuals. In GP there is strong evidence of the link between small size and good generalisation capabilities [15, 17]. So we need some kind of parsimony pressure. This is achieved by a Pareto tournament where individuals can win either by being good or being short. Our hope is that the parsimony pressure will replace the bulky if-then-else clauses with more elegant, short and general expressions.

Seeding of initial populations is not a very well studied technique in GP. It has been used in some financial applications [2], game playing [5], information retrieval [6, 9], speeding evolution [11, 6.9], scheduling [11, C.9.4], image compression [14] and planning [1]. There has also been research in genetic algorithms for instance [3, 4, 18]. Here seeding is used not to find a good starting point for learning but to give a perfect individual and a good starting point for generalisation.

In Sect. 2 we describe our experiments on the programmatic image compression problem [7, 14], where the objective is to evolve a program that reproduces a bitmap image when run. This is an extremely hard problem and although our method is possible, we had to abandon these experiments in favour of problems which need less computational power. We chose (Sect. 3) the North American Pima Indians diabetes problem. Despite starting with a very specific individual, GP was able to quickly boil the genetic soup down to very simple expressions. In one instance, to a program with only one input, see Fig. 1. This is a remarkable data mining and data selection feat. We suggest GP has an inherent capability to perform *variable selection* where by non-relevant variables are discarded during evolution. Finally we also report results from the Wisconsin Breast Cancer Database (9 attributes, Sect. 4) and from a real-world customer profiling problem with 85 attributes (Sect. 5).

Fig. 1. Short program evolved by the second seeded Pima Diabetes run. It means *if Plasma glucose concentration at two hours in an oral glucose tolerance test exceeds 155.5 then the Diabetes test will be positive.* (All values in the training and verification sets are actually integers). This very simple program scores approximately as well as sophisticated machine learning techniques [12].

2 Programmatic Compression

We can view any system which creates programs which create data as a compression system if the programs are smaller than the data they create. In GP terms, the data to be compressed are fitness cases for symbolic regression. GP tries to evolve an individual program that outputs the uncompressed data, to a certain degree of precision. If the evolved program solution can be expressed by fewer bits than the target data, then we have achieved a compression. Here the data are the individual pixels of a picture (Fig. 2).

Fig. 2. Target image (256 × 167). Note variety of repeating patterns.

After 50 generations in a GP run (using crossover only) with Pareto selection and a population of 100, the seed program proved to be extremely robust. On average across the whole population each pixel was less than 1 grey level out, with 99.41% being exactly correct. However on average the seed had only shrunk from 275,083 to 273,259 i.e. by 0.7%. (Mean depth has increased from 18 to 19.9).

3   Pima Diabetes

This real-world medical classification problem is taken from the UCI Machine Learning repository http://www.ics.uci.edu/~mlearn/MLSummary.html. The target is to predict whether the patient shows signs of diabetes. Five sets of runs where conducted (with five independent runs for each experiment, details in Table 1). Three experiments used three different seeds and in another two the initial population was created at random.

All the seeds were created deterministically by creating a function composed of IF statements, one per training case (less one). Each IF tests all attributes and if they all match returns this training case's class. If not the else branch contains an IF relating to the next training case. And so on. If none of the IF's match, the program returns the class of the last training case.

Terminals containing each attribute's value are compared with constant expressions using the APPROX primitive. APPROX returns true if its two arguments are within 10% of each other. The outputs of the eight APPROX functions per training case are combined using 7 AND functions. The seed generation program chooses the closest available constant to the actual attribute value in the training record. If none are within 10% then an expression combining two constants with ADD, SUB, MUL or DIV is used instead. Typically such expressions will occur more than once in the seed but no effort is made to avoid creation of duplicate code. As we expect these expressions are large, asymmetric, have high fitness on the training cases but fail to generalise.

Table 1. GP parameters for Pima Indians Diabetes (defaults are as [8, page 655])

Objective:	Find a program that predicts Diabetes.
Terminal set:	One terminal per data attribute (8). For each attribute, create unique random constants between its minimum and maximum value (in the training set). Use integers where the attribute has integer only values (total 255 primitives).
Function set:	IF IFLTE MUL ADD DIV SUB AND OR NAND NOR XOR EQ APPROX NOT
Fitness cases:	Training 576 (194 positive). 192 (74 positive) for verification only.
Fitness:	number correct (hits).
Selection:	2 objectives, hits and size, combined in Pareto tournament group size 7. Non-elitist, generational. Fitness sharing (comparison set 81) [11].
Wrapper:	Value ≥ 0.5 taken as true
Pop Size:	500
Max prog. size:	no limit
Initial pop:	100% seeded or created using "ramped half-and-half" [7, pages 92–93] (no duplicate checking).
Parameters:	90% one child crossover, 2.5% function point mutation (rate 100/1024), 2.5% constant mutation (rate 100/1024), 2.5% shrink mutation, 2.5% size fair mutation (subtree size$\leq$30) [10].
Termination:	Maximum number of generations $G = 50$

The first and smallest seed was generated using the first ten negative training case and the first ten positive. These twenty cases alternate, starting with the first negative training case. The second seed was generated using the first half, i.e. 288 cases, of the training set. While the last was generated from all 576 training cases.

Table 2 summarises the verification performance of the five approaches to the Pima problem. For each approach there are two rows. The top relates to the mean of the highest verification score found in each of the five runs, while the lower relates to the first occurrence of programs which score 137 or more. (We chose 137 to indicate satisfactory performance since it is one standard deviation below the best reported results on this problem). For each we report its actual performance "hits", its size (reporting the smallest where multiple programs have the same performance in the same generation) and the first generation when it evolved. In each category: the mean (of five runs), the standard deviation (in brackets) and the minimum and maximum are given.

GP starting from 500 random programs (i.e. no seed) and using a Pareto fitness measure did not perform well. Possibly because it converged too readily to small programs. When we removed the small size objective and just used scalar fitness, 40% of runs achieved reasonable performance. However 100% of runs starting with seeded populations did. It appears there is little actual information in the problem data as a seed constructed from only 20 training records contains enough information to start the GP process off in the right direction. The solutions found when using the two larger seeds are very much bigger. While we anticipate further evolution over more generations would eventually lead to a

Table 2. Highest verification score in run. 5 Pima runs with each seed

Seed size	Hits	Size		Generation	% runs ≥ 137
none (Pareto)	121.0 (4.0) 119–129	25.4 (23.8)	3–56	8.0 (6.6) 1–17	0
none (scalar)	131.4 (7.6) 122–143	30.6 (20.8)	6–64	9.6 (4.8) 4–18	
	138.5 (1.5) 137–140	4.5 (1.5)	3–6	17.0 (3.0) 14–20	40
20	143.4 (0.5) 143–144	34.2 (30.2)	3–91	35.4 (6.0) 28–43	
	139.4 (2.9) 137–143	3.0 (0.0)	3–3	26.0 (12.3) 14–42	100
288	144.0 (1.3) 142–145	1603.0 (2462.9)	21–6489	39.8 (6.9) 29–47	
	137.4 (0.8) 137–139	4.0 (2.0)	3–8	22.4 (6.5) 17–34	100
576	141.8 (2.4) 139–146	6369.4 (4603.7)	294–13479	40.6 (6.5) 31–50	
	137.0 (0.0) 137–137	62.4 (118.8)	3–300	32.8 (13.2) 18–50	100

reduction in size, we use GPQUICK where run time is $O(|\text{programs}| \times |\text{test set}|)$ [11, 8.8]. Thus runs starting from larger seeds are much slower than runs using shorter seeds or starting from random populations.

Figures 3, 4 and 5 plot the evolution of the first run with each of these three seeds. Figure 6 refers to a run starting from random. The graphs on the left hand side refer to training set performance. The two smaller seeds do not generalise to the remainder of the training set and consequently initially fitness is low (cf. gen 0), while the largest seed matches the whole training set and so scores 100%. All three fail to generalise and so have low scores on the verification set (right hand graphs).

The top graphs refer to the Pareto front, i.e. those individuals in the population which are not dominated by others. (Note we have two objectives, training set hits and reducing size). The Pareto front always spreads rapidly away from the initial seed as shorter programs are produced. With the two smaller seeds, there is also a rapid initial increase in fitness in the first generation but slow change after that. With the small seed (20) the population shows less signs of over training and the programs in it shrink. By the end of the run they are on average less than a tenth of the size of the initial seed. With the two bigger seeds, the Pareto front does not collapse in this way but instead the population continues to retain large programs which score well on the training set.

In the case of the two smaller seeds the best validation score in the whole population remains near (but slightly above) the best validation score of the individuals reported on the Pareto front. In the case of the biggest seed, these individuals are large and actually perform much worse on the validations tests than the best program in the population. (The size of these is plotted in the x-y plane of the top right hand graphs of Figs. 3, 4 and 5).

In each case within a few generations programs which score above 70% on the verification set are found. Slow improvements continue so that, by the end of the run, scores (75%) comparable with the best models produced by many machine learning techniques on this problem have been found.

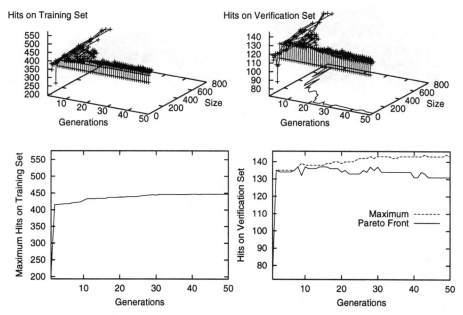

Fig. 3. Evolution of Pareto front in first seeded (20) Pima run

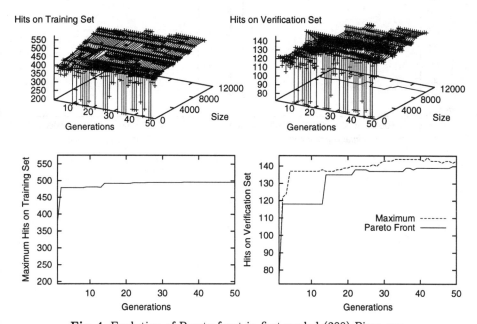

Fig. 4. Evolution of Pareto front in first seeded (288) Pima run

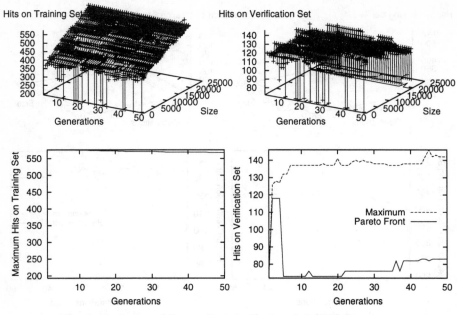

Fig. 5. Evolution of Pareto front in first seeded (576) Pima run

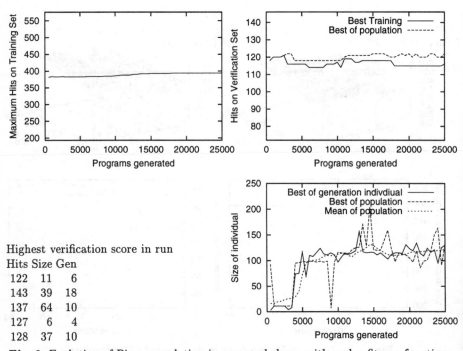

Highest verification score in run

Hits	Size	Gen
122	11	6
143	39	18
137	64	10
127	6	4
128	37	10

Fig. 6. Evolution of Pima population in non-seeded run with scalar fitness function

Table 3. Wisconsin Breast Cancer (as Table 1 except were given)

Objective:	Find a program that predicts Breast Cancer.
Terminal set:	One terminal per data attribute (9). Constants -1 to 10
Fitness cases:	351 (130 positive). 348 (111 positive) verification tests.

Table 4. Cancer. Highest verification score on Pareto front (5 runs on each seed)

Seed size	Hits	Size			Generation		% runs ≥ 326
none	333.2 (4.4) 326–338	22.6 (	21.7)	5–65	28.4 (17.0)	3–50	
(Pareto)	328.2 (2.4) 326–332	6.2 (	1.6)	5–9	14.2 (11.1)	3–35	100
20	329.8 (1.2) 329–332	101.4 (	122.2)	19–342	34.0 (10.1)	20–49	
	327.0 (1.3) 326–329	65.0 (	95.8)	5–256	32.0 (10.9)	16–47	100
128	327.0 (2.9) 323–332	46.4 (	38.1)	20–122	38.8 (8.1)	29–47	
	327.2 (1.1) 326–329	42.0 (	29.1)	20–92	38.0 (9.2)	29–50	80
351	327.4 (2.4) 323–330	6422.0 (2610.8)	3703–11257		41.8 (2.7)	37–45	
	326.5 (0.5) 326–327	2621.8 (2197.1)	871–6390		46.5 (6.1)	36–50	80

4 Wisconsin Breast Cancer

Like the Pima Diabetes, Wisconsin Breast Cancer is a real-world binary medical classification problem often used as a machine learning benchmark. Again three seeds were constructed in the same way as in Sect. 3. Details of the GP parameters are given in Table 3, while Table 4 summarises the verification results of five independent runs starting with each seed. We choose a score of 326 out of 348 on the verification set as indicating satisfactory performance. (State of the art ML performance is about 96% on this problem, so 326 is 2σ below it). Table 4 indicates GP can obtain satisfactory performance without seeding. Indeed the large size of programs in the initial population created from the two bigger seeds appears to hamper GP in this problem.

5 Insurance Customer Profiling

Given 85 attributes relating to a customer, such as age, number of children, number of cars, income, other insurance policies they hold, predict if they want caravan insurance. 5922 records (of which 343 are positive and the rest negative) are available as training data from http://www.swi.psy.uva.nl/benelearn99/comppage.html as part of Benelearn'99. The task is to find 800 records of a further 4000 records which contain as many positive examples as possible.

We conducted two experiments: 1) GP starting from a random population and 2) the initial population was filled with copies of a seed created by C4.5 release 8 [16]. (The unpruned C4.5 trees produced using default parameter settings were used). The 5922 records were randomly split in half. One half (with 179 positive examples) was used as the training set and the other half was used as a verification set. The details are given in Table 5.

Table 5. Insurance Customer Profiling (as Table 1 except were given)

Objective:	Find a program that predicts the most likely $1/5^{th}$ of people to become caravan insurance customers.
Terminal set:	One terminal per data attribute, 0..41 and 110 different random numbers uniformly selected from 0..10 (total 255 primitives)
Function set:	IF IFLTE MUL ADD DIV SUB AND OR NAND NOR XOR EQ APPROX GTEQ LTEQ GT LT NOT
Fitness cases:	2911 (179 positive)
Hits:	number of positive cases predicted
Fitness:	At end of each generation each positive fitness case given a weight equal to the reciprocal of the number of individuals which correctly predicted it. Fitness given by sum of weights of positive cases predicted.
Wrapper:	2911 values sorted, top 1/5 (583) treated as positive predicted
Pop Size:	100, 1000, 5000, 20000
Max prog. size:	no limit
Initial pop:	"ramped half-and-half" depth 5–9 Or 100% seeded with C4.5 decision IFLTE tree
Parameters:	90% one child crossover, 5% point mutation (rate 10/1024), 5% size fair mutation
Termination:	Maximum number of generations G = 50

Figure 7 shows the distribution of performance on both the training and verification sets in the final population. The population changed dramatically from the initial seed. As expect the initial seed performs reasonably well on the training set (from which was created) but only slightly better than random guessing on the verification set. By the end of the run the population is widely spread. On the training set, most of the population lies close to the Pareto front (solid line). Since GP has discovered programs that are shorter and/or fitter than the seed, the front is well to the left of it. After 50 generations the population has bloated as is indicated by the cluster of points at the top left. These programs are long and have high training scores but they do not score markedly more than shorter programs in earlier generations. In fact while almost all programs longer than the seed (681) score better than it on the training case, they are worse than it on the verification set, scarcely better than random guessing. That is the population contains long programs which are heavily over trained.

The performance on the verification set of the best members on the training set is also plotted (lower solid line). Above size $\approx$ 200 many programs perform above the line on the verification set. I.e. many programs do better on the verification set than the best (on the training set) programs.

The lower dotted line shows the best 1% of the population on the verification set. This line is almost flat, Showing that bigger programs give little if any real performance advantage. Indeed in this population programs as short as five appear to be the best. The upper dotted line shows the performance on the training set of the same 50 programs. Not surprisingly it is more erratic than the lower curve and climbs with programs size, again indicating over training.

It is clear that GP has been able to generalise the initial seed program and thereby considerably improve its performance. However there are also clear signs of over training and this increases with size of programs in the population.

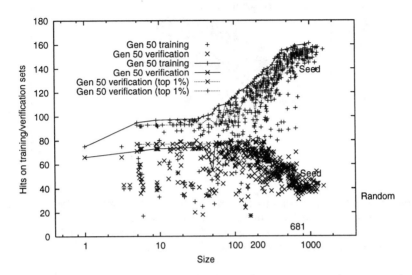

Fig. 7. Distribution of training (+) and verification (×) performance in generation 50 of the first seeded run of the customer profiling problem. Solid lines indicate individuals on the Pareto (training, size) front, while dotted lines indicate the best 1% of the population on the verification set. Note log scale.

Figure 8 show the final population of an unseeded GP run. The best performance on the training set is much lower than for the seeded run. However the population shows only a little sign of over training, and there is less variation with size. The verification performance is approximately the same as the seeded run, Fig. 7.

6 Conclusions

We have seen that it is indeed possible to start GP with non-random populations constructed from perfect individuals and use evolution under parsimony as a way to find good generalisers.

However constructing programs which memorise the training data (as used in Sects. 2, 3, 4 and 5) gives rise to large and unwieldy seeds if the whole training set is used. In our approach GP took many generations to reduce these to general solutions. Sections 3 and 4 show it is not necessary to use the whole training set and sometimes a very small sample of it will do.

GP is unique compared to other soft-computing technique in that it relies on symbolic structures in individuals and population. It is this property that enables

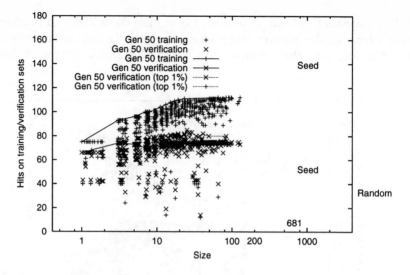

Fig. 8. Distribution of training (+) and verification (×) performance in the final generation of a non-seeded run of the customer profiling problem. Solid lines indicate individuals on the Pareto (training, size) front, while dotted lines indicate the best 1% of the population on the verification set. Note log scale.

seeding of perfect or near perfect starting points. As we saw in Sect. 5, the symbolic nature of GP-individuals also makes the integration of results from other techniques possible and opens possibilities for all sorts of hybrid approaches.

Our studies of populations evolved using Pareto multi-objective size versus performance selection clearly shows the often assumed relationship between long programs and poor generalisation.

In the light of the large amount of previous work trying to make GP memorise examples from Boolean functions or complex multidimensional data, it might seem a bit provocative to have a perfect individual from the start, but the approach turns out to be feasible. We think that this is because GP has a built in ability to generalise [11, 13, 15].

Acknowledgments

Funded by in part by the Wennergren foundation, NUTEK and TFR.

References

1. Ricardo Aler, Daniel Borrajo, and Pedro Isasi. Genetic programming and deductive-inductive learning: A multistrategy approach. In Jude Shavlik, editor, *Proceedings of the Fifteenth International Conference on Machine Learning, ICML'98*, pages 10–18, Madison, Wisconsin, USA, July 1998. Morgan Kaufmann.

2. Shu-Heng Chen and Chia-Hsuan Yeh. Using genetic programming to model volatility in financial time series: The case of nikkei 225 and S&P 500. In *Proceedings of the 4th JAFEE International Conference on Investments and Derivatives (JIC'97)*, pages 288–306, Aoyoma Gakuin University, Tokyo, Japan, July 29-31 1997.

3. E. William East. Infrastructure work order planning using genetic algorithms. In Wolfgang Banzhaf *et. al.*, editors, *Proceedings of the Genetic and Evolutionary Computation Conference*, pages 1510–1516, 1999. Morgan Kaufmann.

4. Jacob Eisenstein. Genetic algorithms and incremental learning. In John R. Koza, editor, *Genetic Algorithms and Genetic Programming at Stanford 1997*, pages 47–56. Stanford Bookstore, Stanford, California, 94305-3079 USA, 17 March 1997.

5. Gabriel J. Ferrer and Worthy N. Martin. Using genetic programming to evolve board evaluation functions for a boardgame. In *1995 IEEE Conference on Evolutionary Computation*, volume 2, page 747, Perth, Australia. IEEE Press.

6. Paul Holmes. The odin genetic programming system. Tech Report RR-95-3, Computer Studies, Napier University, Craiglockhart, Edinburgh, EH14 1DJ, UK, 1995.

7. John R. Koza. *Genetic Programming: On the Programming of Computers by Means of Natural Selection*. MIT Press, 1992.

8. John R. Koza. *Genetic Programming II: Automatic Discovery of Reusable Programs*. MIT Press, 1994.

9. D. H. Kraft, F. E. Petry, W. P. Buckles, and T. Sadasivan. The use of genetic programming to build queries for information retrieval. In *Proceedings of the 1994 IEEE World Congress on Computational Intelligence*, pages 468–473, 1994.

10. W. B. Langdon. The evolution of size in variable length representations. In *1998 IEEE International Conference on Evolutionary Computation*, pages 633–638.

11. William B. Langdon. *Data Structures and Genetic Programming: Genetic Programming + Data Structures = Automatic Programming!* Kluwer, 1998.

12. Tom M. Mitchell. *Machine Learning*. McGraw-Hill, 1997.

13. Peter Nordin and Wolfgang Banzhaf. Complexity compression and evolution. In L. Eshelman, editor, *Genetic Algorithms: Proceedings of the Sixth International Conference (ICGA95)*, pages 310–317, 1995. Morgan Kaufmann.

14. Peter Nordin and Wolfgang Banzhaf. Programmatic compression of images and sound. In John R. Koza *et. al.*, editors, *Genetic Programming 1996: Proceedings of the First Annual Conference*, pages 345–350, 1996. MIT Press.

15. Peter Nordin, Wolfgang Banzhaf, and Frank D. Francone. Compression of effective size in genetic programming. In Thomas Haynes *et. al.*, editors, *Foundations of Genetic Programming*, GECCO'99 workshop, 13 July 1999.

16. J. Ross Quinlan. *C4.5: Programs for Machine Learning*. Morgan Kaufmann, 1993.

17. Justinian Rosca. Generality versus size in genetic programming. In John R. Koza *et. al.*, editors, *Genetic Programming 1996: Proceedings of the First Annual Conference*, pages 381–387, 1996. MIT Press.

18. Amir M. Sharif and Anthony N. Barrett. Seeding a genetic population for mesh optimisation and evaluation. In John R. Koza, editor, *Late Breaking Papers at the Genetic Programming 1998 Conference*, 1998. Stanford University Bookstore.

Distributed Java Bytecode Genetic Programming with Telecom Applications

Eduard Lukschandl[1], Henrik Borgvall[2], Lars Nohle[2], Mats Nordahl[2], Peter Nordin[2]

[1] EHPT Sweden AB, Gothenburg, Sweden
Eduard.Lukschandl@ieee.org
[2]Chalmers University of Technology, Gothenburg, Sweden
(tfemn,Nordin)@fy.chalmers.se

Abstract. This paper describes a method for evolutionary program induction of binary Java bytecode. Like many other machine code based methods it uses a linear genome. The genetic operators are adapted to the stack architecture and preserve stack depth during crossover. In this work we have extended a previous system to run in a distributed manner on several different physical machines. We call our new system Distributed Java Bytecode Genetic Programming (DJBGP). We use the Voyager package for migration of Java individuals. The system's feasibility is demonstrated on a telecom routing problem.

1 Introduction

All a computer can do is to process machine language and all we do with computers, including genetic programming, will in the end be executed as machine code [1] [2] [3] [5]. In some cases it is advantageous to directly address the machine code with genetic programming and to *evolve* machine code. Machine code programming is often used when there is a need for very efficient solutions, e.g. in applications with hard constraints on execution time or memory usage. In general the reasons for evolving machine code - in contrast to higher level languages - are similar to the reasons for programming by hand in machine code or assembler: It is very efficient in terms of speed and space. It is also *close to the machine* ensuring relevance in size and speed between evolved individual and its use in applications [11].

Java is a platform independent open language with accelerating popularity. Most Java programs today are run on a processor with another machine code. In order to execute a binary Java program the system either has to interpret it in a Java virtual machine or compile it to the host machine code language,e for instance using a *just in time* (JIT) compiler [8] [12]. Processors designed to directly execute Java code are just beginning to appear on the market.

The binary *machine code* of Java is called *bytecode*. The interesting feature of this code is that almost all instructions occupy only a single byte. This contrasts with other trends in CPU design with long RISC instructions and even longer compound instructions. However, the short instruction length enables very compact programs, important for distribution, for instance of applets on the Internet.

We have previously presented our evolutionary program induction system for Java: Java Bytecode Genetic Programming (JBGP) and showed how it implements all of the relevant instructions in Java bytecode [9] [11].

In this paper we present enhancements to JBGP. The most important change is the parallelisation of JBGP. The population is divided into demes, which can be run on different physical machines. We call our new system Distributed Java bytecode Genetic Programming (DJBGP). Related work includes:

- AIM-GP: evolutionary induction of binary machine code for a number of platforms [12].
- JAPHET: a Java GP system with a linear genome where the individual is represented as a set of Java classes [6].
- bcGP: A system with similarities to DJBGP but with less annotation information and a smaller function set [4]

2 Method

The objectives of DJBGP is to implement a GP method with the following properties:

1. Evolutionary program induction of true Java bytecode executed directly on a virtual machine or Java chip.
2. The use of a standard Java virtual machine and not our own special implementation. In this way the system will run on many platforms and benefit from the ongoing refinement of the virtual machines.
3. Mastering the built-in difficulties with the Java code verifier and classloader.
4. An extensive function set with push and pops as well as jumps and conditionals.
5. Easy integration of the GP component with the environment using the evolved component.
6. Distributed execution on several machines.

The structure and relationship between different modules of the DJBGP-system is illustrated in figure 1 below.

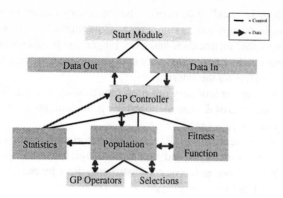

Figure 1: Structure of the JBGP system

The most important data structures are the *population* containing the *individuals*. In the JBGP-system the phenotype is not equal to the genotype. The genotype manipulated by the genetic operators contains additional information to guide crossover and mutation, for instance stack depth. However, the genotype contains true Java bytecode as part of the structure. The phenotype used for execution is created by stripping and concatenating *blocks* from the genotype into an executable consisting of a consecutive array of bytecode.

An instruction is an array of opcodes and parameters describing how it affects the stack and flow. Parameters specifying the number of pop and push operations the instruction performs on the stack, are necessary to avoid stack over- and underflow errors. For instance when the opcode array includes an opcode sequence that first pushes something onto the stack and then pops it, the *local maximum stack* variable is increased by one. JBGP needs this parameter to calculate the correct maximum stack value used by the individual.

The system also gives an opportunity to include new instructions in the instruction set. It is possible to invoke three different kinds of instructions: the general one word opcode, complex opcode instructions with arguments, and a macro containing a mixture of both of them.

An individual consists of a header and a body. The header in turn consists of two parameters, describing the fitness and the maximum stack depth used by the method. The body has two phases, one execution (phenotype) phase and one evolution (genotype) phase. In the evolution phase, the body consists of an instruction array. During the execution phase, it is translated into bytecode as a Java method. Transforming an individual from the evolution phase to the execution phase is accomplished by concatenating the opcode arrays of the instructions, and adding a method *header* and *footer*. Since the only part needed when executing the method is the actual opcode, it is not necessary to copy the entire instruction to each individual, thus saving memory.

To evaluate the population, it is turned into a Java class, passing it to the population output stream. This class is then loaded by a class loader, hence enabling the fitness function to invoke each of the methods of the Java class.

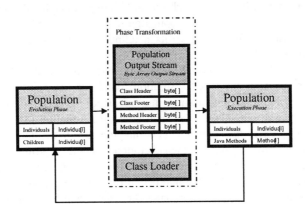

Figure 2: The population data structure

We use both crossover and mutation in our system. Both operators preserve stack depth. Crossover is performed as *two-point string crossover*. However, in our system we also make sure that the switched blocks have the same effect on the stack depth.

Mutation is simply performed by exchanging a single instruction with a random instruction, which *has the same effect on stack depth*. This also guarantees consistency of the stack.

DJBGP is the enhanced version of the JBGP-system that has a graphical user interface (GUI) and support for parallelization with demes. It was created to improve the usability of the system by providing an easy way to alter the parameters in the GP runs and to speed it up with parallel computation.

To implement parallelization we used the Voyager platform from Object Space Inc. It includes both basic features such as the ability to create remote objects and accessing them through an Object Request Broker (ORB) and more advanced, like mobile autonomous agents. Its syntax is pure Java. Once you have imported the relevant Voyager packages you can use them as you would ordinary Java classes.

2.1 Differences between JBGP and DJBGP

Besides the major differences between JBGP and DJBGP concerning the GUI and the demes, there are several smaller differences.. Using only float opcodes is not a large limitation and it results in a less complex system. Using simple opcodes instead of advanced is sensible as a *one action per invocation architecture*.

Demes are created in the Creator of Worlds class by using Voyager's Factory.create() method.

The demes have one population each. The populations are evaluated and propagated in a similar way to the JBGP system. The main difference is that in DJBGP each deme waits for a signal from the Synchronizer before propagating the population.

When a deme has propagated its population it selects a number of individuals, and distributes the selected individuals to its neighbors. It then waits for individuals to arrive from all its neighbors.

When all individuals have arrived, the deme checks if a solution was found in the last generation. It sends the result ("solution found" or "no solution found") to the Synchronizer. When the Synchronizer has received messages from all demes, it checks the messages to see if any of them contained "solution found". If that is the case a new run is started by sending a "new run" message to the demes. Otherwise the current run should proceed and a "new generation" message is sent to the demes. This procedure goes on until the number of solutions found are equal to the number of good runs, the number of runs equals the number of runs or a stop signal is received from the GUI. Then the Synchronizer terminates the experiment by sending a "stop" message to the demes.

The demes communicate with each other using standard Java listeners. The Creator of Worlds registers a deme as a listener to all its neighbors after creating the deme. Which demes that are its neighbors are given by the topology.

3. Evaluations and Results

DJBGP has been evaluated on a number of test problems from toy-problems such as the Santa Fe Trail [7] to more real-world type problems such as the telecom problem presented here. The problem below is included to demonstrate the feasibility in a real-world problem.

3.1 Dynamic Routing in Telecommunication Networks

A telecommunication network can be seen as a graph where the nodes represent telephone switches or exchanges and the edges represent communication links. Telephone calls may originate in any node and may be destined to any other node. A routing algorithm defines the selection of routes for every origin-destination pair of nodes.

A network simulator developed at EHPT has been used for building a graph which corresponds to the northern part of BT's Synchronous Digital Hierarchy Network. The following assumptions about links and nodes have been made:
- Links have unlimited transmission capacity.
- Links have zero transmission time.
- There is only one scarce resource in the nodes. It can be ports, buffers etc. The maximum number of units of this resource available in each node is called the *max load*. All nodes have the same max load.

- One unit of resource is used in a node for each call that passes through it, originates from it or is destined to it.

Each node has a price associated with it. The price is a measurement of how much money the network operator collects for each call that is destined to the node. This feature can be used to simulate local, long distance and international calls.

DJBGP and the network simulator cooperate in the sense that the simulator is used as an environment to the DJBGP-system. DJBGP evolves an individual and passes it to the simulator. The simulator inserts the individual into the nodes of the network and runs a predefined set of calls through it, measuring both the number of calls lost and the amount of lost money. Depending on the experiment, it then returns either the number of lost calls or the amount of money lost to DJBGP as the fitness of the individual.

An individual's return value is used for deciding where to route a call. When the simulator routes a call from a node, it executes the individual in each of its neighbor nodes. The input given to the individual is the load and the distance to the destination node. When finished, the simulator routes the call to the neighbor that returned the smallest value.

A call can become lost in two ways. Either if the call arrives at a node where it has already been handled or if it arrives at a node that operates at max load. In the first case the call is said to be *circular*, in the other *blocked*.

Three different experiments (A, B, and C) were carried out. All three experiments used a set of 250 calls with an average duration of 10 time units. On average, one call is started every time unit for a period of 250 units. In the first two experiments the input to the individuals were the load of the node (L) and the distance to the destination node (d). In the third experiment the individuals also got the price of the node as input.

In experiment A the fitness was the number of lost calls, in experiment B and C the amount of lost money. The number of circular, blocked and lost calls in experiments A, B and C are represented graphically in figure 3, 4 and 5.

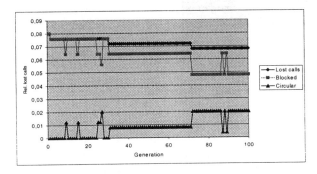

Figure 3: Experiment A: Maximizing the number of lost calls, 2 parameters.

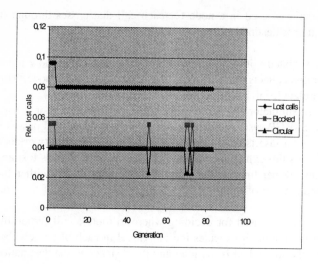

Figure 4: Experiment B: Maximizing the revenue, 2 parameters.

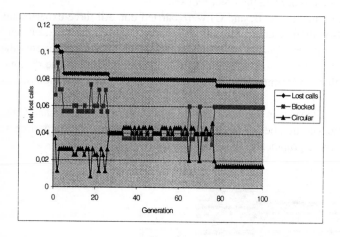

Figure 5: Experiment C: Maximizing the revenue, 3 parameters.

In the graphs above the individuals in experiment A loose the least calls. This is what could be expected because in experiment A, we optimized with respect to the amount of lost calls. The relative amount of lost calls for the best individual in each experiment are 6.8% for experiment A, 8.0% for experiment B and 7.6% for experiment C.

If the revenue collected is plotted instead we get a different result. See figure 6.

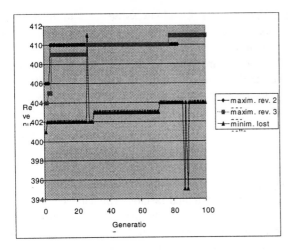

Figure 6: The Revenue from A-C

Here experiment C gives us the best results, closely followed by experiment B. This is also what we expect given the fact that the fitness function in these experiments returns the revenue lost. That experiment C performs a little better than experiment B is not strange either. It has access to more relevant information (the price of the node).

The maximum revenue collected for each experiment was 404 for A, 410 for B and 411 for C. The best individual in experiment A consists of the following opcodes:

```
fload_1
fconst_2
fdiv
fconst_2
fadd
fload_1
fconst_2
fdiv
fload_1
fload_0
fload_1
fload_0
fload_0
fmul
fsub
fdiv
fadd
fmul
fadd
fload_0
fmul
fload_1
fadd
freturn
```

The program can be translated into the expression:

$$R = L * (0.5 * d + 2 + 0.5 * d * (d + L / (d - L^2))) + d \qquad (1)$$

Here R is the return value, L the load and d the distance to the destination node.

324

We see that if the load is zero, the result value will always be equal to the distance d. This is quite reasonable. If no node has any load, the routing algorithm should route the call the shortest way to the destination.

The other terms in the expression above are harder to interpret. It is however interesting to compare the performance of the evolved individuals to other individuals that are constructed manually. Table 3 summarizes the performance of some of these.

4. Conclusion

In this paper we have presented an evolutionary system for induction of Java byte-code that can be used over a network in a distributed fashion. This enables the access to a very powerful computer environment - and with the increase in networking and connectivity - an enhanced performance of the system. Given the fact thhat we base the system on Java enables the use of heterogeneous networks and hosts. We have also demonstrated the feasibility with test runs on a complex real world problem from the telecom world.

Acknowledgement

Peter Nordin gratefully acknowledges support from the Swedish Research Council for Engineering Sciences.

References

1. Banzhaf, W., Nordin, P. Keller, R. E., and Francone, F. D. (1997). *Genetic Programming — An Introduction. On the automatic evolution of computer programs and its applications.* Morgan Kaufmann, Germany
2. Cramer N.L. (1985) A representation for adaptive generation of simple sequential programs. In Proceedings of the 1st International Conference on Genetic Algorithms and Their Applications, J. Grefenstette (ed.), p.183-187.
3. Crepeau R.L. (1995) Genetic Evolution of Machine Language Software, In: Proceeding of the GP workshop at Machine Learning 95, Tahoe City, CA, J. Rosca(ed.), University of Rochester Technical Report 95.2, Rochester, NY, p. 6-22.
4. Harvey, B., J.A.Foster, and D.Frinke (1998). Byte Code Genetic Programming. In: *Late Breaking Papers at the Genetic Programming 1998 Conference*, University of Wisconsin – Madison.
5. Huelsberger L. (1996) Simulated Evolution of Machine Language Iteration, In proceeding of: The First International conf. on Genetic Programming, Stanford,USA.
6. Klanhold,S., S.Frank, R.Keller, and Banzhaf (1998). Exploring the Possibilities and Restrictions of Genetic Programming in Java Bytecode. In: *Late Breaking Papers at the Genetic Programming 1998 Conference*, University of Wisconsin – Madison.

7. Langdon, W.B., and Poli, R. (1998). *Why Ants are Hard.* School of Computer Science, The University of Birmingham, UK., Technical report: CSRP-98-4.

8. Lindholm, T. and Yellin, F. (1996). *The Java Virtual Machine Specification.* Sun Microsystems, Inc., USA.

9. E.Lukschandl, M.Holmlund, E.Modén (1998) *Automatic Evolution of Java Bytecode: First experience with the Java Virtual Machine.* In R.Poli, W.B.Langdon, M.Schoenauer, T.Fogarty, W.Banzhaf (editors). Late Breaking Papers at EuroGP'98: the First European Workshop on Genetic Programming (Paris, 14-15 April 1998)

10. E.Lukschandl, M.Holmlund, E.Modén, M.Nordahl, P.Nordin (1998) *Induction of Java Bytecode with Genetic Programming.* In Koza, John R. (editor). Late Breaking Papers at the Genetic Programming 1998 Conference,University of Wisconsin

11. Nordin, J.P. (1997), Evolutionary Program Induction of Binary Machine Code and its Application. Krehl Verlag, Muenster, Germany

12. Venners, B. (1998). *Inside the Java Virtual Machine.* McGraw-Hill, New York, NY., USA.

Fighting Program Bloat with the Fractal Complexity Measure

Vili Podgorelec, Peter Kokol

Laboratory for System Design, University of Maribor - FERI
Smetanova 17, SI-2000 Maribor, Slovenia
Vili.Podgorelec@uni-mb.si

Abstract. The problem of evolving decision programs to be used for medical diagnosis prediction brought us to the problem, well know to the genetic programming (GP) community – the tendency of programs to grow in length too fast. While searching for a solution we found out that an appropriately defined fractal complexity measure can differentiate between random and non-random computer programs by measuring the fractal structure of the computer programs. Knowing this fact, we introduced the fractal measure α in the evaluation and selection phase of the evolutionary process of decision program induction, which resulted in a significant program bloat reduction.

1 Introduction: Medical Diagnosis Prediction using GP

Decision support systems that help physicians are becoming a very important part of medical decision making, especially in medical diagnostic processes and particularly in those where decision must be made effectively and reliably [Kok94a, Kok98a]. They are based on different models and the best of them are providing an explanation together with an accurate, reliable, and quick response. Since conceptual simple decision-making models with the possibility of automatic learning should be considered for performing such tasks [Kok95], according to recent reviews [Kok94a, Quin93], decision-trees are a very suitable candidate. Decision-trees have been already successfully used for many medical decision making purposes (Fig. 1). Although effective and reliable, the traditional decision-tree construction approach still contains several deficiencies. Therefore we decided to develop a decision support model, that uses genetic programming techniques [Koza92, Banz98] to overcome the deficiencies of the traditional decision-tree-induction method. Using GP techniques we constructed a tool [Podg99] that helps to predict a correct diagnosis of a new patient, based on the knowledge it retrieves from a set of already solved cases.

Our main two propositions for this work were the following: 1) an evolutionary approach to the induction of decision-tree-like decision models should provide better results than traditional induction methods, and 2) automatic programming is the appropriate technique for performing such a task. To prove (or disprove) these propositions we developed an automatic program construction tool, called *proGenesys* (*pro*gram *gene*ration tool based on *gene*tic *sys*tems), that uses the methods, well known from genetic programming and which can develop programs in an arbitrary programming language, described with a context-free grammar [Podg99].

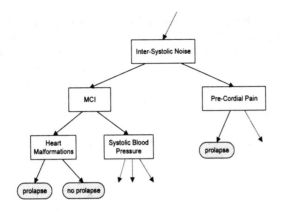

Fig 1. A part of a decision tree for the MVP classification.

First results, obtained with our tool, showed some very bad results (in comparison to the traditional decision trees and according to our expectations). But this was merely the consequence of the fact, very well known to the genetic programming community – program bloat [Tack95, Nord95, Soul96]. Namely, in program bloat, some parts of decision programs that actually had no influence on the result regarding the evaluation function – parts of a program that were never executed (Fig. 2), started to grow in size rapidly. To avoid this problem, we need a tool that would measure in some way how "meaningful" the evolved programs are. Regarding the proposition that the "meaning of the program" is in some way related to its complexity, we decided to include the fractal measure alpha in the evaluation function.

```
if (SystolicBloodPressure<=97) then
  if (MCI=true) then
    if (InjuryMother=true) then return "silent prolapse"
    else return "no prolapse"
    endif
  else
    if (SystolicBloodPressure>107) then
      // this part is never executed
    endif
  ...
```

Fig 2. A part of a decision program for the MVP classification. There we can see a possible situation where a part of a program is never executed because of the logical condition (a value of systolic blood pressure in our case).

2 Evolving Computer Programs in an Arbitrary Programming Language

Evolutionary algorithms have been used with much success for the automatic generation of programs. In particular, Koza's [Koza92] GP has enjoyed considerable popularity and widespread use. Koza's method originally employed Lisp as its target language, and others still generate Lisp code. Lisp enjoys much popularity for a number of reasons, not least of which is the property of Lisp of not having a distinction between programs and data. Hence, the structures being evolved can directly be evaluated and programs may be safely crossed over with each other and still remain syntactically correct.

However, recently there have been several attempts to evolve programs in other languages, by using a grammar to describe the target language. Researchers such as Whigham [Whi95] and Wong [Wong95] used context free languages in conjunction with GP to evolve code. Horner [Hor96] introduced a system called Genetic Programming Kernel, which, similar to standard GP, employs trees to generate code genes. Each tree is a derivation tree made up from the BNF definition. An approach which generates C programs directly, was described by Paterson [Pat97]. This method was applied to the area of evolving caching algorithms in C with some success, but with a number of problems, most notably the problem of the chromosomes containing vast amounts of introns.

We also developed an automatic program construction tool called *proGenesys* [Podg99] that uses the methods from GP and can develop programs in an arbitrary programming language described with a context-free grammar. In our approach an individual (genotype) is represented with a syntax tree (derivation tree); when presenting the solution it must be mapped to the actual program (phenotype).

2.1 Language Definition: AutoDecide

With the *proGenesys* tool, programs can be evolved in an arbitrary programming language, defined with the BNF notation. For the problem of medical diagnosis prediction first we have to decide whether an existing programming language will be used, or a new one will be defined that most suits the selected problem. We decided on the second option and defined a programming language AutoDecide (Fig. 3).

As can be seen in the Fig. 3, the presented BNF is only a language template, which is finalised into a language in accordance with the decision problem (based on the used attributes, possible decisions, etc). In that manner the non-terminal <DISCR_ATTR> can be expanded to any discrete attribute (each discrete attribute is one terminal symbol), <NUM_ATTR> can be expanded to any numeric attribute (each numeric attribute is one terminal symbol), and <DECISION> can be expanded to any of the possible decisions (each possible decision is one terminal symbol). Two special cases are also the non-terminals <INT> and <REAL>; the first can be replaced by a random integer number and the second can be replaced by a random real number.

```
<S>            ::= <PROGRAM>
<PROGRAM>      ::= <ST_LIST> <LAST_ST>
<ST_LIST>      ::= <ST> <ST_LIST> | <ST>
<ST>           ::= if ( <COND> ) then <PROGRAM> endif
                   | if ( <COND> ) then <PROGRAM> else
                     <PROGRAM> endif
                   | return <DECISION>
                   | skip
<LAST_ST>      ::= return "Unable to decide!"
<COND>         ::= <DISCR_COND> | <NUM_COND>
<DISCR_COND>   ::= <DISCR_ATTR> <DISCR_REL_OP> <INT>
<DISCR_ATTR>   ::= all discrete attributes; each represents one terminal
<DISCR_REL_OP> ::= = | <>
<NUM_COND>     ::= <EXPR> <REL_OP> <EXPR>
<EXPR>         ::= <EXPR> <OP> <EXPR> | ( <EXPR> ) |
                   <NUM_ATTR> | <REAL>
<REL_OP>       ::= < | <= | > | >=
<OP>           ::= + | - | / | *
<NUM_ATTR>     ::= all numeric attributes; each represents one terminal
<DECISION>     ::= all possible decisions; each represent one terminal
<INT>          ::= random integer number
<REAL>         ::= random real number
```

Fig 3. AutoDecide programming language for evolving decision programs. The presented BNF is only a language template, which is finalised into a language in accordance with the decision problem (used attributes, possible decisions, etc.)

3 The First Attempt: A Failure

We used the proGenesys tool together with the AutoDecide language for predicting the mitral valve prolapse (MVP) syndrome in children patients. But after several attempts it became clear that we will not be able to obtain some good results; namely, the problem of program bloat appeared which prevented the possibility of a successful decision programs evolution. The evolved programs start to grow in length rapidly after only a few generations (Fig. 4).

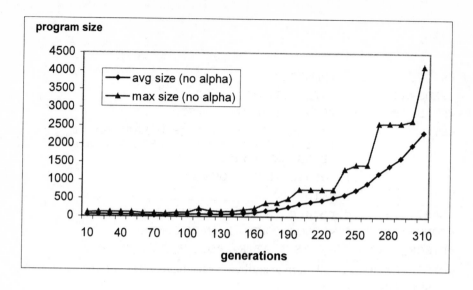

Fig 4. Size of the evolved programs as measured in the first attempt of generating the decision programs for the MVP prediction.

The tendency for programs in GP populations to grow in length has been widely reported [Tack93, Tack94, Ang94, Lang95, Nord95, Soul96]. The principal explanation advanced for bloat has been the growth of introns or redundancy, i.e. code which has no effect on the operation of the program which contains it (see Fig. 2). In some cases such introns are said to protect the program containing them from crossover [Blic94, Nord95, Nord96]. In our case this effect causes more troubles than profit because the system was not able to process such vast programs in a reasonable time. Therefore we needed to introduce some measure that would favour evolved programs without too large bloated parts.

4 Fractal Complexity Measure

The majority of experts agree that complexity is one of the most relevant characteristics of computer programs. For example Brooks states that computer software is the most complex entity among human made artifacts [Broo87]. But what is complexity and how can we measure it? There are two possible, not completely distinct, viewpoints:
- the classical computational complexity [Coh86, Weg95] and
- the recent "science of complexity" [Pin88, Mor95].

The first viewpoint is well know and researched. Thereafter it is much more interesting to concentrate on the second view and the aim of this section is to present some ideas and results concerning it.

4.1 Complexity

According to Morowitz [Mor95], the complex systems share certain features like having a large number of elements, possessing high dimensionality and representing an extended space of possibilities. Such systems are hierarchies consisting of different levels each having its own principles, laws, and structures. The most powerful approach for studying such systems was reductionism, the attempt to understand each level in terms of the next lower level. The big weakness of the reductionistic approach is that it is unable to explain how properties at one level emerge from the next lower level and how to understand the emergence at all. The problem is not the poverty of prediction, but its richness. The operations applied to the entities of one level generate so enormously many possibilities at the next level that it is very difficult to conclude anything. This demands radical pruning and the main task of a complexity as a discipline is to find out the common features of pruning (or more generally selection) algorithms across hierarchical levels and diverse subject matters.

4.2 Quantitative Properties of Complexity

Great many quantities have been proposed as measures of complexity. Gell-Mann [Gell95] suggests there have to be many different measures to capture all our intuitive ideas about what is meant by complexity. Some of the quantities are computational complexity, information content, algorithmic information content, the length of a concise description of a set of the entity's regularities, logical depth [Gell95], etc., (in contemplating various phenomena we frequently have to distinguish between effective complexity and logical depth - for example some very complex behaviour patterns can be generated from very simple formula like Mandelbrot's fractal set, energy levels of atomic nuclei, the unified quantum theory, etc.- that means that they have little effective complexity and great logical depth). A more concrete measure of complexity, based on the generalisation of the entropy, is correlation [Schen93], which can be relatively easy to calculate for special kinds of systems, namely the systems which can be represented as strings of symbols.

4.3 Complexity and Computer Programs

Computer programs, including popular information systems, usually consist of a number of entities like subroutines, modules, functions, etc., on different hierarchical levels. Concerning "laws of software engineering" or the concepts of programming languages [Watt90], the emergent characteristics of the above entities must be very different from the emergent characteristics of the program as a whole. Indeed, programming techniques such as stepwise refinement, top-down design, bottom-up design, or more modern object-oriented-programming, are only meaningful if different hierarchical levels of a program have distinguishable characteristics.

Computer programs are conventionally analysed using the computational complexity or measured using complexity metrics [Cont86, Fent91, Schne94]. Another way to asses complexity is, for example, to use Fractal Metrics [Kok94b]. But, as we can see from above, we can regard computer programs from the viewpoint of "complexity as a discipline" and according to that apply various possible complexity measures. The fact that a computer program is a string of symbols, introduces an elegant method to assess the complexity - namely to calculate long range correlations between symbols, an approach which has been successfully used in the DNA decoding [Buld94] and recently on human writings [Schen93].

4.4 Long Range Power Law Correlations

Long range power law correlations (LRC) have been discovered in a wide variety of systems. Recognising a LRC is very important for understanding the system's behaviour, since we can quantify it with a critical exponent. Quantification of this kind of scaling behaviour for apparently unrelated systems allows us to recognise similarities between different systems, leading to underlying unifications. For example, the recent research has shown that DNA sequences and human writings can be analysed using very similar techniques, and we are interested if a like approach can be applied to computer software, too.

In order to analyse the long range correlations of a string of symbols we must first map the string into a random walk model [Buld94]. The advantage of this method over the more traditional power spectrum method, or direct correlation of string's correlation, is that it yields high quality scaling data [Schen93].

4.5 Fractal Software Complexity Measure: α metric

Alpha metric [Kok98b, Kok99] is based on the long range correlation calculation. In this paper we will use the so-called CHAR method described by Kokol [Kok98b]. A character is taken to be the basic symbol of a computer program. Each character is then transformed into a six bit long binary representation according to a fixed **code table** (i.e. we have 64 different codes for the six bit representation – in our case we assigned 56 codes for the letters and the remaining codes for special symbols like period, comma, mathematical operators, etc) are used. The resulting binary string is then transformed into a two dimensional Brownian walk model (Brownian walk in the text which follows) using each bit as one move - the 0 as a step down and the 1 as a step up.

An important statistical quantity characterising any walk is the root of mean square fluctuation F about the average of the displacement. In a two-dimensional Brownian walk model the F is defined as:

$$F^2(l) \equiv \overline{[\Delta y(l,l_0)]^2} - \overline{[\Delta y(l,l_0)]}^2$$

where

$$\Delta y(l,l_0) \equiv y(l_0 + l) - y(l_0)$$

l is the distance between two points of the walk on the X axis
l_0 is the initial position (beginning point) on the X axis where the calculation of $F(l)$ for one pass starts
y is the position of the walk – the distance between the initial position and the current position on the Y axis

and the bars indicate the average over all positions l_0.

The $F(l)$ can distinguish two possible types of behaviour:

- if the string sequence is uncorrelated (normal random walk) or there are local correlations extending up to a characteristic range i.e Markov chains or symbolic sequences generated by regular grammars, then

$$F(l) \approx l^{0.5}$$

- if there is no characteristic length and the correlations are "infinite" then the scaling property of $F(l)$ is described by a power law

$$F(l) \approx l^{\alpha} \text{ and } \alpha \neq 0.5.$$

The power law is most easily recognised if we plot $F(l)$ and l on a double logarithmic scale. If a power law describes the scaling property then the resulting curve is linear and the slope of the curve represents α. In the case that there are long range correlations in the program analysed, α should not be equal to 0.5.

4.6 Fractal Structure of Computer Programs

Our proposition is that alpha metric measures the fractal structure of computer programs. This can be shown with the comparison of real and random program samples with some simple statistical methods like average, minimal and maximal values; distribution of values; and finally with the Student t-test - in the case that we encounter some differences in α between the two samples, then α is able to measure the fractal structure of programs and is thereafter the metric of fractal structure of programs.

Based on this proposition we would like to test the following hypothesis: the random generated programs have different fractal structure than ordinary programs written by human programmers.

We would like to test the above hypothesis based on the fact that random generated programs (at least in our experiments) are syntactically correct but meaningless - that means that they have a syntactic structure but no semantic structure, in short their structure differs significantly from the structure of ordinary human written programs. To show this difference we had to have an adequate metric. Based on our previous work α is such a metric.

4.7 Difference in fractal structure between random and real programs

We randomly selected 20 Pascal programs written by the first year students during their course Programming I held at our Faculty. Then we generated 20 various random Pascal programs using our generator of random programs approximately the same size (LOC) as student programs. Next we calculated the α for each program. The distribution of α values is shown in table 1.

Table 1. The distribution of α values.

$\lvert \alpha\text{-}0.5 \rvert$	Number of real programs	Number of random programs
0 – 0.02	1	16
0.02 – 0.04	5	3
> 0.04	14	1

From the above table it can be concluded (which was also confirmed with further testing) that programs, which α value differs from 0.5 by less than 0.04 can be almost uniquely identified as random, while programs, which α value differs from 0.5 by more than 0.04 can be almost uniquely identified as "normal" – real programs.

5 The Second Attempt: A Success

Now, we were able to change the proGenesys tool; we implemented the possibility of measuring alpha in the phase of evaluation – besides the existing fitness function, alpha was used to select the individuals for further evolution. This time the evolved programs did not grow in length so much anymore (Fig. 5). The comparison between the size of evolved programs in the first attempt (without measuring the alpha values) and the evolved programs in the second attempt (with the fractal measure alpha included) shows a significant reduction of program bloat effect when the fractal measure alpha is included in the evaluation and selection phase of the program generation process (Fig. 6).

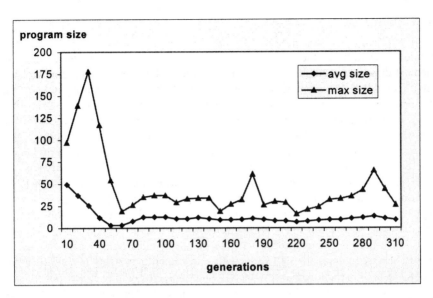

Fig 5. Size of the evolved programs as measured in the second attempt of generating the decision programs for the MVP prediction.

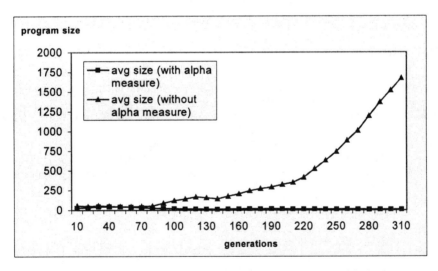

Fig 6. The comparison of the sizes of the evolved programs as measured in both attempts of generating the decision programs for the MVP prediction.

6 Conclusion

While trying to solve the problem of automatic program construction for the use of medical diagnosis prediction we encountered the problem of the programs growing in length too much. Based on our experiences from the software engineering field, where we defined a new fractal software complexity metric, we introduced this fractal measure in the evaluation and selection phase of program evolution. With the fractal measure alpha included the programs evolved with significant bloat reduction.

In the case of automatic program induction in an arbitrary programming language, where the language is defined by BNF notation, sizes of the derivation trees representing programs can grow very fast with the program length. When, additionally, there is also a considerable program bloat effect present, the evolution process needs too much computational resources to finish the construction of the final solution in a reasonable time. Therefore the reduction of program bloat problem is very important. When evaluating whether or not to pursue our fractal approach we should be able to compare our results with some other approaches to limiting bloat. In this manner we performed a few tests with simple parsimony based on number of nodes, which did not perform as well as the fractal approach. But the number of experiments we performed is too small to make some very general conclusions. However, so far our approach showed very promising results and in our opinion this work is worth further investigations by all means.

References

[Ang94]   Angeline, P.J.: Genetic programming and the emergent intelligence, In Advances in Genetic Programming, MIT Press 75-98 (1994)

[Banz98]   Banzhaf, W., Nordin, P, Keller R.E. and Francone, F.D.: Genetic Programming – An Introduction. Morgan Kaufmann, (1998)

[Blic94]   Blickle, T., Thiele, L.: Genetic programming and redundancy, In Genetic Algorithms within the Framework of Evolutionary Computation 33-38 (1994)

[Broo87]   Brooks, P.F.: No silver bullet: essence and accidents of software engineering, IEEE Computer, 20(4) 10-19 (1987)

[Buld94]   Buldyrev, S.V. et al.: Fractals in Biology and Medicine: From DNA to the Heartbeat, Fractals in Science (Eds. Bunde, A, Havlin, S), Springer Verlag (1994)

[Coh86]   Cohen, B, Harwood, W.T., Jackson, M.I.: The specification of complex systems, Addison Wesley (1986)

[Cont86]   Conte S.D., Dunsmore, H.F., Shen, V.Y.: Software engineering metrics and models, Benjamin/Cummings, Menlo Park (1986)

[Fent91]   Fenton, N.E.: Software Metrics: A Rigorous Approach, Chapman & Hall (1991)

[Gell95]   Gell-Mann, M.: What is complexity, Complexity 1(1) 16-19 (1995)

[Hor96]   Horner, H.: A C++ class library for GP, Vienna University of Economics (1996)

[Kok94a]   Kokol, P., et al.: Decision Trees and Automatic Learning and Their Use in Cardiology. Journal of Medical Systems 19:4 (1994)

[Kok94b]   Kokol, P.: Searching For Fractal Structure in Computer Programs, SIGPLAN 29(1) (1994)

[Kok95]   Kokol, P., Stiglic, B. and Zumer, V.: Metaparadigm: a soft and situation oriented MIS design approach. International Journal of Bio-Medical Computing 39 243-256 (1995)

[Kok98a] Kokol, P., Podgorelec, V. and Malcic, I.: Diagnostic Process Optimisation with Evolutionary Programming. In Proceedings of the 11th IEEE Symposium on Computer-based Medical Systems CBMS'98, 62-67 (1998)

[Kok98b] Kokol, P., Podgorelec, V., Brest, J.: A wishful complexity metric, In Proceedings of FESMA (Eds: Combes H et al). Technologish Institut 235–246 (1998)

[Kok99]   Kokol, P., Podgorelec, V., Zorman, M., Pighin, M.: Alpha - a generic software complexity metric, Project control for software quality (Eds: Rob J. Kusters et al.), Maastricht : Shaker Publishing BV 397-405 (1999)

[Koza92] Koza, J.R.: Genetic Programming: On the Programming of Computers by Natural Selection. Cambridge, MA: MIT Press (1992)

[Lang95] Langdon, W.B.: Evolving data structures using genetic programming, In Proceedings of the Sixth International Conference on Genetic Algorithms, Morgan Kaufmann 295-302 (1995)

[Mor95]   Morowitz, H.: The Emergence of Complexity, Complexity 1(1) 4 (1995)

[Nord95] Nordin, P., Banzhaf, W.: Complexity compression and evolution. In Proceedings of the Sixth International Conference on Genetic Algorithms ICGA95, Morgan Kaufmann 310-317 (1995)

[Nord96] Nordin, P., Francone, F., Banzhaf, W.: Explicitly defined introns and destructive crossover in genetic programming, In Advances in Genetic Programming 2, MIT Press 111-134 (1996)

[Pat97]   Paterson, N., Livesey, M.: Evolving caching algorithms in C by GP, In Genetic Programming 1997, MIT Press 262-267 (1997)

[Pin88]   Pines, D. (Ed.): Emerging syntheses in science, Addison Wesley (1988)

[Podg99] Podgorelec, V.: proGenesys – Program Generation Tool Based on Genetic Systems. In Proceedings of IC-AI'99 Conference 299-302 (1999)

[Quin93] Quinlan, J.R.: C4.5: Programs for Machine Learning. Morgan Kaufmann, (1993)

[Schen93] Schenkel, A., Zhang, J., Zhang, Y.: Long range correlations in human writings, Fractals 1(1) 47-55 (1993)

[Schne94] Schneidewind, N.F.: Methodology for Validating Software Metrics, IEEE Trans Soft Eng 18(5) 410-422 (1994)

[Soul96] Soule, T., Foster, J.A. and Dickinson, J.: Code growth in genetic programming. In Proceedings of the First Annual Conference on Genetic Programming. MIT Press 213-223 (1996)

[Tack94] Tackett, W.A.: Recombination, Selection, and the Genetic Construction of Computer Programs, PhD thesis, University of Southern California (1994)

[Tack95] Tackett, W.A.: Greedy recombination and genetic search on the space of computer programs. In Foundations of Genetic Algorithms 3, Morgan Kaufmann 271-297 (1995)

[Watt90] Watt, D.A.: Programming Language Concepts and Paradigms, Prentice Hall (1990)

[Weg95]   Wegner, P., Israel, M (Eds.): Symposium on Computational Complexity and the Nature of Computer Science, Computing Surveys 27(1) 5-62 (1995)

[Whi95]   Whigham, P.: Search Bias, Language Bias and Genetic Programming, In Genetic Programming 1996 230-237 (1996)

[Wong95] Wong, M., Leung, K.: Applying logic grammars to induce subfunctions in genetic programming, In Proceedings of the 1995 IEEE conference on Evolutionary Computation, IEEE Press 737-740 (1995)

Paragen - The First Results

Conor Ryan and Laur Ivan
Department of Computer Science and Information Systems
University of Limerick
Ireland
{Conor.Ryan|Laur.Ivan}@ul.ie

Abstract. The Paragen System is an automatic parallelisation system which takes a serial program and, using a combination of Genetic Programming and Genetic Algorithms, transforms it into a functionally identical parallel program. This paper gives an overview of the system, describing the novel way in which GP and GA are integrated, before giving the first substantial results for the system. The programs that are processed by the system illustrate that it can handle programs which contain a combination of loops and atomic instructions, and we show the advantage of applying the system to a program as a whole, rather than simply to one kind of instruction.

1 Introduction

This paper describes the current state of the Paragen system, a Genetic Programming tool which automatically converts serial programs into functionally equivalent parallel ones. The system is a combination of GP and GAs, as we have found that certain kinds of code are easier to manipulate with one or the other. We describe how the system divides a program into various sections and then decides which type of evolution to use.

1.1 Software Re-Engineering

A significant portion of the time, effort and expense of many Data Processing departments is consumed by software maintenance. Up to 70% of total expenditure is directed towards this activity [Lucia et al., 1996]. These tasks vary from meeting the changing needs and desires of users, to the improvement of the control structure of a program, e.g. from removing *GOTOs* to make subsequent modifications easier. Other re-engineering tasks include the ubiquitous year 2000 problem, as well as that of Euro-conversion, which involves adapting software to display monetary values in both local currency and the new European currency.

Those involved in software maintenance are first faced with the task of understanding the code. This task is made more difficult as it is often the case that those involved in the initial writing of that code are long since departed, and is further complicated by sparse or nonexistent documentation. Given the scarcity of information available, it is not surprising that a major part of the effort in re-engineering is spent trying to understand the code. It has been estimated [Rajlich, 1994] that up to 50% of the costs associated with re-engineering a piece of code can be attributed to code comprehension.

The difficulty and level of concern, which often bordered on panic with the recent portents of doom accompanying the new millennium, have caused many to look for a third party to re-engineer their code. The scale of their problems are evidenced by the existence of re-engineering companies whose sole service was to provide Year 2000 solutions. The most successful re-engineering companies are those that have developed tools that automate, or at the very least semi-automate, the re-engineering process. In general, the greater the level of automation, the greater the success rate and thus, the less testing required.

Certain re-engineering tasks can be executed by carrying out standard transformations on the original code, i.e. transform two digit dates to four digits, remove sections of code which are clones, etc. These tasks can only be automated if there are some rules that govern which transformations can legally be applied and, if order dependent, the order they should occur in.

It is reasonable to expect that automated software re-engineering tools will be increasingly more important to the software industry as a whole, as the shortage of trained computer personnel increases. This is because the greater the level of automation, the less the programmers need be concerned with code comprehension. This paper is concerned with applying GP to re-engineering in general, and to auto-parallelization in particular. It is shown that is is possible to develop a system which meets the above demand, to the extent that both re-engineering and testing are fully automated. It is shown that GP is particularly suitable for the generation of parallel code, because it eagerly embraces the maze of transformations required for re-engineering. Furthermore, the often lateral approach required for parallel program design, while a foreign way of thinking for many programmers, is tailor made for the bottom up approach GP takes.

1.2 Auto-Parallelization

The auto-parallelization of serial programs is an important issue due to the enormous advances that have been made in parallel architectures in the past few years. Until recently, parallel programming tended to be restricted to either purely academic activities or to exotic super computer systems. The extension of systems such as PVM [Geist, 1993] (Parallel Virtual Machine) and MPI (Message Passing Interface) to clusters of workstations has changed this, however. For it is now possible to treat even the most modest network of (possibly heterogeneous) computers as though each were a node in a parallel computer.

Despite the apparent ease with which one can adopt parallel architectures, they have yet to enjoy widespread use. One of the main causes of this is that the kind of users who stand to benefit most from parallel processing are often the least likely to have the knowledge necessary to extract the optimal, or even near optimal, performance from their machines. Moreover, they tend to have large legacy systems running on serial machines, and re-writing this legacy code can represent an enormous cost.

Difficulties with the production of good parallel code are not restricted to re-writing, however. The generation of parallel code is an arduous task that requires a substantial mind-shift and no small amount of training, particularly if the code is to be optimized. Persons or institutions wishing to produce parallel code would stand to benefit from a tool that would allow them to develop their code in a traditional manner, and subsequently convert it to parallel code. Of course, programmers who take this route would, by necessity, demand proof that the newly converted code is equivalent to their original code.

2 Parallel Compilers

In line with the increased accessibility of parallel computers, so too has the demand for parallel compilers increased. Unfortunately, most of them were designed for specific architectures which, although providing maximum performance, do so at the cost of minimum portability.

The vast majority of parallel compilers are, by necessity, only concerned with parallel code. That is, they can take advantage of code writen using some extensions of the standard language (for example, compiling directives in PGI Fortran). If normal sequential code is passed as input for a such compiler, the improvements are minor.

Other types of compilers are concerned with trying to parallelise normal sequential code. These kind of compilers are extremely useful as many the companies already have the software writen in form of sequential programs. Most of these compilers have been designed to tackle a specific set of problems with a specific set of transformations, in order to reduce the data dependency checking. If a different style of problem is given as input, the result can have fery few optimizations. Paragen is included the latter class of compilers, as its purpose is to combine the proven parallelization techniques and laws with evolutionary algorithms in order to create a more general parallelization tool. Its current design stresses the coarse grain parallelisation for a PVM/MPI based output. It can be eventually combined with a finer grain parallelisation for more specific architectures (vector processing in SIMD computers, pipe processing and so on).

3 Paragen Architecture

Like other systems that evolve sequences of transformations, e.g. Koza's electrical circuits [Koza et al., 1999] and Gruau's [Gruau, 1994] neural nets, Paragen can be looked upon as an *embryonic* one. That is, it starts with a serial program, and progressive application of the transformations modify it, until eventually the system produces a parallel version of the program. It is only after this embryonic stage that an individual is tested.

All the transformations employed are the standard parallel type, and syntax preserving with one caveat; that is, that the area of the program which a transformation affects does not contain any data dependencies. A description of these transformations is beyond the scope of this paper, but the interested reader is directed to [Ryan, 1999] and [Ryan and Ivan, 1998] for further information. If a transformation violates that condition while being applied it may change the semantics, and any transformation that runs this risk causes an individual's fitness to be reduced.

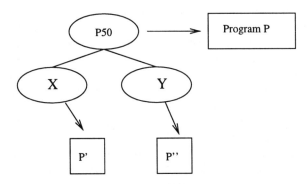

Fig. 1. Applying transformations to a program.

We introduce the notion of a *program segment*. A program segment is the part of a program (or function) that is currently being modified by a transformation. Notice how in figure 1 the size of the program segment decreases, this is because of the nature of the transformations being employed. Typically they schedule one or more instructions, and pass the remaining ones on as the program segment to the next transformation. However, by far the greatest parallelism can be extracted from loops, as it tends to be within these that the largest amount of processing is carried out. There are a large amount of standard transformations, such as loop skewing, fusion etc. which are peculiar to loops.

To reflect the dual nature of the transformations, individuals must be capable of employing both, and applying the appropriate type depending on the nature of the code encountered in the embryonic stage. Each type of transformation is stored separately, the *atom* type, which affect single instructions, are manipulated by standard tree structures in a similar fashion to standard GP. The second type of transformation, the *loop* type are stored in a linear genome.

Essentially, an individual is evaluated in "atom mode", that is, the tree is traversed in the normal manner, with the transformations being applied to the serial program. However, if a loop structure is encountered, the system enters "loop mode" and a transformation is instead read from the linear part of the genome, which is then applied. After this, the system returns to atom mode and applies the next available transformation to the code within the loops, thus, not only is it possible to parallelize the loop itself, but also the code contained within the loop.

3.1 The client-server architecture

Paragen employs a MPI/OOMPI (Beowulf) based client/server architecture, which greatly reduces the time required for compilation. Evolutionary Algorithms are often criticised for their time demands, but with this approach we are able to spread instances of atom and loop modes across processors and dramatically reducing the compilation time.

Notice that unlike many EA uses of parallel equipment, we do not simply run several copies of the program in parallel, rather this version implements a nesting level based *divide et impera* tactics for breaking up the program.

We look upon a program as a set of nested logical blocks, with each block being a set of one or more instructions. We define a logical block as an atom and we also define an atom as the basic operating unit for paragen. This nesting level view is used to spread a program's sequences across the network of processors: for each logical block, a Paragen instance is created with that block contents as input parameter. A special case is the loop atom which is one or more consecutive loops. When a loop atom is encountered, Paragen seeks for other consecutive loop atoms in order to create a metaloop atom. This is done regardless of the outcome of any previous atom mode execution.

3.2 Example

Consider the code below:

```
void function2()
    {
    if(a==b)
        {
        a++;
        b--;
        }
    else
        {
        a--;
        b++;
        }
    }
```

Notice that although the code in the example above is C, Paragen actually operates at a language independant level. This is because the only information it requires about an instruction are the used and modified variables.

The first step is to separate the two nesting levels present in this sequence. The first nesting level contains the "if" statement, while the second one represents the instructions within the "if"'s branches (the compound statements). For each set of instructions within paragen generates a file containing the modified/used variables, transmitting the non-local variables (both used and modified) to the upper nesting level. The result is three files: one containing the equivalent of the function2 with the collapsed "if" instruction, and two files, one for each "if" branch. The resulting representation is similar to that in figure 2. The only data required for each instruction are those variables which are read and those that are modified, as this is the information required to test if two or more instructions are dependant on each other.

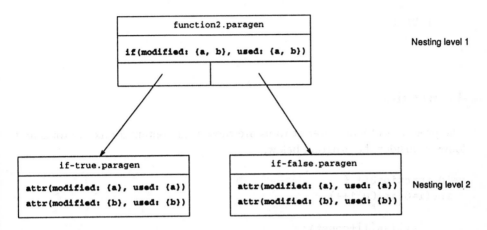

Fig. 2. Information required for each part of the program. If two items on the same level are to be executed in parallel, the level that caused this execution is passed back the information for each atom. This information is then used to test if there are any data dependencies.

The Paragen server sends the first file (**function2.paragen**) to a client. which then executes "atom mode" using this file as an input parameter. In this case, there will be no optimisation because this file contains only one instruction.

The client then advances to the second step of processing by parsing the nested files fields. If a loop is encountered, it seeks out any consecutive loops (regardless of being executed in parallel or not) and generates a metaloop atom. For each file (or metaloop atom), the client requests another free processor from the server. The request is blocking, so the client will wait until a processor is available. When a processor is found, the client sends to that processor (client) the filename to be processed and the mode (default is atom mode).

This is an asynchronous approach which permits maximum processor usage with a monolithic atom and loop modes.

The final result for the source is:

```
void function2() {
   if(a==b)
      {
      PAR
         a++;
         b--;
```

```
     ENDPAR
     }
   else
     {
     PAR
       a--;
       b++;
     ENDPAR
     }
}
```

4   Results

The programs of most interest to us are those that contain a mix of atoms and loops. Consider the program below:

```
void function3() {
  for(i=0;i<100;i++)
     {
       a[i]=a[i]+const1;
     }
for(i=0;i<100;i++)
     {
       b[i]=b[i]+const1;
     }
for(i=0;i<100;i++)
     {
       c[i]=c[i]+const1;
       d[i]=d[i]+const1;
     }
}
```

Firstly, an "atom mode" is launched with the three collapsed loops as parameters. The variable i is a loop counter and it does not affect the data dependencies, this "pseudo-dependency" is detected by a simple initial parse of the code. This indicates that the loops can be executed in parallel.

The second step is launching a "loop mode" with one metaloop atom having three loops. The iterations can be executed in parallel, because there is no cross-iteration data dependency. If a "split" operator is called for the last loop, the outcome is a metaloop atom having for loops, while if a "fusion" operator is called, then the outcome is a parallel loop having four data independent instructions. Furthermore, another set of "atom mode" sessions can be launched to obtain the final atom execution topology.

There are several functionally equivalent combinations that were produced by the system. We present two of the more interesting ones here:

- four parallel loops (parfor) being executed in parallel (this can be obtained by splitting the last loop into two loops, each with one instruction),
- one parallel loop where all the instructions within can be executed in parallel.

The first version is suitable for a MPI based machine rather than a multi-processor. Having a machine with 4 processors, we can assign one processor to each loop and transfer the data vectors accordingly. The second version would imply the transfer of one element at time, increasing dramatically the communication overhead. With an SMP architecture, The first version would launch four threads, one for each loop, while the second one would launch four threads for 100 timesteps, each thread executing only one instruction at a time.

The second version is suitable for SIMD architectures: the instruction is addition and the data is provided by the four attribution instructions.

Another interesting situation arises with the following code sequence:

```
void function4()
{
        for(i=0;i<100;i++)
        {
            a[i]=a[i]+1;
}

        b=b+m;
        for(i=0;i<100;i++)
        {
            a[i]=a[i]+2;
        }
        for(i=0;i<100;i++)
        {
            c[i]=c[i]+1;
        }
}
```

The first nesting level contains all the collapsed loops and the attribution instruction "b=b+m;":

```
loop1    loop (modified: {counter1, a[counter1]}, used: {counter1, a[counter1]})
attr     attr (modified: {b}, used: {b,m})
loop2    loop (modified: {counter2, a[counter2]}, used: {counter2, a[counter2]})
loop3    loop (modified: {counter3, c[counter3]}, used: {counter3, c[counter3]})
```

Fig. 3. Used and modified variables for the second level of nesting. The information extracted for each individual is the kind of instruction, e.g. loop, attribution etc. and the variables modified, and those used (read).

As can be seen from figure 3, the first and third instructions cannot be executed in parallel, because a[] is used/changed by both. Therefore, if only atom mode is executed, the result will be two timesteps. There are several equivalent "best trees", but two of them are particularly interesting, as illustrated in figure 4.

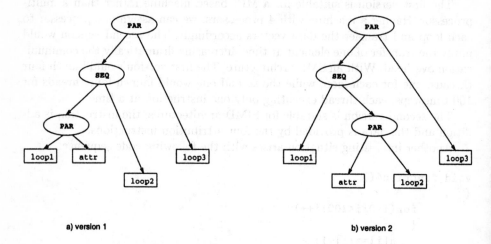

a) version 1   b) version 2

Fig. 4. Two equivalent individuals for the example individual.

The first version will still be executed in at least two timesteps even after a loop mode is applied, because the "loop2" has to be executed after "loop1". The second tree is more interesting, because it states that the attibutive instruction and the loop2 can be executed in parallel. This means that the two instructions can be swapped (altering the original instruction sequence). and the code would become:

```
loop1    loop (modified: (counter1, a[counter1]), used: (counter1, a[counter1]))
loop2    loop (modified: (counter2, a[counter2]), used: (counter2, a[counter2]))
attr     attr (modified: (b), used: (b,m))
loop3    loop (modified: (counter3, c[counter3]), used: (counter3, c[counter3]))
```

Fig. 5. The used and modified variables for function4().

Launching a loop mode for the first two loops will generate (with a fusion transformation) a single loop with two instructions, which can be eventually collapsed with a trivial speed optimizer compiler. Applying atom mode again, the result will be executed in only one timestep.

```
void function4()
{
    PAR
```

```
     PAR
        parfor(i=0;i<100;i++)
            {
                a[i]=a[i]+1;
                a[i]=a[i]+2;
            }
            b=b+m;
     ENDPAR
     parfor(i=0;i<100;i++)
     {
         c[i]=c[i]+1;
     }
     ENDPAR
}
```

4.1 Time Considerations

All the experiments used a population of 100 with steady state replacement. Experiments ran for ten generations. In those experiments that didn't involve loops an optimal solution was found on average after just two generation equivalents.

Those experiments that involved loops as well as atoms had similar parameters, but also included a loop chromosome for each loop, consisting of three transformations for each loop. On average, an optimal solution was generated after four generation equivalents.

5 Conclusions

We have demonstrated that Paragen can process programs which are made up of a combination of loops and atomic instructions. Unlike other parallelising compilers of which the authors are aware, which tend to deal with loops *in situ*, the dual nature of the system permits it to move various segments of a program around to allow it apply more powerful transformations.

The examples described in this paper indicate how general this approach is, even with a relatively small number of transformations available to the system. The next step in the work is to increase the number loop-specific transformations available to the system to enable it to be applied to more complex benchmark problems.

A perenniel problem of GP, and indeed, all EA based systems, is the difficulty with scaling them up to larger problems. Encouraged by our first results, we hope this will not be quite such a problem for Paragen, particularly due to the divide and conquer approach used to evaluate individuals.

References

[Geist, 1993] Geist, G. (1993). Pvm 3 beyond network computing. In *Lecture Notes in Computer Science 734, 2nd International Confernce of the Austrian Center for Parallel Computation.*

[Gruau, 1994] Gruau, F. (1994). *Neural Network Synthesis using Cellular Encoding and The Genetic Algorithm.* PhD thesis, Centre d'etude nucleaire de Grenoble.

[Koza et al., 1999] Koza, J. R., Bennett, F. H., and Andre, D. (1999). *Genetic Programming III : Darwinian Invention and Problem Solving.* Morgan Kaufmann.

[Lucia et al., 1996] Lucia, A., Fasolino, A., and Munro, A. (1996). Understanding function behaviors through program slicing. In *Proceedings of the Fourth Workshop on Program Comprehension.*

[Rajlich, 1994] Rajlich, V. (1994). Program reading and comprehension. In *Proceedings of Summer School on Engineering of Existing Software,* pages 161–178.

[Ryan, 1999] Ryan, C. (1999). *Automatic Re-engineering of Software Using Genetic Programming.* Kluwer.

[Ryan and Ivan, 1998] Ryan, C. and Ivan, L. (1998). Automatic parallelization of loops in sequential programs using genetic programming. In Koza, J. R., Banzhaf, W., Chellapilla, K., Deb, K., Dorigo, M., Fogel, D. B., Garzon, M. H., Goldberg, D. E., Iba, H., and Riolo, R., editors, *Genetic Programming 1998: Proceedings of the Third Annual Conference,* pages 344–349, University of Wisconsin, Madison, Wisconsin, USA. Morgan Kaufmann.

Multi-robot Cooperation and Competition with Genetic Programming

Kai Zhao, Jue Wang

Institute of Automation, Chinese Academy of Sciences,
P.O.Box 2728, Beijing 100080, China
{wjst, wangj}@sunserver.ia.ac.cn

Abstract. In this paper, we apply Genetic Programming(GP) on multi-robot cooperation and competition problem. GP is taken as a real time planning method in stead of learning method. Robot all use GP to make a plan and then walk according to the plan. The environment is composed of two parts, natural environment, which is the obstacles, and social environment that refers to other robots. The cooperation process is accomplished by robot's adaptation to both of them. In spite of the fact that there is no communication among robots and little knowledge about how to cooperate well, the adaptive capability in dynamic environment enable robots to complete a common task or solve the competition. Several experiments are taken and the results are shown.

1   Introduction

Traditionally, Genetic Programming is used as a learning method and automatic program generation is its major application[8]. In this paper, GP is applied as a real time planning method in multi-robot cooperation and competition solution.

Genetic Programming has been applied in single robot path planning[15] in which the robot doesn't know the distribution of the obstacles. Robot makes and optimizes plans with GP and then walks according to the plan. Robot makes a new plan when it senses some new obstacles that invalidate the old plan, and then it continues walking according to the new plan. This process repeats until the robot reaches the destination.

In the above case, the environment is dynamic because of the ignorance of the distribution of obstacles. Compared to the knowledge methods[2][11][12], one of the advantages of GP is that it is not necessary to equip the robot with much knowledge about how to look for path under various cases, which is complex and not robust. Instead, a robot reaches its destination by adapting to the dynamically changing environment. We call it a kind of adaptive method. Here, "adaptive" means the robot alters its plan automatically when environment changes.

In multi-robot cooperating problem, robots are required to complete task that can't be achieved by any one alone. In this circumstance, we'd like to divide the environment into two parts: natural environment and social environment. Natural environment includes the obstacles, while social environment is the composition of all the other robots. Between the two parts, there are some similarities. First, they are both dynamic, in sense that obstacles are unknown as prior and robots move all the time. Second, each robot must reach its own destination without colliding with obstacles and other robots. Third, reaching destination or cooperating with others, a path has to be planned. Therefore, the adaptive method in single robot case is extendable to multi-robot case.

In the adaptive method, the two parts of environment are considered the same, that is, at planning time, other robots are considered as moving obstacles. As in single robot case, we don't equip robots with knowledge about how to cooperate with others. To complete its task, a robot adapts to the social environment as well as the natural part.

Cooperation among multi-robot has attracted much attention in recent years [7][10][13]. In this field, there are mainly two types of control: model-based control and behavior-based control, which is the so-called reactive control[1]. In model-based control, the environment is assumed stable. The behavior of robots is carefully designed so that the results can be predicted precisely before they really run. Compared with the model-based control, the reactive control does not have an exact model. Instead, it emphasizes the interaction with environment, usually taking perception-action rules[4]. In general, the reactive control system is easier to design and more robust than model-based one. However, it is relatively difficult for optimizing the performance of the robot. Compared to the above methods, adaptive method keeps optimized performance as well as robust property.

2   Path planning with GP

The environment is described with grid representation, which divides the environment into some cells [6](Fig. 1). Every robot occupies one cell, while obstacles do more. A robot doesn't know the distribution of obstacles as prior. When it walks in the environment, the obstacles in view are sensed and labeled on its own environment map.

To make and optimize plans for a robot, we use Chromosome-Protein Scheme (CPS) with multiple-tree[14][15], which is based on GP. The original purpose of CPS is to make the planning course easier when field changes. This objective is realized by introducing amino acid library, which stores fundamental elements of the field. Thus, when field changes, the main work is changing amino acid library in stead of redesigning the function set.

In the specific path planning field, there are four elements stored in amino acid library, each of which stands for a road sign(Table 1). In population, every individual contains a chromosome composing several separated hierarchical structures(multiple trees). The functions in the trees are Get, Follow, Change, Insert, and Put, which are responsible for getting out road signs and arranging them on the map(Table 2). To generate a real plan, the program(chromosome) is executed and the road signs are got out from amino acid library, combined and then put into the map. All the road signs on the map together indicates a specific path. Still, not every path is suitable for the robot to walk along. Some of them may corrupt when collide with obstacles, while others may be circuitous from the destination. Differences in the paths decide the fitness of the individual. The faster a robot arrives at destination, the better.

Fig. 1. The environment representation(R: robot, *: obstacle)

Table 1. Amino Acid Library

Index	Amino Acid
0	• →
1	←
2	↓
3	↑

Table 2. Functions in chromosome structure

(1)Get(index): fetching the amino acid with the index from the amino acid library.
(2)Follow(ele1, ele2, ele3): arranging the elements in turn. The result is

ele1	ele2	Ele3

(3)Insert(ele1, ele2, ele3): inserting ele3 between the first two elements. The result is

ele1	ele3	Ele2

(4)Change(ele1, ele2): altering two elements' order. The result is

Ele2	ele1

(5)Put(ele,arg1,arg2): putting the element in the map at the special position referred by "arg1" and "arg2". In path planning, the "arg1" and "arg2" indicate the row and column index of the map.

The following is an example. Fig. 2 shows a chromosome that is composed of three trees.

        1. (Put (Change (Get 2) (Get 0)) 0 5)
        2. (Put (Get 0) 0 0)
        3. (Put (Get 0) 7 6)

Fig. 2. Chromosome with multiple-tree

Fig. 3(1) shows the plan generated according to the chromosome in Fig. 2. Fig. 3(2) shows the path indicated by the plan in real environment. Along the path, the robot stops at 11'th step because an obstacle stands in front of it. Although the road sign "→" at the bottom of the map leads to the destination, the plan is infeasible in real environment.

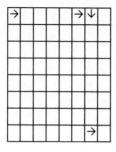   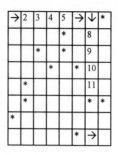

Fig. 3. (1) the plan generated according to the chromosome in Fig. 2   (2) the real path a robot walks according to the plan in (1)

3 Adaptive method for cooperation and competition

In multi-robot domain, the most attractive subjects are cooperation and competition. With cooperation, some robots are able to accomplish tasks that can't be done by any single one. Competition is also an important issue, for it must be solved before robots finish their own or shared tasks.

A typical cooperation problem is pursuit game. The original version of the game is introduced by Benda et al[3]. In the game, there are two types of robots, predators and prey(Fig. 4). Both of them move freely in the world. At each turn, robot can move one cell horizontally or vertically, but not diagonally. The prey moves randomly. The predators aim at completely surrounding the prey, by occupying the grid positions immediately north, south, east, and west of the prey. In this paper, we explain the prey as a ball, so that robots try to enclose and capture the ball.

Fig. 4. Cooperation problem: pursuit game. r1~r4: predators. B: prey.

About competition, we introduce bridge passing game(Fig. 5). In the game, there are two robots, r1 and r2. They are required to go to diagonal destinations respectively, @1 and @2. In the environment, there is a river and a bridge. To reach destination, robot must cross the river through the bridge. But the bridge is only accessible for one robot at any time. Therefor, sometimes the two robots collide on the bridge. This competition must be solved.

r1								@2
*	*	*	*		*	*	*	
			*		*			
			*		*			
*	*	*	*		*	*	*	
*								
r2								@1

Fig. 5. Competition problem: bridge passing. r1~r2: robots. @1~@2: destination.

We propose that some kinds of cooperation and competition may be solved by adapting to the environment. In multi-robot problem, the environment is divided into two parts: natural one and social one. The natural environment is the composition of obstacles, while all the robots make up the social environment. Both environments are uncertain, because the obstacles are unknown as prior and robots are moving all the time. Therefore a robot needs to modify its behavior plan according to the variation of the whole environment.

In traditional multi-robot planning, social environment is always thought of as a special part which is totally different from the natural one. Specific organization methods, central or distributed, with communication or not, are carefully designed to enable robots to cooperate with others[5][9]. But from robot's point of view, natural and social environment have a high degree of similarity. First, they are both dynamic and variable all the time. A robot does not aware of the distribution of all the obstacles, neither can it predict next position of any other robot. In fact, when moving in the environment, a robot keeps sensing new obstacles and new positions of robots. Second, to arrive at a special position, a robot must avoid obstacles as well as other robots. Third, reaching destination or cooperating with others, a path has to be planned. Therefore, in multi-robot case, a robot may consider the social and natural environment as the same, that is, regard other robots as moving obstacles. Thus, the adaptive method employed in single robot case extends to multi-robot case.

To explain the idea clearly, let's take the pursuit game as an example. When a robot(predator) tries to arrive at a position just around the ball(prey), it must avoid the obstacles as well as other robots. Any plan colliding with obstacles or robots will get a bad fitness and being abandoned. Therefore, the robot will select an optimized path leading to the destination. For instance, at a special time, three robots, r1~r3, have occupied north, south and west cells of the ball, and only the east cell is vacant. As for r4, because r1~r3 are considered as obstacles, only one way is accessible, which leads the robot to the ball from east direction. Since this is the general case for every robot, from the group point of view, all the robots tend to surround and enclose the ball.

With the above discussion, it is clear that for some kinds of cooperation and competition, with the adaptive method, it is not necessary to organize robots, design strategy for surrounding, or keep communication among them. The current position of the ball is the shared destination of all the robots. If some capturing positions (positions just beside the ball) have been occupied, robots will automatically surround the ball in other positions.

Competition solving is similar to cooperation case. It is discussed concretely in experiments in section 5.

4 Algorithm and fitness function for multi-robot problem

4.1 Algorithm for multi-robot problem

1) All the robots are initialized in turn. Each of them observes the environment and makes an initial plan.
2) In turn, each robot:
   Stops if arrives at destination, or stopping condition is satisfied;
   Observes the environment;
   Re-evaluates the individuals of population if environment changes;
   if no obstacle or robot stays on the optimized path, or thinking time exceeds time limit,
       Walks one step according to the plan;
   else,
       optimizes plan by CPS with multiple-tree.
   turns to 2).

Fig. 6. Multi-robot algorithm

The following points are some illustrations of the above algorithm.
1) When a robot observes the environment, it records the positions of obstacles and robots in view.
2) If environment changes, the old plans need to be re-evaluated, for obstacles or robots may occupy the path.
3) To prevent staying and making plan all the time, a time limit is set. When thinking time exceeds the limit, robot is obliged to walk a step according to the plan, even if the plan is not yet optimized.
4) When plans are optimized with CPS, the generation of evolution is rather little, usually 3~10. Other parameters and information of experiment computer are listed in Table 3.
5) Robots move in turn, that is, $r_1, r_2, \ldots, r_n$.

4.2 Fitness funciton

Some experiments have been made in section 5. They run under CPS with multiple-tree, with 266MHz PC Pentium type computer. The following table shows the fitness function and other parameters.

Table 3. Fitness function and other parameters

Objective:	To plan the shortest path for a robot to reach the goal while avoiding the obstacles and other robots.
Terminal Set:	Random constants from 0 to 7.
Function Set:	Put,Follow,Insert,Change,Get.
Raw Fitness:	SumFitness-Weight1*DistanceToGoal-Weight2* CollideObstacleNumber SumFitness:65, Weight1:7,Weight2:30 DistanceToGoal:|GoalX-NowX|+|GoalY-NowY|
Standardized Fitness:	Not reach the goal: Row Fitness Have reached the goal: Row Fitness+Award-WalkStep Award: 50.
Parameters:	• Population Size,M,is 200 or 300. • Maximum number of generation to be run,G,is 1500. • 1% reproductions, 14% crossover,10% muta- tion, 60% gene-recombination, 10% gene-copy, 5% gene-lose Zhao,1997b) • Tree's initial depth is less than 5. • Rank selection method. • View: 6.
Time:	Several minutes(to find the shortest path).

5 Experiments and results

In this section, we do three types of experiments:
1)cooperation in pure social environment without natural one.
2)cooperation in both social and natural environment.
3)competition solving in both social and natural environment.

5.1 Experiment 1: cooperation in pure social environment without natural one

This experiment is a simplified case of the pursuit game, in which only two robots are involved. The robots start at position r1 and r2 respectively, and the ball starts at "B" (Fig. 7). The experiment succeeds when the two robots clip the ball from symmetrical directions(north and south, or east and west), or enclose the ball in the corner.

In this experiment, the robot1, robot2 and the ball move in turn. The view of a robot is a rectangle whose size is 5*5, taking current position as center. All the obstacles and robots out of this range can't be sensed. But the ball is always sensate.

Fig. 8 shows one of the running processes. The running time is less than 1 minute.

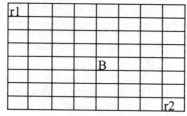

Fig. 7. Cooperation in pure social environment without natural part.

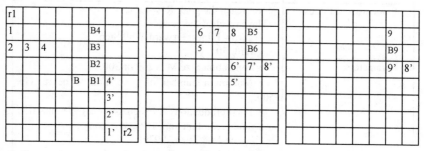

Fig. 8. 1)steps 1~4   2) step 5~8, B7=B5, B8=B6   3)final result

Illustration:

At the beginning, the ball moves upward. The two robots pursue the ball from two sides. They change directions at step 3 and 2 respectively to track the ball(Fig. 8(1)). Then, the ball jumps back and forth in step 5~8. The robots seize the opportunity and come closer to the ball rapidly(Fig. 8(2)). At step 9, the ball is clipped.

In this experiment, the two robots take the position of the ball as shared destination. They make plan with CPS to close to the destination. They have no rules designed prior like in reactive control, but just make plan and move according to the plan. They have no attention and knowledge to clip the ball from symmetrical positions either. No communication helps them cooperate. What the robots know is that the closer it is to the ball, the better, and the faster it come close to the ball, the better. But the adaptive ability enables robots to cooperate well.

In this experiment, robot2 finally moves left(8'~9' in Fig. 8(3)) to clip the ball and ends up the game. Please note that this is not a necessity. At position 8', robot2 tries to close to the ball. It has two selections, move left, or upward. In this experiment, it randomly selects to move left and the game ends up coincidently. Robot2 doesn't know it should move left because it has little knowledge about how to cooperate with robot1. If the experiment runs again, maybe robot2 selects to move upward, which resumes the running process.

5.2 Experiment 2: cooperation in both social and natural environment

In this experiment, the environment is more complex than it in the above experiment, because natural environment is appended. It is more difficult for robots to clip the ball, because the obstacles restrict the movement of robots.

As shown in Fig. 9, some obstacles are distributed like walls.

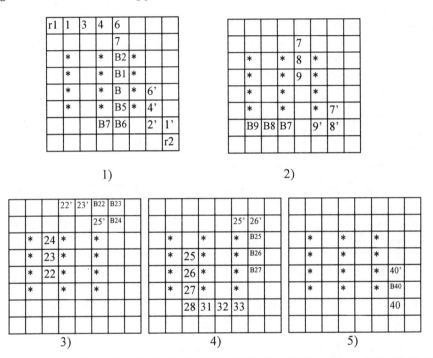

Fig. 9. Environment and starting position

1)   2)

3)   4)   5)

Fig. 10. 1) B3=B1,B4=B,2=1,5=4,8=B2, 3'=2',5'=4',7'=4'   3) 24'=B22   4) 29=27 30=28
5)final position

Illustration:

At the beginning, the ball hides between the walls and moves up and down. The two robots try to pursue the ball from upside and downside respectively. Since new obstacles are sensed continuously, robots make new plan from time to time. It is noted that robot1 re-plans at step 2 and 5, and robot2 does so at step 3 and 5(the robot view is 5*5). At step 7, robot1 moves downward, and occupies the way between the right two walls, which is a necessary part of robot2's plan(Fig. 10(2)). Therefore, robot2 has to make a new plan. The new plan leads robot2 downward, and tracks the ball in another way(step 7',8',9' in Fig. 10(2)). The similar case occurs again at step 25. But this time, it is robot2 who occupies the path of robot1(position 25' is on the way of robot1). The consequence is that robot1 alters its plan (step 25~33, in Fig. 10(4)). At step 40, the ball is clipped.

In this experiment, we see that with the adaptive ability, robots do not follow the ball in group. Instead, sometimes they take intelligent group behavior, like outflanking (Fig. 10(4)), as animals and human do.

5.3 Experiment 3: competition solving in both social and natural environment

The competition problem is described in section 3(Fig. 5). Although the environment is identical to the two robots, they do not always pass the bridge at the same time, because of the randomness in evolutionary course. If one robot crosses the river after the other one, no competition occurs. But a usual case is that they collide with each other on the bridge. To solve the competition, one of them must quit the bridge and let the other one pass first.

Fig. 11 shows one of the running results.

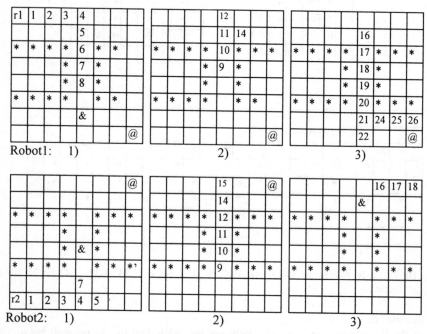

Fig. 11. Robot1 quits the bridge. The upper three maps are paths of robot1, while the lower ones of robot2.   @: destination &: the other robot.   1) robot2: 8=7.   2) robot1: 13=11, 15=14; robot2: 13=12   3) robot1: 23=21

Illustration:

Robot1 gets on the bridge first(step 1~8 in Fig. 11(1)). Robot2 tries to reach the lower-right corner first, and then go upward directly to the destination. However, at step 5, robot2 finds this plan infeasible because the obstacle labeled with "* " occupies the path(this obstacle comes into view at step 5). Now it learns that only the bridge is a feasible way. Therefore it changes its plan and tries to cross the river through the bridge(step 6~7 in Fig. 11(1)). But robot1 has occupied the bridge in advance, so that robot2 finds no way to cross the river and stays at position 7 making new plans. At the same time, robot1 sensed robot2(at "&") and think this way is infeasible. It replans and tries to reach the destination through the right way. Here, robot1 doesn't know the obstacle at the most right side, because it is out of view all along. Therefore robot1 quits the bridge(step 9~14 in Fig. 11(2)). However, at step 14, the obstacle at the most right side is sensed. Robot1 stays at position 14 and makes plans again. Robot2 takes the chance and passes the bridge successfully(step 9~15 in Fig. 11(2)). Finally the two robots reach their own destination(Fig. 11(3)).

6 Conclusion

In this paper, we suggest that adaptive method like GP is powerful in some kinds of cooperation and competition problem. Unlike model-based control and reactive control, adaptive method solves the problem by adapting to the environment. The environment is usually composed of two parts, natural part and social part. The obstacles make up the natural environment, while other robots build the social one. In the adaptive method, the two parts are considered as the same, that is, at planning time, other robots are considered as obstacles. Through the evolutionary course, plans are optimized for robots. Taking these plans, robots accomplish cooperation or solve competition.

References

1. Arkin,R.C.: Motor Schema-based Mobile-Robot Navigation. The International Journal of Robotics Research (1989) 8(4): 92-112
2. Baltsan.A,and Sharir.M.: On the Shortest Paths Between Two Convex Polyhedral. ACM Journal (1988) 35(2): 267-287
3. Benda, M., Jagannathan, V., Dodhiawalla, R.: On Optimal Cooperation of Knowledge Sources. Technical Report BCS-G2010-28, Boeing AI Center (1985)
4. Brooks,R.A.: A Robust Layered Control System For a Mobile Robot. IEEE Journal of Robotics and Automation (1986) 2(1): 14-23
5. Donald, B.R., Jennings, J., Rus D.: Analyzing Teams of Cooperation Mobile Robots. Proceedings of the 1994 IEEE International Conference on Robotics and Automata. San Deigo, CA, (1994) 1896-1903
6. Huang,Y.K, and Ahuja,N.: Gross Motion Planning: a Survey, ACM Computing Survey, (1992) 24(3): 219-291
7. Johnson, P.J., and Bay, J.S.: Distributed Control of Simulated Autonomous Mobile-Robot Collective in Payload Transportation. Autonomous Robots (1995) 2(1): 43-63
8. Koza.J.R.: Genetic Programming II : Automatic Discovery of Reusable Programs. Cambridge, MA, MIT Press (1994)
9. Kuniyoshi, Y., Rougeaux, S., Ishii, M., Kita, N., Sakane, S., and Kakikura, M.: Cooperation by Observation: The Framework and Basic Task Patterns. Proceedings of the 1994 IEEE International Conference on Robotics and Automata, San Deigo, CA, (1994) 767-774

10.   Mataric, M. J., Nilsson, M., and Simsarin, K, T.: Cooperative Multirobot Box Pushing. Proceedings of the 1995 IEEE/RSJ International Conference on Intelligent Robots and System(IROS'95). Washington, DC, (1995) 556-561

11.   Mount, M.D.: The Number of Shortest Paths on the Surface of a Polyhedron, SIAM Journal of Computing. (1990) 19(4): 593-611

12.   Schwartz.J.T, Sharir.M.: On the "Piano Movers" Problem I. the Case of a Two-Dimensional Rigid Polygonal Body Moving Amidst Polygonal Barriers, Communications on Pure and Applied Mathematics, (1983) 36: 345-398

13.   Yamaguchi, H.: A Cooperative Hunting Behavior by Mobile-Robot Troops. The International Journal of Robotics Research, (1999) 18(8): 931-940

14.   Zhao kai, Wang Jue: Chromosome-Protein: a Representation Scheme, Proceedings of the Second Annual Conference on Genetic Programming, Stanford, CA, Morgan Kaufmann, (1997)

15.   Zhao kai,Wang Jue: Path planning in Computer Animation Employing Chromosome-Protein Scheme, Proceedings of the Third Annual Conference on Genetic Programming, Madison, Wisconsin, Morgan Kaufmann, (1998) 439-444

Author Index

Lecture Notes in Computer Science

For information about Vols. 1–1713
please contact your bookseller or Springer-Verlag

Vol. 1745: P. Banerjee, V.K. Prasanna, B.P. Sinha (Eds.), High Performance Computing – HiPC'99. Proceedings, 1999. XXII, 412 pages. 1999.

Vol. 1746: M. Walker (Ed.), Cryptography and Coding. Proceedings, 1999. IX, 313 pages. 1999.

Vol. 1747: N. Foo (Ed.), Adavanced Topics in Artificial Intelligence. Proceedings, 1999. XV, 500 pages. 1999. (Subseries LNAI).

Vol. 1748: H.V. Leong, W.-C. Lee, B. Li, L. Yin (Eds.), Mobile Data Access. Proceedings, 1999. X, 245 pages. 1999.

Vol. 1749: L. C.-K. Hui, D.L. Lee (Eds.), Internet Applications. Proceedings, 1999. XX, 518 pages. 1999.

Vol. 1750: D.E. Knuth, MMIXware. VIII, 550 pages. 1999.

Vol. 1751: H. Imai, Y. Zheng (Eds.), Public Key Cryptography. Proceedings, 2000. XI, 485 pages. 2000.

Vol. 1752: S. Krakowiak, S. Shrivastava (Eds.), Advances in Distributed Systems. VIII, 509 pages. 2000.

Vol. 1753: E. Pontelli, V. Santos Costa (Eds.), Practical Aspects of Declarative Languages. Proceedings, 2000. X, 327 pages. 2000.

Vol. 1754: J. Väänänen (Ed.), Generalized Quantifiers and Computation. Proceedings, 1997. VII, 139 pages. 1999.

Vol. 1755: D. Bjørner, M. Broy, A.V. Zamulin (Eds.), Perspectives of System Informatics. Proceedings, 1999. XII, 540 pages. 2000.

Vol. 1757: N.R. Jennings, Y. Lespérance (Eds.), Intelligent Agents VI. Proceedings, 1999. XII, 380 pages. 2000. (Subseries LNAI).

Vol. 1758: H. Heys, C. Adams (Eds.), Selected Areas in Cryptography. Proceedings, 1999. VIII, 243 pages. 2000.

Vol. 1759: M.J. Zaki, C.-T. Ho (Eds.), Large-Scale Parallel Data Mining. VIII, 261 pages. 2000. (Subseries LNAI).

Vol. 1760: J.-J. Ch. Meyer, P.-Y. Schobbens (Eds.), Formal Models of Agents. Poceedings. VIII, 253 pages. 1999. (Subseries LNAI).

Vol. 1761: R. Caferra, G. Salzer (Eds.), Automated Deduction in Classical and Non-Classical Logics. Proceedings. VIII, 299 pages. 2000. (Subseries LNAI).

Vol. 1762: K.-D. Schewe, B. Thalheim (Eds.), Foundations of Information and Knowledge Systems. Proceedings, 2000. X, 305 pages. 2000.

Vol. 1763: J. Akiyama, M. Kano, M. Urabe (Eds.), Discrete and Computational Geometry. Proceedings, 1998. VIII, 333 pages. 2000.

Vol. 1764: H. Ehrig, G. Engels, H.-J. Kreowski, G. Rozenberg (Eds.), Theory and Application of Graph Transformations. Proceedings, 1998. IX, 490 pages. 2000.

Vol. 1765: T. Ishida, K. Isbister (Eds.), Digital Cities. IX, 444 pages. 2000.

Vol. 1767: G. Bongiovanni, G. Gambosi, R. Petreschi (Eds.), Algorithms and Complexity. Proceedings, 2000. VIII, 317 pages. 2000.

Vol. 1768: A. Pfitzmann (Ed.), Information Hiding. Proceedings, 1999. IX, 492 pages. 2000.

Vol. 1769: G. Haring, C. Lindemann, M. Reiser (Eds.), Performance Evaluation: Origins and Directions. X, 529 pages. 2000.

Vol. 1770: H. Reichel, S. Tison (Eds.), STACS 2000. Proceedings, 2000. XIV, 662 pages. 2000.

Vol. 1771: P. Lambrix, Part-Whole Reasoning in an Object-Centered Framework. XII, 195 pages. 2000. (Subseries LNAI).

Vol. 1772: M. Beetz, Concurrent Ractive Plans. XVI, 213 pages. 2000. (Subseries LNAI).

Vol. 1773: G. Saake, K. Schwarz, C. Türker (Eds.), Transactions and Database Dynamics. Proceedings, 1999. VIII, 247 pages. 2000.

Vol. 1774: J. Delgado, G.D. Stamoulis, A. Mullery, D. Prevedourou, K. Start (Eds.), Telecommunications and IT Convergence Towards Service E-volution. Proceedings, 2000. XIII, 350 pages. 2000.

Vol. 1776: G.H. Gonnet, D. Panario, A. Viola (Eds.), LATIN 2000: Theoretical Informatics. Proceedings, 2000. XIV, 484 pages. 2000.

Vol. 1777: C. Zaniolo, P.C. Lockemann, M.H. Scholl, T. Grust (Eds.), Advances in Database Technology – EDBT 2000. Proceedings, 2000. XII, 540 pages. 2000.

Vol. 1778: S. Wermter, R. Sun (Eds.), Hybrid Neural Systems. IX, 403 pages. 2000. (Subseries LNAI).

Vol. 1780: R. Conradi (Ed.), Software Process Technology. Proceedings, 2000. IX, 249 pages. 2000.

Vol. 1781: D.A. Watt (Ed.), Compiler Construction. Proceedings, 2000. X, 295 pages. 2000.

Vol. 1782: G. Smolka (Ed.), Programming Languages and Systems. Proceedings, 2000. XIII, 429 pages. 2000.

Vol. 1783: T. Maibaum (Ed.), Fundamental Approaches to Software Engineering. Proceedings, 2000. XIII, 375 pages. 2000.

Vol. 1784: J. Tiuryn (Ed.), Foundations of Software Science and Computation Structures. Proceedings, 2000. X, 391 pages. 2000.

Vol. 1785: S. Graf, M. Schwartzbach (Eds.), Tools and Algorithms for the Construction and Analysis of Systems. Proceedings, 2000. XIV, 552 pages. 2000.

Vol. 1786: B.H. Haverkort, H.C. Bohnenkamp, C.U. Smith (Eds.), Computer Performance Evaluation. Proceedings, 2000. XIV, 383 pages. 2000.

Vol. 1790: N. Lynch, B.H. Krogh (Eds.), Hybrid Systems: Computation and Control. Proceedings, 2000. XII, 465 pages. 2000.

Vol. 1793: O. Cairo, L.E. Sucar, F.J. Cantu (Eds.), MICAI 2000: Advances in Artificial Intelligence. Proceedings, 2000. XIV, 750 pages. 2000. (Subseries LNAI).

Vol. 1795: J. Sventek, G. Coulson (Eds.), Middleware 2000. Proceedings, 2000. XI, 436 pages. 2000.

Vol. 1794: H. Kirchner, C. Ringeissen (Eds.), Frontiers of Combining Systems. Proceedings, 2000. X, 291 pages. 2000. (Subseries LNAI).

Vol. 1801: J. Miller, A. Thompson, P. Thomson, T.C. Fogarty (Eds.), Evolvable Systems: From Biology to Hardware. Proceedings, 2000. X, 286 pages. 2000.

Vol. 1802: R. Poli, W. Banzhaf, W.B. Langdon, J. Miller, P. Nordin, T.C. Fogarty (Eds.), Genetic Programming. Proceedings, 2000. X, 361 pages. 2000.

Vol. 1803: S. Cagnoni et al. (Eds.), Real-World Applications and Evolutionary Computing. Proceedings, 2000. XII, 396 pages. 2000.